T0291857

CAMBRIDGE LIBRARY COLLECTION

Books of enduring scholarly value

Physical Sciences

From ancient times, humans have tried to understand the workings of the world around them. The roots of modern physical science go back to the very earliest mechanical devices such as levers and rollers, the mixing of paints and dyes, and the importance of the heavenly bodies in early religious observance and navigation. The physical sciences as we know them today began to emerge as independent academic subjects during the early modern period, in the work of Newton and other 'natural philosophers', and numerous sub-disciplines developed during the centuries that followed. This part of the Cambridge Library Collection is devoted to landmark publications in this area which will be of interest to historians of science concerned with individual scientists, particular discoveries, and advances in scientific method, or with the establishment and development of scientific institutions around the world.

Cosmos

Polymath Alexander von Humboldt (1769-1859), a self-described 'scientific traveller', was one of the most respected scientists of his time. Humboldt's wanderlust led him across Europe and to South America, Mexico, the U.S., and Russia, and his voyages and observations resulted in the discovery of many species previously unknown to Europeans. Originating as lectures delivered in Berlin and Paris (1827–8), his two-volume *Cosmos: Sketch of a Description of the Universe* (1845–60) represented the culmination of his lifelong interest in understanding the physical world. As Humboldt writes, 'I ever desired to discern physical phenomena in their widest mutual connection, and to comprehend Nature as a whole, animated and moved by inward forces'. Volume 2 (1848) reviews poetic descriptions of nature as well as landscape painting from antiquity through to modernity, before using the same time-span to examine a 'History of the Physical Contemplation of the Universe'.

Cambridge University Press has long been a pioneer in the reissuing of out-of-print titles from its own backlist, producing digital reprints of books that are still sought after by scholars and students but could not be reprinted economically using traditional technology. The Cambridge Library Collection extends this activity to a wider range of books which are still of importance to researchers and professionals, either for the source material they contain, or as landmarks in the history of their academic discipline.

Drawing from the world-renowned collections in the Cambridge University Library, and guided by the advice of experts in each subject area, Cambridge University Press is using state-of-the-art scanning machines in its own Printing House to capture the content of each book selected for inclusion. The files are processed to give a consistently clear, crisp image, and the books finished to the high quality standard for which the Press is recognised around the world. The latest print-on-demand technology ensures that the books will remain available indefinitely, and that orders for single or multiple copies can quickly be supplied.

The Cambridge Library Collection will bring back to life books of enduring scholarly value (including out-of-copyright works originally issued by other publishers) across a wide range of disciplines in the humanities and social sciences and in science and technology.

Cosmos

Sketch of a Physical Description of the Universe
of the Universe

VOLUME 2

ALEXANDER VON HUMBOLDT
EDITED BY EDWARD SABINE

CAMBRIDGE UNIVERSITY PRESS

Cambridge, New York, Melbourne, Madrid, Cape Town, Singapore,
São Paolo, Delhi, Dubai, Tokyo

Published in the United States of America by Cambridge University Press, New York

www.cambridge.org
Information on this title: www.cambridge.org/9781108013642

This edition first published 1848
This digitally printed version 2010

ISBN 978-1-108-01364-2 Paperback

COSMOS:

SKETCH

OF A

PHYSICAL DESCRIPTION OF THE UNIVERSE.

BY

ALEXANDER VON HUMBOLDT.

VOL. II.

Naturæ vero rerum vis atque majestas in omnibus momentis fide caret, si quis modo partes ejus ac non totam complectatur animo.—PLIN. H. N. lib. vii. c. 1.

TRANSLATED UNDER THE SUPERINTENDENCE OF

LIEUT.-COL. EDWARD SABINE, R.A., FOR. SEC. R.S.

LONDON:

PRINTED FOR

LONGMAN, BROWN, GREEN, AND LONGMANS,

PATERNOSTER ROW; AND

JOHN MURRAY, ALBEMARLE STREET.

1848.

CONTENTS.

INCITEMENTS TO THE STUDY OF NATURE.

II. Landscape Painting.

III. Culture of characteristic Exotic Plants.

HISTORY OF THE PHYSICAL CONTEMPLATION OF THE UNIVERSE.

*** *See* NOTICE *at back.*

*** A notice is appended by M. de Humboldt at the close of the second volume of "Kosmos," stating, that the first portion of that volume, viz. "On the Incitements to the Study of Nature," was printed in July 1846; and that the printing of the second portion, viz. "The History of the Physical Contemplation of the Universe," was completed in the month of September 1847.

From page 100 to the conclusion of the text, the Translation, in its progress through the press, has had the advantage of being compared with the original by the Chevalier Bunsen.

February 21, 1848.

COSMOS.

COSMOS:

A PHYSICAL DESCRIPTION OF THE UNIVERSE.

INCITEMENTS TO THE STUDY OF NATURE.

Action of the external world on the imaginative faculty, and the reflected image produced—Poetic descriptions of nature—Landscape painting—Cultivation of those exotic plants which determine the characteristic aspect of the vegetation in the countries to which they belong.

WE now pass from the domain of objects to that of sensations. The principal results of observation, in the form in which, stripped of all additions derived from the imagination, they belong to a pure scientific description of nature, have been presented in the preceding volume. We have now to consider the impression which the image received by the external senses produces on the feelings, and on the poetic and imaginative faculties of mankind. An inward world here opens to the view, into which we desire to penetrate, not, however, for the purpose of investigating—as would be required if the philosophy of art were our aim—

what in æsthetic performances belongs essentially to the powers and dispositions of the mind, and what to the particular direction of the intellectual activity,—but that we may trace the sources of that animated contemplation which enhances a genuine enjoyment of nature, and discover the particular causes which, in modern times especially, have so powerfully promoted, through the medium of the imagination, a predilection for the study of nature, and for the undertaking of distant voyages.

I have alluded, in the preceding volume, to three (1) kinds of incitement more frequent in modern than in ancient times; 1st, the æsthetic treatment of natural scenery by vivid and graphical descriptions of the vegetable and animal world, which is a very modern branch of literature; 2d, landscape painting, so far as it pourtrays the characteristic aspect of vegetation; and, 3d, the more extended cultivation of tropical plants, and the assemblage of contrasted exotic forms. Each of these subjects might be historically treated and investigated at some length; but it appears to me better suited to the spirit and object of my work, to unfold only a few leading ideas relating to them,—to recal how differently the contemplation of nature has acted on the intellect and the feelings of different races of men, and at different periods of time,—and to notice how, at epochs when there has been a general cultivation of the mental faculties, the severe pursuit of exact knowledge, and the more delicate workings of the imagination, have tended to interpenetrate and blend with each other. If we would describe the full majesty of nature, we must not dwell solely on her external phænomena, but we must also regard her in her reflected image—at one time filling the visionary land of physical myths with graceful phantoms, and at another developing the noble germs of presentive art.

I here limit myself to the consideration of incitements to a scientific study of nature; and, in so doing, I would recal the lessons of experience, which tell us how often impressions received by the senses from circumstances seemingly accidental, have so acted on the youthful mind as to determine the whole direction of the man s course through life. Childish pleasure in the form of countries and of seas, as delineated in maps (2); the desire to behold those southern constellations which have never risen in our horizon (3); the sight of palms and of the cedars of Lebanon, figured in a pictorial bible, may have implanted in the spirit the first impulse to travels in distant lands. If I might have recourse to my own experience, and say what awakened in me the first beginnings of an inextinguishable longing to visit the tropics, I should name George Forster's descriptions of the islands of the Pacific—paintings, by Hodge, in the house of Warren Hastings, in London, representing the banks of the Ganges—and a colossal dragon tree in an old tower of the Botanic Garden at Berlin. These objects, which I here cite as exemplifications taken from fact, belong respectively to the three classes above noticed, viz. to descriptions of nature flowing from a mind inspired by her contemplation, to imitative art in landscape painting, and to the immediate view of characteristic natural objects. Such incitements are, however, only influential where general intellectual cultivation prevails, and when they address themselves to dispositions suited to their reception, and in which a particular course of mental development has heightened the susceptibility to natural impressions.

INCITEMENTS TO THE STUDY OF NATURE.

I.—Description of natural scenery, and the feelings associated therewith at different times and among different races and nations.

It has often been said, that if delight in nature were not altogether unknown to the ancients, yet that its expression was more rare and less animated among them than in modern times. Schiller (4), in his considerations on naïve and sentimental poetry, remarks, that "when we think of the glorious scenery which surrounded the ancient Greeks, and remember the free and constant intercourse with nature in which their happier skies enabled them to live, as well as how much more accordant their manners, their habits of feeling, and their modes of representation, were with the simplicity of nature, of which their poetic works convey so true an impress, we cannot but remark with surprise how few traces we find amongst them of the sentimental interest with which we moderns attach ourselves to natural scenes and objects. In the description of these, the Greek is indeed in the highest degree exact, faithful, and circumstantial, but without exhibiting more warmth of sympathy than in treating of a garment, a shield, or of a suit of armour. Nature appears to interest his understanding rather than his feelings; he does not cling to her with intimate affection and sweet melancholy, as do the moderns." Much as there is that is true and excellent in these remarks, they are far from being applicable to all antiquity, even in the sense ordinarily attached to the term; I cannot, moreover, but regard as far too limited, the restriction of antiquity (as opposed to modern times), exclusively to the Greeks and Romans: a profound feeling of nature speaks forth in the earliest poetry of the Hebrews and of the Indians;—in nations, therefore, of very different descent, Semitic, and Indo-Germanic.

We can only infer the feeling with which the ancients regarded nature from the portions of its expression which have reached us in the remains of their literature; we must therefore seek for such passages the more diligently, and pronounce upon them the more circumspectly, as they present themselves but sparingly in the two great forms of epical and lyrical poetry. In Hellenic poetry, at that flowery season of the life of mankind, we find, indeed, the tenderest expression of the love and admiration of nature mingling with the poetic representation of human passion, in actions taken from legendary history; but specific descriptions of natural scenes or objects appear only as subordinate; for in Grecian art all is made to concenter within the sphere of human life and feeling.

The description of nature in her manifold diversity, as a distinct branch of poetic literature, was altogether foreign to the ideas of the Greeks. With them the landscape is always the mere background of a picture, in the foreground of which human figures are moving. Passion breaking forth in action rivetted their attention almost exclusively; the agitation of politics, and a life passed chiefly in public, withdrew men's minds from enthusiastic absorption in the tranquil pursuit of nature. Physical phænomena were always referred to man (5) by supposed relations or resemblances either of external form or of inward spirit. It was almost exclusively by such applications that the consideration of nature was thought worthy of a place in poetry in the form of comparisons or similitudes, which often present small detached pictures, full of objective vividness and truth.

At Delphi, pæans to spring (6) were sung—probably to express men's joy that the privations and discomforts of winter were past. A natural description of winter has been interwoven (may it not be by a later Ionian rhapsodist?) with the "Works and Days" of Hesiod (7). This poem, full of a noble simplicity, but purely didactic in its form, gives advice respecting agriculture, and directions for different kinds of work and profitable employment, together with ethical exhortations to a blameless life. Its tone rises to a more lyrical character when the poet clothes the miseries of mankind, or the fine allegorical mythus of Epimetheus and Pandora, with an anthropomorphic garb. In Hesiod's Theogony, which is composed of various ancient and dissimilar elements, we find repeatedly (as, for example, in the enumeration of the Nereides (8)), natural descriptions veiled under the significant names of mythic personages. In the Bœotian bardic school, and generally in all ancient Greek poetry, the phænomena of the external world are introduced only by personification under human forms.

But if it be true, as we have remarked, that natural descriptions, whether of the richness and luxuriance of southern vegetation, or the portraiture in fresh and vivid colours of the habits of animals, have only become a distinct branch of literature in very modern times, it was not that sensibility to the beauty of nature was absent (9), where the perception of beauty was so intense,—or the animated expression of a contemplative poetic spirit wanting, where the creative power of the Hellenic mind produced inimitable master works in poetry and in the plastic arts. The deficiency which appears to our modern ideas in this department of antiquity, betokens not so much a want of sensibility, as the absence of a prevailing impulse to disclose in words the feeling of natural beauty. Directed less to the inanimate world of phænomena than to that of human action, and of the internal spontaneous emotions, the earliest and the noblest developments of the poetic spirit were epical and lyrical. These were forms in which natural descriptions could only hold a subordinate, and, as it were, an accidental place, and could not appear as distinct productions of the imagination. As the influence of antiquity gradually declined, and as its blossoms faded, both descriptive and didactic poetry became more and more rhetorical; and the latter, which, in its earlier philosophical and semi-priestly character, had been severe, grand, and unadorned, as in Empedocles' "Poem of Nature," gradually lost its early simple dignity.

I may be permitted to illustrate these general observations by a few particular instances. Conformably to the character of the Epos, natural scenes and images, however charming, appear in the Homeric songs always as mere incidental adjuncts. "The shepherd rejoices in the calm of night, when the winds are still; in the pure ether, and in the bright stars shining in the vault of heaven; he hears from afar the rushing of the suddenly-swollen forest torrent, bearing down earth and trunks of uprooted oaks" [10]. The fine description of the sylvan loneliness of Parnassus, and of its dark, thickly-wooded rocky valleys, contrasts with the smiling pictures of the many-fountained poplar groves of the Phæacian Islands, and especially with the land of the Cyclops, "where swelling meads of rich waving grass surround the hills of undressed vines" (11). Pindar, in a vernal dithyrambus recited at Athens, sings "the earth covered with new flowers, what time in Argive Nemea the first opening shoot of the palm announces the approach of balmy spring;" he sings of Etna, "the pillar of heaven, the nurse of enduring snows;" but he quickly hastens to turn from the awful form of inanimate nature, to celebrate Hiero of Syracuse, and the Greeks' victorious combats with the powerful Persian nation.

Let us not forget that Grecian scenery possesses the peculiar charm of blended and intermingled land and sea; the breaking waves and changing brightness of the resounding ocean, amidst shores adorned with vegetation, or picturesque cliffs richly tinged with aerial hues. Whilst to other nations the different features and the different pursuits belonging to the sea and to the land appeared separate and distinct, the Greeks, not only of the islands, but also of almost all the southern portion of the mainland, enjoyed the continual presence of the greater variety and richness, as well as of the higher character of beauty, given by the contact and mutual influence of the two elements. How can we imagine that a race so happily organised by nature, and whose perception of beauty was so intense, should have been unmoved by the aspect of the wood-crowned cliffs of the deeply-indented shores of the Mediterranean, the varied distribution of vegetable forms, and, spread over all, the added charms dependent on atmospheric influences, varying by a silent interchange with the varying surfaces of land

and sea, of mountain and of plain, as well as with the varying hours and seasons? Or how, in the age when the poetic tendency was highest, can emotions of the mind thus awakened through the senses have failed to resolve themselvos into ideal contemplation? The Greeks, we know, imagined the vegetable world connected by a thousand mythical relations with the heroes and the gods: avenging chastisement followed injury to the sacred trees or plants. But while trees and flowers were animated and personified, the prevailing forms of poetry in which the peculiar mental development of the Greeks unfolded itself, allowed but a limited space to descriptions of nature.

Yet, a deep sense of the beauty of nature breaks forth sometimes even in their tragic poets, in the midst of deep sadness, or of the most tumultuous agitation of the passions. When Œdipus is approaching the grove of the Furies, the chorus sings, "the noble resting-place of glorious Colonos, where the melodious nightingale loves to dwell, and mourns in clear and plaintive strains:" it sings "the verdant darkness of the thick embowering ivy, the narcissus bathed in the dews of heaven, the golden beaming crocus, and the ineradicable, ever fresh-springing olive tree" (12). Sophocles, in striving to glorify his native Colonos, places the lofty form of the fate-pursued, wandering king, by the side of the sleepless waters of the Cephisus, surrounded by soft and bright imagery. The repose of nature heightens the impression of pain called forth by the desolate aspect of the blind exile, the victim of a dreadful and mysterious destiny. Euripides (13) also takes pleasure in the picturesque description of "the pastures of Messenia and Laconia, refreshed by a thousand fountains, under an ever mild sky, and through which the beautiful Pamisus rolls his stream."

Bucolic poetry, born in the Sicilian fields, and popularly inclined to the dramatic, has been called, with reason, a transitional form. These pastoral epics on a small scale depict human beings rather than scenery: they do so in Theocritus, in whose hands this form of poetry reached its greatest perfection. A soft elegiac element is indeed every where proper to the idyll, as if it had arisen from "the longing for a lost ideal;" or as if in the human breast a degree of melancholy were ever blended with the deeper feelings which the view of nature inspires.

When the true poetry of Greece expired with Grecian liberty, that which remained became descriptive, didactic, instructive;—astronomy, geography, and the arts of the hunter and the fisherman, appeared in the age of Alexander and his successors as objects of poetry, and were indeed often adorned with much metrical skill. The forms and habits of animals are described with grace, and often with such exactness that our modern classifying natural historians can recognise genera and even species. But in none of these writings can we discover the presence of that inner life—that inspired contemplation—whereby to the poet, almost unconsciously to himself, the external world becomes a subject of the imagination. The undue preponderance of the descriptive element shews itself in the forty-eight cantos of the Dionysiaca of the Egyptian Nonnus, which are distinguished by a very artfully constructed verse. This poet takes pleasure in describing great revolutions of nature; he makes a fire kindled by lightning on the wooded banks of the Hydaspes burn even the fish in the bed of the river; he tells how ascending vapours produce the meteorological processes of storm and electric rain. Nonnus of Panopolis is inclined to romantic poetry, and is remarkably unequal; at times spirited and interesting, at others verbose and tedious.

A more delicate sensibility to natural beauty shews itself occasionally in the Greek Anthology, which has been handed down to us in such various ways, and from such different periods. In the pleasing translation by Jacobs, all that relates to plants and animals is collected in one section: these passages form small pictures, most commonly, of only single objects. The plane tree, which "nourishes among its boughs the grape swelling with rich juice," and which, in the time of Dionysius the Elder, reached the banks of the Sicilian Anapus from Asia Minor, through the Island of Diomedes, occurs perhaps but too often; still, on the whole, the antique mind shews itself in these songs and epigrams as more inclined to dwell on animal than on vegetable forms.

The vernal idyll of Meleager of Gadara in Cœlo-Syria is a noble and more important composition ([14]). I am unwilling, were it only for the ancient renown of the locality, to omit all notice of the description of the wooded Vale of Tempe given by Ælian ([15]), probably from an earlier notice by Dicearchus. It is the most detailed description of natural scenery by a Greek prose writer which we possess; and, although topographic, is at the same time picturesque. The shady valley is enlivened by the Pythian procession (theoria), "which gathers from the sacred laurel the reconciling bough."

In the latest Byzantine epoch, towards the end of the

fourth century, we find descriptions of scenery frequently introduced in the romances of the Greek prose writers; as in the pastoral romance of Longus (16), in which, however, the author is much more successful in the tender scenes taken from life, than in the expression of sensibility to the beauties of nature.

It is not the object of these pages to introduce more than such few references to particular forms of poetic art, as may tend to illustrate general considerations respecting the poetic conception of the external world; and I should here quit the flowery circle of Hellenic antiquity, if, in a work to which I have ventured to give the name of "Cosmos," I could pass over in silence the description of nature, with which the pseudo Aristotelian book of the Cosmos (or "Order of the Universe") commences. This description shews us "the terrestrial globe adorned with luxuriant vegetation, abundantly watered, and, which is most worthy of praise, inhabited by thinking beings" (17). The rhetorical colouring of this rich picture of nature, so unlike the concise and purely scientific manner of the Stagirite, is one of the many indications by which it has been judged not to have been his composition. Conceding this point, and ascribing it to Appuleius (18), or to Chrysippus (19), or to any other author, its place is fully supplied by a brief but genuine fragment which Cicero has preserved to us from a lost work of Aristotle (20). "If there were beings living in the depths of the earth, in habitations adorned with statues and paintings, and every thing which is possessed in abundance by those whom we call fortunate, and if these beings should receive tidings of the dominion and power of the gods, and should then be brought from their hidden dwelling places to the surface which we inhabit, and should suddenly behold the earth, and the sea, and the vault of heaven; should perceive the broad expanse of the clouds and the strength of the winds; should admire the sun in his majesty, beauty, and effulgence; and, lastly, when night veiled the earth in darkness, should gaze on the starry firmament, the waxing and waning moon, and the stars rising and setting in their unchanging course, ordained from eternity, they would, of a truth, exclaim, 'there are gods, and such great things are their work.'" It has been justly said, that these words would alone be sufficient to confirm Cicero's opinion of "the golden flow of the Aristotelian eloquence" ([21]), and that there breathes in them somewhat of the inspired genius of Plato. Such a testimony as this to the existence of heavenly powers, from the beauty and infinite grandeur of the works of creation, is indeed rare in classical antiquity.

That which we miss with regard to the Greeks, I will not say in their appreciation of natural phænomena, but in the direction which their literature assumed, we find still more sparingly among the Romans. A nation which, in conformity with the old Siculian manners, manifested a marked predilection for agriculture and rural life, might have justified other hopes; but with all their capacity for practical activity, the Romans, in their cold gravity, and measured sobriety of understanding, were, as a people, far inferior to the Greeks in the perception of beauty, and far less sensitive to its influence; and were much more devoted to the realities of everyday life, than to an idealising poetic contemplation of nature.

These inherent differences between the Greek and Roman mind are faithfully reflected, as is always the case with national character, in their respective literatures; and I must add to this consideration, that of the acknowledged difference in the organic structure of the two languages, notwithstanding the affinity between the races. The language of ancient Latium is regarded as possessing less flexibility, a more limited adaptation of words, and "more of realistic tendency" than of "ideal mobility." The predilection for the imitation of foreign Greek models in the Augustan age, might, moreover, have been unfavourable to the free outpourings of the native mind and feelings in reference to nature; but yet, powerful minds, animated by love of country, have effectually surmounted these varied obstacles, by creative individuality, by elevation of ideas, and by tender grace in their presentation. The great poem which is the fruit of the rich genius of Lucretius, embraces the whole Cosmos: it has much affinity with the works of Empedocles and Parmenides; and the grave tone in which the subject is presented is enhanced by its archaic diction. Poetry and philosophy are closely interwoven in it; without, however, falling into that coldness of composition, which, as contrasted with Plato's views of nature so rich in imagination, is severely blamed by the rhetor Menander, in the sentence passed by him on the "hymns to nature" (22). My brother has pointed out, with great ingenuity, the striking analogies and diversities produced by the interweaving of metaphysical abstraction with poetry in the ancient Greek didactic poems, in that of Lucretius, and in the Bhagavad-Gita episode of the Indian epic Mahabharata (23). In the great physical picture of the universe traced by the Roman poet, we find contrasted with his chilling atomic doctrine, and his often extravagantly wild geological fancies, the fresh and animated description of mankind exchanging the thickets of the forest for the pursuits of agriculture, the subjugation of natural forces, the cultivation of the intellect and of language, and the formation of civil society (24).

When, in the midst of the busy and agitated life of a statesman, and in a mind excited by political passions, an animated love of nature and of rural solitude still subsists, its source must be sought in the depths of a great and noble character. Cicero's writings shew the truth of this assertion. Although it is generally recognised that in the book De Legibus, and in that of the Orator, many things are imitated from the Phædrus of Plato (25), yet the picture of Italian nature does not lose its individuality and truth. Plato, in more general characters, praises the dark shade of the lofty plane tree, the luxuriant abundance of fragrant herbs and flowers, the sweet summer breezes, and the chorus of grasshoppers." In Cicero's smaller pictures, we find, as has been recently well remarked (26), all those features which we still recognise in the actual landscape: we see the Liris shaded by lofty poplars; and in descending the steep mountain side to the east, behind the old castle of Arpinum, we look on the grove of oaks near the Fibrenus, as well as on the island now called Isola di Carnello, which is formed by the division of the stream, and into which Cicero retired, as he says, to "give himself up to his meditations, to read, or to write." Arpinum, on the Volscian Mountains, was the birthplace of the great statesman; and his mind and character were doubtless influenced in his boyhood by the grand scenery of the vicinity. In the mind of man, the reflex action of the external aspect of surrounding nature is early and unconsciously blended with that which belongs to the original tendencies, capacities, and powers of his own inner being.

In the midst of the stormy and eventful period of the year 708 (from the foundation of Rome), Cicero found consolation in his villas, alternately at Tusculum, Arpinum, Cumæ, and Antium. "Nothing," he writes to Atticus (27), "can be more delightful than this solitude; more pleasing than this country dwelling, the neighbouring shore, and the prospect over the sea. In the lonely island of Astura, at the mouth of the river of the same name, and on the shore of the Tyrrhenian sea, no human being disturbs me; and when, early in the morning, I hide myself in a thick wild forest, I do not leave it until the evening. Next to my Atticus, nothing is so dear to me as solitude, in which I cultivate intercourse with philosophy; but this intercourse is often interrupted with tears. I strive against these as much as I can, but I have not yet prevailed." It has been repeatedly remarked, that in these letters, and in those of the younger Pliny, expressions resembling those so common amongst the sentimental writers of modern times may be unequivocally recognised; I find in them only the accents of a mind deeply moved, such as in every age, and every nation or race, escape from the heavily-oppressed bosom.

From the general diffusion of Roman literature, the master works of Virgil, Horace, and Tibullus, are so widely and intimately known, that it would be superfluous to dwell on individual instances of the delicate and ever wakeful sensibility to nature, by which many of them are animated. In the Æneid, the epic character forbids the appearance of descriptions of natural scenes and objects otherwise than as subordinate and accidental features, limited to a very small space; individual localities are not pourtrayed (28), but an intimate understanding and love of nature manifest themselves occasionally with peculiar beauty. Where have the soft play of the waves, and the repose of night, ever been more happily described? and how finely do these mild and tender images contrast with the powerful representations of the gathering and bursting tempest in the first book of the Georgics, and with the descriptions in the Æneid of the navigation and landing at the Strophades, the crashing fall of the rock, and of Ætna with its flames (29). We might have expected from Ovid, as the fruit of his long sojourn in the plains of Tomi in Lower Mæsia, a poetic description of the aspect of nature in the steppes; but none such has come down to us from antiquity, either from him or from any other writer. The Roman exile did not indeed see that kind of steppe which in summer is thickly covered by rich herbage and flowering plants from four to six feet high, which, as each breeze passes over them, present the pleasing picture of an undulating many-coloured sea of flowers and verdure. The place of his banishment was a desolate marshy district. The broken spirit of the exile, which yielded to unmanly lamentations, was filled with recollections of the social pleasures and the political occurrences of Rome, and had no place for the contemplation of the Scythian desert by which he was surrounded. On the other hand, this richly-gifted poet, so powerful in vivid representation, has given us, besides general descriptions of grottos, fountains, and silent moonlight nights, which are but too frequently repeated, an eminently-characteristic, and even geologically-important description of the volcanic eruption at Methone between

Epidaurus and Trœzene, which has been referred to in the "General View of Nature" contained in the preceding volume (30).

It is especially to be regretted that Tibullus should not have left us any great composition descriptive of natural scenery, general or individual. He belongs to the few among the poets of the Augustan age who, being happily strangers to the Alexandrian learning, and devoted to retirement and a rural life, full of feeling and therefore simple, drew from their own resources. In many of his elegies (31), indeed, the landscape forms only the background of the picture; but the Lustration of the Fields and the 6th Elegy of the first book shew what might have been expected from the friend of Horace and Messala.

Lucan, the grandson of the rhetor Marcus Annæus Seneca, is indeed only too nearly related to his progenitor in the rhetorical ornateness of his style; yet we find among his writings a fine description of the destruction of a Druidic forest (32) on the now treeless shore of Marseilles, which is thoroughly true to nature: the severed oaks, leaning against each other, support themselves for a time before they fall; and, denuded of their leaves, admit the first ray of light to penetrate the awful gloom of the sacred shade. Those who have lived long in the forests of the New Continent, feel how vividly the poet has depicted, with a few traits, the luxuriant growth of trees whose giant remains are still found buried in turf bogs in France (33).

In a didactic poem entitled Ætna, written by Lucilius the Younger, a friend of L. Annæus Seneca, the phænomena of a volcanic eruption are described, not inaccurately, but yet in a far less animated and characteristic manner than in the "Ætna Dialogus" (34) of the youthful Bembo, mentioned with praise in the preceding volume.

When, after the close of the fourth century, poetry in its grander and nobler forms faded away, as if exhausted, poetic attempts, deprived of the magic of creative imagination, were occupied only with the drier realities of knowledge and description: and a certain rhetorical polish of style could ill replace the simple feeling for nature, and the idealising inspiration, of an earlier age. We may name as a production of this barren period, in which the poetic element appears only as an accidental and merely external ornament, a poem on the Moselle, by Ausonius, a native of Aquitanian Gaul, who had accompanied Valentinian in his campaign against the Allemanni. The "Mosella," which was composed at ancient Trèves (35), describes sometimes not unpleasingly the already vine-covered hills of one of the loveliest rivers of Germany; but the mere topography of the country, the enumeration of the streams which flow into the Moselle, and the characters, in form, colour, and habits, of some of the different kinds of fish which are found in the river, are the principal objects of this purely didactic composition.

In the works of Roman prose writers, among which we have already referred to some remarkable passages by Cicero, descriptions of natural scenery are as rare as in those of Greek writers of the same class; but the great historians—Julius Cæsar, Livy, and Tacitus—in relating the conflicts of men with natural obstacles and with hostile forces, are sometimes led to give descriptions of fields of battle, and of the passage of rivers, or of difficult mountain passes. In the Annals of Tacitus, I am delighted with the description of Germanicus's unsuccessful navigation of the Amisia, and with the grand geographical sketch of the mountain chains of Syria and of Palestine (36). Curtius (37) has left us a fine natural picture of a forest wilderness to the west of Hekatompylos, through which the Macedonian army had to pass in entering the humid province of Mazanderan; to which I would refer more in detail, if, in a writer whose period is so uncertain, we could distinguish with any security between what he has drawn from his own lively imagination, and what he has derived from historic sources.

The great encyclopædic work of the elder Pliny, which, as his nephew, the younger Pliny, has finely said, is "varied as nature herself," and which, in the abundance of its contents, is unequalled by any other ancient work, will be referred to in the sequel, when treating of the "History of the Contemplation of the Universe." This work, which exerted a powerful influence on the whole of the middle ages, is a most remarkable result of the disposition to comprehensive, but often indiscriminate collection. Unequal in style—sometimes simple and narrative, sometimes thoughtful, animated, and rhetorically ornate—it has, as, indeed, might be expected from its form, few individual descriptions of nature; but wherever the grand concurrent action of the forces in the universe, the well-ordered Cosmos (naturæ majestas), is the object of contemplation, we cannot mistake the evidences of true inward poetic inspiration.

We would gladly adduce the pleasantly-situated villas of the Romans, on the Pincian Mount, at Tusculum, and Tibur, on the promontory of Misenum, and near Puteoli and Baiæ, as evidences of a love of nature, if these spots had not, like those in which were the villas of Scaurus and Mæcenas, Lucullus and Adrian, been crowded with sumptuous buildings—temples, theatres, and race-courses alternating with aviaries and houses for rearing snails and dormice. The elder Scipio had surrounded his more simple country seat at Liturnum with towers like a fortress. The name of Matius, a friend of Augustus, has been handed down to us as that of the individual whose predilection for unnatural constraint first introduced the custom of cutting and training trees into artificial imitations of architectural and plastic models. The letters of the younger Pliny furnish us with pleasing descriptions of two (38) of his numerous villas, Laurentinum and Tuscum. Although buildings, surrounded by box cut into artificial forms, are more numerous and crowded than our taste for nature would lead us to desire, yet these descriptions, as well as the imitation of the Vale of Tempe in the Tiburtine villa of Adrian, shew us that among the inhabitants of the imperial city, the love of art, and the solicitous care for comfort and convenience manifested in the choice of the positions of their country houses with reference to the sun and to the prevailing winds, might be associated with love for the free enjoyment of nature. It is cheering to be able to add, that on the estates of Pliny this enjoyment was less disturbed than elsewhere by the painful features of slavery. The wealthy proprietor was not only one of the most learned men of his period, but he had also those compassionate and truly humane feelings for the lower classes of the people who were not in the enjoyment of freedom, of which the expression at least is most rare in antiquity. At his villas fetters were unused; and he provided that the slave, as a cultivator of the soil, should freely bequeath that which he had acquired (39).

No description of the eternal snows of the Alps, when tinged in the morning or evening with a rosy hue, of the beauty of the blue glacier ice, or of any part of the grandeur of the scenery of Switzerland, have reached us from the ancients, although statesmen and generals, with men of letters in their train, were constantly passing through Helvetia into Gaul. All these travellers think only of complaining of the badness of the roads; the romantic character of the scenery never seems to have engaged their attention. It is even known that Julius Cæsar, when returning to his legions in Gaul, employed his time, while passing over the Alps, in preparing a grammatical treatise "De Analogia" (40). Silius Italicus, who died under Trajan, when Switzerland was already in great measure cultivated, describes the district of the Alps merely as an awful and barren wilderness (41); although he elsewhere loves to dwell in verse on the rocky ravines of Italy, and the wood-fringed banks of the Liris (Garigliano) (42). It is deserving of notice that the remarkable appearance of groups of jointed basaltic columns, such as are seen in several parts of the interior of France, on the banks of the Rhine, and in Lombardy, never engaged the attention of the Romans sufficiently to lead their writers to describe or even to mention them.

At the period when the feelings which had animated classical antiquity, and had directed the minds of men to the active manifestation of human power, almost to the exclusion of the passive contemplation of the natural world, were expiring, a new influence, and new modes of thought, were gaining sway. Christianity gradually diffused itself; and, as where it was received as the religion of the state, its beneficent action on the lower classes of the people favoured the general cause of civil freedom, so also did it render man's contemplation of nature more enlarged and free. The forms of the Olympic gods no longer fixed the eyes of men: the fathers of the church proclaimed, in their æsthetically correct, and often poetically imaginative language, that the Creator shews himself great no less in inanimate than in living nature; in the wild strife of the elements as well as in the silent progress of organic development. But during the gradual dissolution of the Roman Empire, vigour of imagination, and simplicity and purity of diction, declined more and more, first in the Latin countries, and afterwards in the Greek or eastern portion of the empire. A predilection for solitude, for saddened meditation, and for an internal absorption of mind, seems to have influenced simultaneously both the language itself and the colouring of the style.

Where a new element appears to develop itself suddenly and generally in the feelings of men, we may almost always trace earlier indications of a deep-seated germ existing previously in detached and solitary instances. The softness of Mimnermus (43) has often been called a sentimental direction of the mind. The ancient world is not abruptly separated from the modern; but changes in the religious sentiments and apprehensions of men, in their tenderest moral feelings, and in the particular mode of life of those who influence the ideas of the masses, gave a sudden predominance to that which previously escaped notice.

The tendency of the Christian mind was to shew the

greatness and goodness of the Creator from the order of the universe and the beauty of nature; and this desire to glorify the Deity through his works, favoured a disposition for natural descriptions. We find the earliest and most detailed instances of this kind in the writings of Minucius Felix, a rhetorician and advocate living in Rome in the beginning of the third century, and a contemporary of Tertullian and Philostratus. We follow him with pleasure in the evening twilight to the sea shore near Ostia, which, indeed, he describes as more picturesque, and more favourable to health, than we now find it. The religious discourse entitled "Octavius" is a spirited defence of the new faith against the attacks of a heathen friend (44).

This is the place for introducing from the Greek fathers of the church extracts descriptive of natural scenes, which are probably less known to my readers than are the evidences of the ancient Italian love for a rural life contained in Roman literature. I will begin with a letter of the great Basil, which has long been an especial favourite with me. Basil, who was a native of Cesarea in Cappadocia, left the pleasures of Athens when little more than thirty years of age, and, having already visited the Christian hermitages of Cœlo-Syria and Upper Egypt, withdrew, like the Essenes and Therapeuti before Christianity, into a wilderness adjacent to the Armenian river Iris. His second brother, Naucratius (45), had been drowned there while engaged in fishing, after leading for five years the life of a rigid anchorite. Basil writes to his friend Gregory of Nazianzum, "I believe I have at last found the end of my wanderings: my hopes of uniting myself with thee—my pleasing dreams, I should rather say, for the hopes of men have been justly called waking dreams,—have remained unfulfilled. God has caused me to find a place such as has often hovered before the fancy of us both; and that which imagination shewed us afar off, I now see present before me. A high mountain, clothed with thick forest, is watered towards the north by fresh and ever flowing streams; and at the foot of the mountain extends a wide plain, which these streams render fruitful. The surrounding forest, in which grow many kinds of trees, shuts me in as in a strong fortress. This wilderness is bounded by two deep ravines; on one side the river, precipitating itself foaming from the mountain, forms an obstacle difficult to overcome; and the other side is enclosed by a broad range of hills. My hut is so placed on the summit of the mountain, that I overlook the extensive plain, and the whole course of the Iris, which is both more beautiful, and more abundant in its waters, than the Strymon near Amphipolis. The river of my wilderness, which is more rapid than any which I have ever seen, breaks against the jutting precipice, and throws itself foaming into the deep pool below—to the mountain traveller an object on which he gazes with delight and admiration, and valuable to the native for the many fish which it affords. Shall I describe to thee the fertilising vapours rising from the moist earth, and the cool breezes from the broken water? shall I speak of the lovely song of the birds, and of the profusion of flowers? What charms me most of all is the undisturbed tranquillity of the district: it is only visited occasionally by hunters; for my wilderness feeds deer and herds of wild goats, not your bears and your wolves. How should I exchange any other place for this! Alcmæon, when he had found the Echinades, would not wander farther" (46). In this simple description of the landscape and of the life of the forest, there speak feelings more intimately allied to those of modern times than any thing that Greek and Roman antiquity have bequeathed to us. From the lonely mountain hut to which Basilius had retired, the eye looks down on the humid roof of foliage of the forest beneath; the resting-place for which he and his friend Gregory of Nazianzum (47) have so long panted is at last found. The sportive allusion at the close to the poetic mythus of Alcmæon sounds like a distant lingering echo, repeating in the Christian world accents belonging to that which had preceded it.

Basil's Homilies on the Hexæmeron also bear witness to his love of nature. He describes the mildness of the constantly serene nights of Asia Minor, where, according to his expression, the stars, "those eternal flowers of heaven," raise the spirit of man from the visible to the Invisible (48). When, in speaking of the creation of the world, he desires to praise the beauty of the sea, he describes the aspect of the boundless plain of waters in its different and varying conditions—"how, when gently agitated by mildly-breathing airs, it gives back the varied hues of heaven, now in white, now in blue, and now in roseate light; and caresses the shore in peaceful play!"

We find in Gregory of Nyssa, the brother of Basil, the same delight in nature, the same sentimental and partly melancholy vein. "When," he exclaims, "I behold each craggy hill, each valley, and each plain clothed with fresh-springing grass; the varied foliage with which the trees are adorned; at my feet the lilies to which nature has given a double dower, of sweet fragrance, and of beauty of colour;

and in the distance the sea, towards which the wandering cloud is sailing,—my mind is possessed with a sadness which is not devoid of enjoyment. When, in autumn, the fruits disappear, the leaves fall, and the branches of the trees, stripped of their ornaments, hang lifeless, in viewing this perpetual and regularly recurring alternation the mind becomes absorbed in the contemplation, and rapt as it were in unison with the many-voiced chorus of the wondrous forces of nature. Whoso gazes through these with the inward eye of the soul feels the littleness of man in the greatness of the universe" ([49]).

While the early Christian Greeks were thus led, by glorifying God in a loving contemplation of nature, to poetic descriptions of her various beauty, they were at the same time full of contempt for all works of human art. We find in Chrysostom many such passages as these: "when thou lookest on the glittering buildings, if the ranges of columns would seduce thy heart, turn quickly to contemplate the vault of heaven and the open fields, with the flocks grazing by the water's side. Who but despises all that art can shew whilst he gazes at early morn, and, in the silence of the heart, on the rising sun pouring his golden light upon the earth; or when seated by the side of a fountain in the cool grass, or in the dark shade of thick foliage, his eye feeds the while on the wide-extended prospect far vanishing in the distance" ([50]). Antioch was at this period surrounded by hermitages, in one of which Chrysostom dwelt: it might have seemed that eloquence had found again her element, freedom, on returning to the bosom of nature in the then forest-covered mountain districts of Syria and Asia Minor.

But when, during the subsequent period, so hostile to all

intellectual cultivation, Christianity spread among the Germanic and Celtic races, who had previously been devoted to the worship of nature, and who honoured under rude symbols its preserving and destroying powers, the close and affectionate intercourse with the external world of phænomena which we have remarked among the early Christians of Greece and Italy, as well as all endeavours to trace the action of natural forces, fell gradually under suspicion, as tending towards sorcery. They were therefore regarded as not less dangerous than the art of the sculptor had appeared to Tertullian, Clemens of Alexandria, and almost all the most ancient fathers of the church. In the twelfth and thirteenth centuries, the Councils of Tours (1163) and of Paris (1209) forbade to monks the sinful reading of writings on physical science ([51]). These intellectual fetters were first broken by the courage of Albert the Great and Roger Bacon; when nature was pronounced pure, and reinstated in her ancient rights.

Hitherto we have sought to depict differences which have shewn themselves in different periods of time; and in two literatures so nearly allied as were those of the Greeks and the Romans. But not only are great differences in modes of feeling produced by time,—by the changes which it brings with it, in forms of government, in manners, and in religious views,—but diversities still more striking are produced by differences of race and of mental disposition. How different in animation and in poetic colouring are the manifestations of the love of nature and the descriptions of natural scenery among the Greeks, the Germans of the north, the Semitic races, the Persians, and the Indians!

An opinion has been repeatedly expressed, that the delight in nature felt by northern nations, and the longing desire for the pleasant fields of Italy and Greece, and for the wonderful luxuriance of tropical vegetation, are principally to be ascribed to the long winter's privation of all such enjoyments. We do not mean to deny that the longing for the climate of palms seems to diminish as we approach the South of France and the Iberian Peninsula; but the now generally employed, and ethnologically correct name of Indo-Germanic races, might alone be sufficient to remind us that we must be cautious lest we generalise too much respecting the influence thus ascribed to northern winters. The richness of the poetic literature of the Indians teaches us, that within and near the tropics south of the great chain of the Himalaya, the sight of ever verdant and ever flowering forests has at all times acted as a powerful stimulus to the poetic and imaginative faculties of the East-Arianic nations, and that these nations have been more strongly inclined to picturesque descriptions of nature than the true Germanic races, who, in the far inhospitable north, had extended even into Iceland. A deprivation, or, at least, a certain interruption of the enjoyment of nature, is not, however, unknown even to the happier climates of Southern Asia: the seasons are there abruptly divided from each other by alternate periods of fertilising rain and of dusty desolating aridity. In the Persian plateau of West Aria, the desert often extends in deep bays far into the interior of the most smiling and fruitful lands. In Middle and in Western Asia, a margin of forest often forms as it were the shore of a widely extended inland sea of steppe; and thus the inhabitants of these hot countries have presented to them the strongest contrasts of desert barrenness and luxuriant vegetation, in the same horizontal plane, as well as in the vertical elevation of the snow-capped mountain chains of India and of Afghanistan. Wherever a lively tendency to the contemplation of nature is interwoven with the whole intellectual cultivation, and with the religious feelings of a nation, great and striking contrasts of season, of vegetation, or of elevation, are unfailing stimulants to the poetic imagination.

Delight in nature, inseparable from the tendency to objective contemplation which belongs to the Germanic nations, shews itself in a high degree in the earliest poetry of the middle ages. Of this the chivalric poems of the Minnesingers during the Hohenstauffen period afford us numerous examples. Many and varied as are its points of contact with the romanesque poetry of the Provençals, yet its true Germanic principle can never be mistaken. A deep felt and all pervading love of nature may be discerned in all Germanic manners, habits, and modes of life; and even in the love of freedom characteristic of the race (52). The wandering Minnesingers, or minstrels, though living much in courtly circles (from which, indeed, they often sprang), still maintained frequent and intimate intercourse with nature, and preserved, in all its freshness, an idyllic, and often an elegiac, turn of thought. I avail myself on these subjects of the researches of those most profoundly versed in the history and literature of our German middle ages, my noble-minded friends Jacob and Wilhelm Grimm. "The poets of our country of that period," says the last named writer, "never gave separate descriptions of nature, their object being solely to represent, in brilliant colours, the impression of the landscape on the mind. Assuredly the eye and the feeling for nature were not wanting in these old German masters; but the only expressions thereof which they have left us are such as flowed forth in lyrical strains, in connection with the occurrences or the feelings belonging to the narrative. To begin with the best and oldest monuments of the popular epos, we do not find any description of scenery either in the Niebelungen or in Gudrun ([53]), even where the occasion might lead us to look for it. In the otherwise circumstantial description of the chase during which Siegfried is murdered, the only natural features mentioned are the blooming heather and the cool fountain under the linden tree. In Gudrun, which shews something of a higher polish, a finer eye for nature seems also discernible. When the king's daughter, with her companions, reduced to slavery, and compelled to perform menial offices, carry the garments of their cruel lord to the sea-shore, the time is indicated as being the season 'when winter is just dissolving, and the birds begin to be heard, vying with each other in their songs; snow and rain still fall, and the hair of the captive maidens is blown by the rude winds of March. When Gudrun, hoping for the approach of her deliverer, leaves her couch, the morning star rises over the sea, which begins to glisten in the early dawn, and she distinguishes the dark helmets and the shields of her friends.' The words are few, but they convey to the fancy a visible picture, suited to heighten the feeling of expectation and suspense previous to the occurrence of an important event in the narrative. In like manner, when Homer paints the island of the Cyclops and the gardens of Alcinous, his purpose is to bring before our eyes the luxuriant fertility and abundance of the wild dwelling-place of the giant monsters, and the magnificent residence of a powerful king. In neither poet is the description of nature a primary or independent object."

"Opposed to these simple popular epics, are the more varied and artificial narrations of the chivalrous poets of the thirteenth century; among whom, Hartmann von Aue, Wolfram von Eschenbach, and Gottfried von Strasburg (54), in the early part of the century, are so much distinguished above the rest, that they may be called great and classical. It would be easy to bring together from their extensive writings sufficient proof of their deep feeling for nature, as it breaks forth in similitudes; but distinct and independent descriptions of natural scenes are never found in their pages; they never arrest the progress of the action to contemplate the tranquil life of nature. How different is this from the writers of modern poetic compositions! Bernardin de St.-Pierre uses the occurrences of his narratives only as frames for his pictures. The lyric poets of the thirteenth century, when singing of love, (which is not, however, their constant theme), speak, indeed, often of 'gentle May,' of the 'song of the nightingale,' and 'the dew glistening on the bells of heather,' but always in connection with sentiments springing from other sources, which these outward images serve to reflect. Thus, when feelings of sadness are to be indicated, mention is made of fading leaves, birds whose songs are mute, and the fruits of the field buried in snow. The same thoughts recur incessantly, not indeed without considerable variety as well as beauty in the manner in which they are expressed. Walther von der Vogelweide, and Wolfram von Eschenbach, the former characterised by tenderness and the latter by deep thought, have left us some lyric pieces, unfortunately only few in number, which are deserving of honourable mention."

"If it be asked whether contact with Southern Italy, and, by means of the crusades, with Asia Minor, Syria, and Palestine, did not enrich poetic art in Germany with new imagery drawn from the aspect of nature in more sunny climes, the question must, on the whole, be answered in the negative. We do not find that acquaintance with the East changed the direction of the minstrel poetry of the period: the crusaders had little familiar communication with the Saracens, and there was much of repulsion even between the warriors of different nations associated for a common cause. Friedrich von Hausen, who perished in Barbarossa's army, was one of the earliest German lyrical poets. His songs often relate to the crusades, but only to express religious feelings, or the pains of absence from a beloved object. Neither he nor any of the writers who had taken part in the expeditions to Palestine, as Reinmar the Elder, Rubin, Neidhart, and Ulrich of Lichtenstein, ever take occasion to speak of the country in which they were sojourning. Reinmar came to Syria as a pilgrim, it would appear, in the train of Duke Leopold VI. of Austria: he complains that the thoughts of home leave him no peace, and draw him away from God. The date-tree is occasionally mentioned, in speaking of the palms which pious pilgrims should bear on their shoulders. Neither do I remember any indication of the loveliness of Italian nature having stimulated the imagination of those minstrels who crossed the Alps. Walther von der Vogelweide, though he had wandered far, had in Italy seen only the Po; but Freidank ([55]) was in Rome, and he merely remarks that 'grass now grows in the palaces of those who once ruled there.'"

The German Thier-epos, which must not be confounded with the oriental "fable," originated in habitual association and familiarity with the animal world; to paint which was not, however, its purpose. This peculiar class of poem, which Jacob Grimm has treated in so masterly a manner, in the introduction to his edition of Reinhart Fuchs, shews a cordial delight in nature. The animals, not attached to the ground, excited by passion, and gifted by the poet with speech, contrast with the still life of the silent plants, and form a constantly active element enlivening the landscape. "The early poetry loves to look on the life of nature with human eyes, and lends to animals, and even to plants, human thoughts and feelings; giving a fanciful and childlike interpretation to all that has been observed of their forms and habits. Plants and flowers, gathered and used by gods and heroes, are afterwards named from them. In reading the old German epic, in which brutes are the actors, we breathe an air redolent as it were with the sylvan odours of some ancient forest" (56).

Formerly we might have been tempted to number among the memorials of Germanic poetry having reference to external nature, the supposed remains of the Celto-Irish poems, which, for half a century, passed as shapes of mist from nation to nation, under the name of Ossian; but the spell has been broken since the complete discovery of the literary fraud of the talented Macpherson, by his publication of the supposed Gaelic original text, now known to have been a retranslation from the English work. There are, indeed, ancient Irish Fingalian songs belonging to the times of Christianity, and perhaps not even reaching as far back as the eighth century; but these popular songs contain little of the sentimental description of nature which gives a particular charm to Macpherson's poems (57).

We have already remarked, that if sentimental and romantic turns of thought and feeling in reference to nature belong in a high degree to the Indo-Germanic races of Northern Europe, it should not be regarded only as a consequence of climate; that is, as arising from a longing desire enhanced by protracted privation. I have noticed, that the literatures of India and of Persia, which have unfolded under the glowing brightness of southern skies, offer descriptions full of charm, not only of organic, but also of inorganic nature; of the transition from parching drought to tropical rain; of the appearance of the first cloud on the deep azure of the pure sky, and the first rustling sound of the long desired etesian winds in the feathered foliage of the summits of the palms.

It is now time to enter somewhat more deeply into the subject of the Indian descriptions of nature. "Let us imagine," says Lassen, in his excellent work on Indian antiquity (58), "a portion of the Arianic race migrating from their primitive seats, in the north-west, to India: they would there find themselves surrounded by scenery altogether new, and by vegetation of a striking and luxuriant character. The mildness of the climate, the fertility of the soil, the profusion of rich gifts which it lavishes almost spontaneously, would all tend to impart to the new life of the immigrants a bright and cheerful colouring. The originally fine organisation of this race, and their high endowments of intellect and disposition, the germ of all that the nations of India have achieved of great or noble, early rendered the spectacle of the external world productive of a profound meditation on the forces of nature, which is the groundwork of that contemplative tendency which we find intimately interwoven with the earliest Indian poetry. This prevailing impression on the mental disposition of the people, has embodied itself most distinctly in their fundamental religious tenets, in the recognition of the divine in nature. The careless ease of outward life likewise favoured the indulgence of the contemplative tendency. Who could have less to disturb their meditations on earthly life, the condition of man after death, and on the divine essence, than the Indian anchorites, the Brahmins dwelling in the forest (59), whose ancient schools constituted one of the most peculiar phænomena of Indian life, and materially influenced the mental development of the whole race?"

In referring now, as I did in my public lectures under the guidance of my brother and of others conversant with Sanscrit literature, to particular instances of the vivid sense of natural beauty which frequently breaks forth in the descriptive portions of Indian poetry, I begin with the Vedas, or sacred writings, which are the earliest monuments of the civilisation of the East Arianic nations, and are principally occupied with the adoring veneration of nature. The hymns of the Rig-Veda contain beautiful descriptions of the blush of early dawn, and the appearance of the "golden-handed" sun. The great heroic poems of Ramayana and Mahabharata are later than the Vedas, and earlier than the Puranas; and in them the praises of nature are connected with a narrative, agreeably to the essential character of epic poetry. In the Vedas, it is seldom possible to assign the particular locality whence the sacred sages derive their inspiration; in the heroic poems, on the contrary, the descriptions are mostly individual, and attached to particular localities, and are animated by that fresher life which is found where the writer has drawn from impressions of which he was himself the recipient. Rama's journey from Ayodhya to the capital of Dschanaka, his sojourn in the primeval forest, and the picture of the hermit life of the Panduides, are all richly coloured.

The name of the great poet Kalidasa, who flourished at the highly polished court of Vikramaditya, contemporaneously with Virgil and Horace, has obtained an early and extensive celebrity among the nations of the west: nearer our own times, the English and German translations of Sacontala have further contributed, in a high degree, to the admiration so largely felt for an author, whose tenderness of feeling, and rich creative imagination, claim for him a distinguished place among the poets of all countries ([60]). The charm of his descriptions of nature is seen also in the lovely drama of "Vikrama and Urvasi," in which the king wanders through the thickets of the forest in search of the nymph Urvasi; in the poem of "The Seasons;" and in "The Meghaduta," or "Cloud Messenger." The last named poem paints, with admirable truth to nature, the joyful welcome which, after a long continuance of tropical drought, hails the first appearance of the rising cloud, which shews that the looked-for season of rains is at hand. The expression, "truth to nature," which I have just employed, can alone justify me in venturing to recal, in connection with the Indian poem, a sketch of the commencement of the rainy season (61) traced by myself, in South America, at a time when I was wholly unacquainted with Kalidasa's Meghaduta, even in Chézy's translation. The obscure meteorological processes which take place in the atmosphere, in the formation of vapour, in the shape of the clouds, and in the luminous electric phænomena, are the same in the tropical regions of both continents; and idealising art, whose province it is to form the actual into the ideal image, will surely lose none of its magic power by the discovery that the analysing spirit of observation of a later age confirms the truth to nature of the older, purely graphical and poetical representation.

We pass from the East Arians, or the Brahminic Indians, and their strongly marked sense of picturesque beauty in nature (62), to the West Arians, or Persians, who had migrated into the northern country of the Zend, and were originally disposed to combine with the dualistic belief in Ormuzd and Ahrimanes a spiritualised veneration of nature. What we term Persian literature does not reach farther back than the period of the Sassanides; the older poetic memorials have perished; and it was not until the country had been subjugated by the Arabs, and the characteristics of its earlier inhabitants in great measure obliterated, that it regained a national literature, under the Samanides, Gaznevides, and Seldschuki. The flourishing period of its poetry, from Firdusi to Hafiz and Dschami, can hardly be said to have lasted four or five centuries, and extends but little beyond the epoch of Vasco de Gama. The literatures of Persia and of India are separated by time as well as by space; the Persian belonging to the middle ages, while the great literature of India belongs strictly to antiquity. In the Iraunian highlands, nature does not present the luxuriance of arborescent vegetation, or the admirable variety of form and colour, which adorn the soil of Hindostan. The Vindhya chain, which was long the boundary of the East Arianic nations, is still within the torrid zone, while the whole of Persia is situated beyond the tropics, and its poetic literature even belongs in part to the northern soil of Balkh and Fergana. The four paradises celebrated by the Persian poets (63), were the pleasant valley of Soghd near Samarcand, Maschanrud near Hamadan, Tcha'abi Bowan near Kal'eh Sofid in Fars, and Ghute the plain of Damascus. Both Iran and Turan are wanting in the sylvan scenery and the hermit life of the forest which influenced so powerfully the imaginations of the Indian poets. Gardens refreshed by springing fountains, and filled with rose bushes and fruit trees, could ill replace the wild and grand scenery of Hindostan. No wonder, therefore, that the descriptive poetry of Persia has less life and freshness, and is even often tame, and full of artificial ornament. Since, in the judgment of the Persians, the highest meed of praise is given to that which we term sprightliness and wit, our admiration must be limited to the productiveness of their poets, and to the infinite variety of forms (64) which the same materials assume under their hands: we miss in them depth and earnestness of feeling.

In the national epic of Persia, Firdusi's Shahnameh, the course of the narrative is but rarely interrupted by descriptions of landscape. The praises of the coast land of Mazanderan, put into the mouth of a wandering bard, and describing the mildness of its climate, and the vigour of its vegetation, appear to me to have much grace and charm, and a high degree of local truth. In the story, the king

(Kei Kawus) is induced by the description to undertake an expedition to the Caspian, and to attempt a new conquest (65). Enweri, Dschelaleddin Rumi (who is considered the greatest mystic poet of the East), Adhad, and the half Indian Feisi, have written poems on spring, parts of which breathe poetic life and freshness, although in other parts our enjoyment is often unpleasingly disturbed by petty efforts in plays on words and artificial comparisons (66). Joseph von Hammer, in his great work on the history of Persian poetry, remarks of Sadi, in the Bostan and Gulistan (Fruit and Rose Gardens), and of Hafiz, whose joyous philosophy of life has been compared with that of Horace, that we find in the first an ethical teacher, and in the love songs of the second, lyrical flights of no mean beauty; but that in both the descriptions of nature are too often marred and disfigured by turgidity and false ornament (67). The favourite subject of Persian poetry, the loves of the nightingale and the rose, is wearisome, from its perpetual recurrence; and the genuine love of nature is stifled in the East under the conventional prettinesses of the language of flowers.

When we proceed northwards from the Iraunian highlands through Turan (in the Zend Tuirja) (68), into the chain of the Ural which forms the boundary between Europe and Asia, we find ourselves in the early seat of the Finnish races; for the Ural is as deserving of the title of the ancient land of the Fins as the Altai is of that of the Turks. Among the Fins who have settled far to the west in European lowlands, Elias Lönnrot has collected, from the lips of the Karelians and the country people of Olonetz, a great number of Finnish songs, in which Jacob Grimm (69) finds, in regard to nature, a tone of emotion and of reverie rarely met with except in Indian poetry. An old epic of nearly three thousand lines, which is occupied with the wars between the Fins and the Lapps, and the fortunes and fate of a godlike hero named Vaino, contains a pleasing description of the rural life of the Fins; especially where the wife of the ironworker, Ilmarine, sends her flocks into the forest, with prayers for their safeguard. Few races present more remarkable gradations in the character of their minds and the direction of their feelings, as determined by servitude, by wild and warlike habits, or by persevering efforts for political freedom, than the race of Fins, with its subdivisions speaking kindred languages. I allude to the now peaceful rural population among whom the epic just mentioned was discovered,—to the Huns, (long confounded with the Monguls,) who overrun the Roman world,—and to a great and noble people, the Magyars.

We have seen that the vividness of the feeling with which nature is regarded, and the form in which that feeling manifests itself, are influenced by differences of race, by the particular character of the country, by the constitution of the state, and by the tone of religious feeling; and we have traced this influence in the nations of Europe, and in those of kindred descent in Asia (the Indians and Persians) of Arianic or Indo-Germanic origin. Passing from thence to the Semitic or Aramean race, we discover in the oldest and most venerable memorials in which the tone and tendency of their poetry and imagination are displayed, unquestionable evidences of a profound sensibility to nature.

This feeling manifests itself with grandeur and animation in pastoral narratives, in hymns and choral songs, in the splendour of lyric poetry in the Psalms, and in the schools of the prophets and seers, whose high inspiration, almost estranged from the past, is wrapped in futurity.

Besides its own inherent greatness and sublimity, Hebrew poetry presents to Jews, to Christians, and even to Mahometans, local reminiscences more or less closely entwined with religious feelings. Through missions, favoured by the spirit of commerce, and the territorial acquisitions of maritime nations, names and descriptions belonging to oriental localities, preserved to us in the writings of the Old Testament, have penetrated far into the recesses of the forests of the new continent, and into the islands of the Pacific.

It is characteristic of Hebrew poetry in reference to nature, that, as a reflex of monotheism, it always embraces the whole world in its unity, comprehending the life of the terrestrial globe as well as the shining regions of space. It dwells less on details of phænomena, and loves to contemplate great masses. Nature is pourtrayed, not as self-subsisting, or glorious in her own beauty, but ever in relation to a higher, an over-ruling, a spiritual power. The Hebrew bard ever sees in her the living expression of the omnipresence of God in the works of the visible creation. Thus, the lyrical poetry of the Hebrews in its descriptions of nature is essentially, in its very subject, grand and solemn, and, when touching on the earthly condition of man, full of a yearning pensiveness. It is deserving of notice, that notwithstanding its grand character, and even in its highest lyrical flights elevated by the charm of music, the Hebrew poetry, unlike that of the Hindoos, scarcely ever appears unrestrained by law and measure. Devoted to the pure contemplation of the Divinity, figurative in language, but clear and simple in thought, it delights in comparisons, which recur continually and almost rhythmically.

As descriptions of natural scenery, the writings of the Old Testament shew as in a mirror the nature of the country in which the people of Israel moved and dwelt, with its alternations of desert, fruitful land, forest, and mountain. They pourtray the variations of the climate of Palestine, the succession of the seasons, the pastoral manners of the people, and their innate disinclination to agriculture. The epic, or historical and narrative, portions are of the utmost simplicity, almost more unadorned even than Herodotus; and from the small alteration which has taken place in the manners, and in the usages and circumstances of a nomade life, modern travellers have been enabled to testify unanimously to their truth to nature. The Hebrew lyrical poetry is more adorned, and unfolds rich and animated views of the life of nature. A single psalm, the 104th, may be said to present a picture of the entire Cosmos:—"The Lord covereth himself with light as with a garment, He hath stretched out the heavens like a canopy. He laid the foundations of the round earth that it should not be removed for ever. The waters springing in the mountains descend to the valleys, unto the places which the Lord hath appointed for them, that they may never pass the bounds which He has set them, but may give drink to every beast of the field. Beside them the birds of the air sing among the branches. The trees of the Lord are full of sap, the cedars of Lebanon which He hath planted, wherein the birds make their nests, and the fir trees wherein the stork builds her house." The great and wide sea is also described, "wherein are living things innumerable; there move the ships, and there is that leviathan whom Thou hast made to sport therein." The fruits of the field, the objects of the labour of man, are also introduced; the corn, the cheerful vine, and the olive garden. The heavenly bodies complete this picture of nature. "The Lord appointed the moon for seasons, and the sun knoweth the term of his course. He bringeth darkness, and it is night, wherein the wild beasts roam. The young lions roar after their prey, and seek their meat from God. The sun ariseth and they get them away together, and lay them down in their dens:" and then "man goeth forth unto his work and to his labour until the evening." We are astonished to see, within the compass of a poem of such small dimension, the universe, the heavens and the earth, thus drawn with a few grand strokes. The moving life of the elements is here placed in opposition to the quiet laborious life of man, from the rising of the sun, to the evening when his daily work is done. This contrast, the generality in the conception of the mutual influence of phænomena, the glance reverting to the omnipresent invisible power, which can renew the face of the earth, or, cause the creature to return again to the dust, give to the whole a character of solemnity and sublimity rather than of warmth and softness.

Similar views of the Cosmos present themselves to us repeatedly in the Psalms (70), (as in the 65th, v. 7—14, and in the 74th, 15—17), and with perhaps most fulness in the ancient, if not premosaic, book of Job. The meteorological processes taking place in the canopy of the clouds, the formation and dissolution of vapour as the wind changes its direction, the play of colours, the production of hail, and the rolling thunder, are described with the most graphic individuality; many questions are also proposed, which our modern physical science enables us indeed to propound more formally, and to clothe in more scientific language, but not to solve satisfactorily. The book of Job is generally regarded as the most perfect example of Hebrew poetry; it is no less picturesque in the presentation of single phænomena than skilful in the didactic arrangement of the whole. In all the various modern languages into which this book has been translated, its imagery, drawn from eastern nature, leaves on the mind a deep impression. "The Lord walks on the heights of the sea, on the ridges of the towering waves heaped up by the storm" (chap. xxxviii. v.16). "The morning dawn takes hold of the ends of the earth, and moulds variously the canopy of clouds, as the hand of man moulds the ductile clay" (chap xxxviii. v. 13—14.) The habits of animals are depicted, of the wild ass and the horse, the buffalo, the river horse of the Nile, the crocodile, the eagle, and the ostrich. We see (chap. xxxvii. v. 18) during the sultry heat of the south wind, "the pure ether spread over the thirsty desert like a molten mirror ([71])." Where the gifts of nature are sparingly bestowed, man's perceptions are rendered more acute, so that he watches every variation in the atmosphere around him and in the clouds above him; and in the desert, as on the billows of the ocean, traces back every change to the signs which foretold it. The climate of the arid and rocky portions of Palestine is particularly suited to give birth to such observations.

Neither is variety of form wanting in the poetic literature of the Hebrews: while from Joshua to Samuel it breathes a warlike tone, the little book of Ruth presents a natural picture of the most naïve simplicity, and of an inexpressible charm. Goethe, at the period of his enthusiasm for the East, said of it, that we have nothing so lovely in the whole range of epic and idyllic poetry. (72)

Even in later times, in the earliest memorials of the literature of the Arabians, we discover a faint reflex of that grandeur of view in the contemplation of nature, which so early distinguished the Semitic race: I allude to the picturesque description of the Bedouin life of the deserts, which the grammarian Asmai has connected with the great name of Antar, and has woven (together with other pre-mohamedan legends of knightly deeds), into a considerable work. The hero of this romantic tale is the same Antar of the tribe of Abs, son of the princely chief Sheddad and of a black slave, whose verses are preserved among the prize poems, (moallakät), which are hung up in the Kaaba. The learned English translator, Terrick Hamilton, has called attention to the biblical tones in the style of Antar. (73). Asmai makes the son of the desert travel to Constantinople, and thus introduces a picturesque contrast of Greek culture with nomadic simplicity. We should be less surprised at finding that natural descriptions of the surface of the Earth occupy only a very small space in the earliest Arabian poetry, since, according to the remark of an accomplished Arabic scholar, my friend Freytag of Bonn, narratives of deeds of arms, and praises of hospitality and of fidelity in love, are its principal themes, and since scarcely any, if any, of its writers were natives of Arabia Felix. The dreary uniformity of sandy deserts or grassy plains is ill fitted to awaken the love of nature, excepting in rare instances and in minds of a peculiar cast.

Where the earth is unadorned by forests, the imagination, as we have already remarked, is the more occupied by the atmospheric phenomena of storm, tempest, and long desired rain. Among faithful natural pictures of this class, I would instance particularly Antar's Moallakat, which describes the pasture fertilised by rain, and visited by swarms of humming insects (74); the fine descriptions of storms, both by Amru'l Kais, and in the 7th book of the celebrated Hamasa (75), which are also distinguished by a high degree of local truth; and lastly, the description in the Nabegha Dhobyani (76) of the swelling of the Euphrates, when its waters roll down masses of reeds and trunks of trees. The eighth book of the Hamasa, which is entitled "Travel and Sleepiness," naturally attracted my attention: I soon found that the "sleepiness" (77) belongs only to the first fragment of the book, and even there is more excusable, as it is ascribed to a night journey on a camel.

I have endeavoured in this section to unfold in a fragmentary manner the different influence which the external world, that is, the aspect of animate and inanimate nature, has exercised at different epochs, and among different races and nations, on the inward world of thought and feeling. I have tried to accomplish this object by tracing throughout the history of literature, the particular characteristics of the vivid manifestation of the feelings of men in regard to nature. In this, as throughout the whole of the work, my aim has been to give not so much a complete, as a general, view, by the selection of such examples as should best display the peculiarities of the various periods and races. I have followed the Greeks and Romans to the gradual extinction of those feelings which have given to classical antiquity in the West an imperishable lustre; I have traced in the writings of the Christian fathers of the Church, the fine expression of a love of nature nursed in the seclusion of the hermitage. In considering the Indo-Germanic nations, (the denomination being here taken in its most restricted sense), I have passed from the poetic works of the Germans in the middle ages, to those of the highly cultivated ancient East Arianic nations (the Indians); and of the less gifted West Arians, (the inhabitants of ancient Irān). After a rapid glance at the Celtic or Gaelic songs, and at a newly discovered Finnish epic, I have described the rich perception of the life of nature which, in races of Aramean or Semitic origin, breathes in the sublime poetry of the Hebrews, and in the writings of the Arabians. Thus I have traced the reflected image of the world of phænomena, as mirrored in the imagination of the nations of the north and the south-east of Europe, of the west of Asia, of the Persian plateaus, and of tropical India. In order to conceive Nature in all her grandeur, it seemed to me necessary to present her under a two-fold aspect; first objectively, as an actual phænomenon; and next as reflected in the feelings of mankind.

After the fading of Aramaic, Greek, and Roman glory—I might say after the destruction of the ancient world—we find in the great and inspired founder of a new world, Dante Alighieri, scattered passages which manifest the most profound sensibility to the aspect of external nature. The period at which he lived followed immediately that of the decline of the minstrelsy of the Suabian Minnesingers, on the north side of the Alps, of whom I have already spoken. Dante, when treating of natural objects, withdraws himself for a time from the passionate, the subjective, and the mystic elements of his wide range of ideas. Inimitably does he paint, for instance, at the close of the first canto of the Purgatorio ([78]), the sweet breath of morning, and the trembling light on the gently agitated distant mirror of the sea, (il tremolar de la marina); in the fifth canto, the bursting of the clouds and the swelling of the rivers, which, after the battle of Campaldino, caused the body of Buonconte da Montefeltro to be lost in the Arno ([79]). The entrance into the thick grove of the terrestrial paradise reminds the poet of the pine forest near Ravenna: "la pineta in sul lito di Chiassi" ([80]), where the early song of birds is heard in the tall trees. The local truth of this natural picture contrasts with the description of the river of light in the heavenly paradise, from which "sparks burst forth, sink amidst the flowers on the banks, and then, as if intoxicated by their perfumes, plunge again into the stream ([81])." It seems not impossible that this fiction may have had for its groundwork the poet's recollection of that peculiar state of the ocean, in which, during the beating of the waves, luminous points dash above the surface, and the whole liquid plain forms a moving sea of sparkling light. The extraordinary conciseness of the style of the Divina Commedia augments the depth and earnestness of the impression produced.

Lingering on Italian ground, but avoiding those frigid compositions, the pastoral romances, I would next name the sonnet in which Petrarch describes the impression which the lovely valley of Vaucluse made on him when Laura was no more; then, the smaller poems of Boiardo, the friend of Hercules of Este; and at a later period some noble stanzas by Vittoria Colonna ([82]).

When the sudden intercourse which took place with Greece in her low state of political depression caused a more general revival of classical literature, we find, as the first example among prose writers, a charming description of nature from the pen of the lover of the arts, the counsellor and friend of Raphael, Cardinal Bembo. His juvenile work, entitled Ætna Dialogus, gives us an animated picture of the geographical distribution of plants on the declivity of the mountain, from the rich corn fields of Sicily to the snow-covered margin of the crater. The finished work of his maturer years, the Historiæ Venetæ, characterises in a still more picturesque manner the climate and the vegetation of the new continent.

At that period every thing concurred to fill the mind at once with views of the suddenly enlarged boundaries both of the earth, and of the powers of man. In antiquity, the march of the Macedonian army to the Paropamisus, and to the forest-covered river-valleys of Western Asia, left impressions derived from the aspect of a richly adorned exotic nature, of which the vividness manifested itself whole centuries afterwards in the works of highly gifted writers; and now, in like manner, the western nations were acted upon a second time, and in a higher degree than by the crusades, by the discovery of America. The tropical world, with all the richness and luxuriance of its vegetation in the plain, with all the gradations of organic life on the declivities of the Cordilleras, with all the reminiscences of northern climates in the inhabited plateaus of Mexico, New Grenada, and Quito, was now first disclosed to the view of Europeans. Imagination, without which no truly great work of man can be accomplished, gave a peculiar charm to the descriptions of nature traced by Columbus and Vespucci. The description of the coast of Brazil, by the latter, is characterised by an accurate acquaintance with the poets of ancient and modern times; that given by Columbus of the mild sky of Paria, and of the abundant waters of the Orinoco, flowing as he imagines from the east of Paradise, is marked by an earnestly religious tone of mind, which afterwards, by the influence of increasing years, and of the unjust persecutions which he encountered, became touched with melancholy, and with a vein of morbid enthusiasm.

In the heroic times of the Portuguese and Castilian races, it was not the thirst of gold alone (as has been asserted, in ignorance of the national character of the period), but rather a general excitement which led so many to dare the hazards of distant voyages. In the beginning of the sixteenth century, the names of Hayti, Cubagua, and Darien, acted on the imagination of men as in more recent times, since Anson and Cook, those of Tinian and Tahiti have done. If the tidings of far distant lands then drew the youth of the Iberian peninsula, of Flanders, Milan, and Southern Germany, under the victorious banners of the great Emperor, to the ridges of the Andes and to the burning plains of Uraba and Coro;—in more modern times, under the milder influence of a later cultivation, and as the earth's surface became more generally accessible in all its parts, the restless longing for distant regions acquired fresh motives and a new direction. The passionate love for the study of nature which proceeded chiefly from the north, inflamed the minds of men; intellectual grandeur of view became associated with the enlargement of material knowledge; and the particular poetic sentimental turn belonging to the period, has embodied itself, since the close of the last century, in literary works under forms which were before unknown. If we once more cast our eyes on the period of those great discoveries which prepared the way for the modern tendency of which we have been speaking, we must in so doing refer preeminently to those descriptions of nature which have been left us by Columbus himself. It is only recently that we have obtained the knowledge of his own ship's journal, of his letters to the treasurer Sanchez, to Donna Juana de la Torre governess of the Infant Don Juan, and to Queen Isabella. In my critical examination of the history of the geography of the 15th and 16th centuries (83), I have sought to show with how deep a feeling and perception of the forms and the beauty of nature the great discoverer was endowed, and how he described the face of the earth, and the "new heaven" which opened to his view, ("viage nuevo al nuevo cielo i mundo que fasta entonces estaba en occulto"), with a beauty and simplicity of expression which can only be fully appreciated by those who are familiar with the ancient force of the language as it existed at the period.

The aspect and physiognomy of the vegetation; the impenetrable thickets of the forests, "in which one can hardly distinguish which are the flowers and leaves belonging to each stem;" the wild luxuriance which clothed the humid shores; the rose-coloured flamingoes fishing at the mouth of the rivers in the early morning, and giving animation to the landscape;—attract the attention of the old navigator while sailing along the coast of Cuba, between the small Lucayan islands and the Jardinillos, which I also have visited. Each newly discovered land appears to him still more beautiful than those he had before described; he complains that he cannot find words in which to record the sweet impressions which he has received. Wholly unacquainted with botany, (although through the influence of Jewish and Arabian physicians some superficial knowledge of plants had at that time extended into Spain), the simple love of nature leads him to discriminate truly between the many strange forms presented to his view. He already distinguished in Cuba seven or eight different kinds of palms "more beautiful and loftier than date-trees," (variedades de palmas superiores a las nuestras en su belleza y altura); he writes to his friend Anghiera, that he has seen on the same plain palms and pines, (palmeta and pineta), wonderfully grouped together; he regards the vegetation presented to his view with a glance so acute, that he was the first to observe that, on the mountains of Cibao, there are pines whose fruits are not fir cones, but berries like the olives of the Axarafe de Sevilla; and, to cite one more and very remarkable example, Columbus, as I have already noticed (84), separated the genus Podocarpus from the family of Abietineæ.

"The loveliness of this new land," says the discoverer, "far surpasses that of the campiña de Cordoba. The trees are all bright with ever-verdant foliage, and perpetually laden with fruits. The plants on the ground are tall and full of blossoms. The breezes are mild like those of April in Castille; the nightingales sing more sweetly than I can describe. At night other small birds sing sweetly, and I also hear our grasshoppers and frogs. Once I came into a deeply enclosed harbour, and saw high mountains which no human eye had seen before, from which the lovely waters (lindas aguas) streamed down. The mountain was covered with firs, pines, and other trees of very various form, and adorned with beautiful flowers. Ascending the river which poured itself into the bay, I was astonished at the cool shade, the crystal clear water, and the number of singing birds. It seemed to me as if I could never quit a spot so delightful,—as if a thousand tongues would fail to describe it,—as if the spell-bound hand would refuse to write. (Para hacer relacion a los Reyes de las cosas que vian, no bastaran mil lenguas a referillo, ni la mano para lo escribir, que le parecia questaba encantado.)" ([85])

We here learn from the journal of an unlettered seaman, the power which the beauty of nature, manifested in her individual forms, may exert on a susceptible mind. Feelings ennoble language; for the prose of the Admiral, especially when, on his fourth voyage, at the age of 67, he relates his wonderful dream on the coast of Veragua ([86]), is, if not more eloquent, yet far more moving than the allegorical pastoral romance of Boccaccio and the two Arcadias of Sannazaro and of Sydney; than Garcilasso's Salicio y Nemoroso; or than the Diana of Jorge de Montemayor. The elegiac idyllic element was unhappily too long predominant in Italian and Spanish literature; it required the fresh and living picture which Cervantes has drawn of the adventures of the Knight of La Mancha, to efface the Galatea of the same author. The pastoral romance, however ennobled in the works of these great writers by beauty of language and tenderness of feeling, is from its nature, like the allegorical artifices of the intellect of the middle ages, cold and wearisome. Individuality of observation alone leads to truth to nature; in the finest descriptive stanzas of the "Jerusalem Delivered," impressions derived from the poet's recollection of the picturesque landscape of Sorrento have been supposed to be recognised ([87].)

That truth to nature which springs from actual contemplation, shines most richly in the great national epic of Portuguese literature; it is as if a perfumed air from Indian flowers breathed throughout the whole poem, written under the sky of the tropics, in the rocky grotto near Macao and in the Moluccas. It is not for me to confirm a bold sentence of Friedrich Schlegel's, according to which the Lusiad of Camoens excels Ariosto in colouring and richness of fancy; ([88]) but as an observer of Nature, I may well add that in the descriptive portion of the Lusiad, the poet's inspiration, the ornaments of language, and the sweet tones of melancholy, never impair the accuracy of the representation of physical phænomena. Rather, as is always the case when art draws from pure sources, they heighten the living impressions of grandeur and of truth in the pictures of nature. Inimitable are the descriptions in Camoens of the never ceasing mutual relations between the air and sea, between the varying form of the clouds above, their meteorological changes, and the different states of the surface of the ocean. He shews us this surface at one time, as, when curled by gentle breezes the short waves glance sparklingly in the play of the reflected sunbeams; and at another, when the ships of Coelho' and Paul de Gama, overtaken by a dreadful tempest, sustain the conflict of the deeply agitated elements ([89]). Camoens is in the most proper sense of the term, a great sea painter. He had fought at the foot of Atlas in the empire of Morocco, in the Red Sea, and in the Persian Gulf; twice he had sailed round the Cape, and for sixteen years watched the phænomena of the ocean on the Chinese and Indian shores. He describes the electric fires of St. Elmo, (the Castor and Pollux of the ancient Greek navigators "the living light, sacred to the mariner" (90). He paints the danger-threatening water-spout in its gradual development; "how the cloud, woven of thin vapour, whirls round in a circle, and sending down a slender tube sucks up the flood as if athirst; and how, when the black cloud has drunk its fill, the foot of the cone recedes, and flying back to the sky, restores to the waves, as fresh water, the salt stream which it had drawn from them with a surging noise" (91). "Let the book-learned," says the poet—and his taunt might almost as well apply to the present time—"try to explain the wonderful things hidden from the world; they who, guided by (so-called) science and their own conceptions only, are so willing to pronounce as false, what is heard from the mouth of the sailor whose only guide is experience."

Camoens shines, however, not only in the description of single phænomena, but also where large masses are comprehended in one view. The third canto paints with a few traits the whole of Europe, from the coldest north, "to the Lusitanian kingdom, and the strait where Hercules accomplished his last labour" (92). The manners and state of civilisation of the different nations are alluded to. From the Prussians, the Muscovites, and the tribes "que o Rheno frio lava," he hastens to the glorious fields of Hellas, "que creastes os peitos eloquentes, e os juizos de alta phantasia." In the tenth canto the view becomes still more extended; Thetys conducts Gama to the summit of a lofty mountain to shew him the secrets of the structure of the universe ("machina do mundo"), and to disclose to him the courses of the planets, (according to the views of Ptolemy). (93) It is a vision in the style of Dante, and as the Earth is the centre of motion, we have in the description of the globe, a review of all the countries then known, and of their productions. (94) Even the "land of the Holy Cross," (Brazil), is named, and the coasts which Magellan discovered "by the act, but not by the loyalty of a son of Lusitania."

When I before extolled Camoens as especially a marine painter, it was to indicate that the aspect of nature on the land seems to have attracted him less vividly. Sismondi has remarked with justice, that the whole poem contains absolutely no trace of graphical description of the vegetation of the tropics, and its peculiar physiognomy and forms. He only notices the spices and other productions which have commercial value. The episode of the magic island (95) does, indeed, present a charming landscape picture, but, as befits an "Ilha de Venus," the vegetation consists of "fragrant myrtles, citrons, lemon trees, and pomegranates;" all belonging to the climates of South Europe. In the writings of the great discoverer of the new world, we find far greater delight in the forests of the coasts seen by him, and far more attention to the forms of the vegetable kingdom; but it should be remarked, that Columbus, writing the journal of his voyage, records in it the living impressions of each day. The epic of Camoens, on the other hand, is written to celebrate the great achievements of the Portuguese. To have borrowed from native languages uncouth names of plants, and to have interwoven them in the descriptions of landscapes forming the background to the actors in his narrative, might have appeared but little attractive to the poet accustomed to harmonious sounds.

By the side of the knightly form of Camoens has often been placed the equally romantic one of a Spanish warrior who served under the banners of the great Emperor in Peru and Chili, and sung in those distant regions the deeds of arms in which he had borne a distinguished part. But in the whole Epic of the Araucana of Don Alonso de Ercilla, the immediate presence of volcanoes clad with external snows, of valleys covered with tropical forests, and of arms of the sea penetrating far into the land, have scarcely called forth any description which can be termed graphical. The excessive praise which Cervantes bestows on Ercilla, on the occasion of the ingenious satirical review of Don Quixote's books, is probably to be attributed only to the vehement rivalry subsisting at that time between Spanish and Italian poetry, though it would appear to have misled Voltaire and several modern critics. The Araucana is, indeed, a work imbued with a noble national feeling; and the description which it contains of the manners of a wild race who perish in fighting for the freedom of their native land, is not without animation; but Ercilla's style is heavy, loaded to excess with proper names, and without any trace of true poetic inspiration. (96)

We recognise this essential element, however, in several strophes of the Romancero Caballeresco (97); we perceive its presence, mixed with a vein of religious melancholy, in the writings of Fray Luis de Leon,—as, for example, where he celebrates the "eternal luminaries (resplandores eternales) of the starry heaven";— (98) and we find it in the great creations of Calderon. The most profound critic of the dramatic literature of different countries, my friend Ludwig Tieck, has remarked the frequent occurrence in Calderon and his cotemporaries of lyrical strains in varied metres, often containing dazzlingly beautiful pictures of the ocean, of mountains, of wooded valleys, and of gardens; but these pictures are always introduced in allegorical applications, and are characterised by a species of artificial brilliancy. In reading them we feel that we have before us ingenious descriptions, recurring with only slight variations, and clothed in well-sounding and harmonious verse; but we do not feel that we breathe the free air of nature; the reality of the mountain scene, and the shady valley, are not made present to our imagination. In Calderon's play of "Life is a Dream," (la vida es sueno), he makes Prince Sigismund lament his captivity in a series of gracefully drawn contrasts with the freedom of all living nature. He paints the birds, "which fly across the wide sky with rapid wing," the fish, which, but just escaped from the sand and shallows where they were brought to life, seek the wide sea, whose boundless expanse seems still too small for their bold range. Even the stream meandering among flowers, finds a free path through the meadow: "and I," exclaims Sigismund despairingly, "who have more life than they, and a spirit more free, must endure an existence in which I enjoy less freedom." In a similar manner, too often disfigured by antitheses, witty comparisons, and artificial turns from the school of Gongora, Don Fernando speaks to the king of Fez in the "Steadfast Prince" (99).

I have referred to particular instances, because they show how in dramatic poetry, which is chiefly concerned with action, passion, and character, "descriptions of natural objects become as it were only mirrors in which the mental emotions of the actors in the scene are reflected. Shakspeare, who amidst the pressure of his animated action has scarcely ever time and opportunity to introduce deliberate descriptions of natural scenes, does yet so paint them by occurrences, by allusions, and by the emotions of the acting personages, that we seem to see them before our eyes, and to live in them. We thus live in the midsummer-night in the wood; and in the latter scenes of the Merchant of Venice we see the moonshine brightening the warm summer night, without direct descriptions. An actual and elaborate description of a natural scene occurs, however, in King Lear, where Edgar, who feigns himself mad, represents to his blind father, Gloucester, while on the plain, that they are mounting to the summit of Dover Cliff. The picture drawn of the downward view into the depths below actually turns one giddy" (100).

If in Shakspeare the inward life of feeling, and the grand simplicity of the language, animate thus wonderfully the individual expression of nature, and render her actually present to our imagination; in Milton's sublime poem of Paradise Lost, on the other hand, such descriptions are, from the very nature of the subject, magnificent rather than graphic. All the riches of imagination and of language are poured forth in painting the loveliness of Paradise; but the description of vegetation could not be otherwise than general and undefined. This is also the case in Thomson's pleasing didactic poem of The Seasons. Kalidasa's poem on the same subject, the Ritusanhara, which is more ancient by above seventeen centuries, is said by critics deeply versed in Indian literature to individualise more vividly the vigorous nature of the vegetation of the tropics; but it wants the charm which, in Thomson, arises from the more varied division of the seasons which is proper to the higher latitudes; the transition from fruit-bringing autumn to winter, and from winter to reanimating spring; and the pictures afforded by the varied laborious or pleasurable pursuits of men belonging to the different portions of the year.

Arriving at the period nearest to our own time, we find that, since the middle of the last century, descriptive prose has more particularly developed itself, and with peculiar vigour. Although the study of nature, enlarging on every side, has increased beyond measure the mass of things known to us, yet amongst the few who are susceptible of the higher inspiration which this knowledge is capable of affording, the intellectual contemplation of nature has not sunk oppressed under the load, but has rather gained a wider comprehensiveness and a loftier elevation, since a deeper insight has been obtained into the structure of mountain masses (those storied cemeteries of perished organic forms), and into the geographical distribution of plants and animals, and the relationship of different races of men. The first modern prose writers who have powerfully contributed to awaken, through the influence of the imagination, the keen perception of natural beauty, the delight in contact with nature, and the desire for distant travel which is their almost inseparable companion, were in France, Jean Jacques Rousseau, Buffon, Bernardin de St.-Pierre, and (to name exceptionally one living writer), my friend Auguste de Chateaubriand; in the British islands the ingenious Playfair; and in Germany, George Forster, who was the companion of Cook on his second voyage of circumnavigation, and who was gifted botlr with eloquence and with a mind peculiarly favourable to every generalisation in the view of nature.

I must not attempt in these pages to examine the characteristics of these different writers; or what it is that, in works so extensively known, sometimes lends to their descriptions of scenery such grace and charm, or at others disturbs the impressions which the authors desire to awaken; but it may be permitted to a traveller who has derived his knowledge principally from the immediate contemplation of nature, to introduce here a few detached considerations respecting a recent, and on the whole little cultivated, branch of literature.

Buffon, with much of grandeur and of gravity,—embracing simultaneously the structure of the planetary system, the world of organic life, light, and magnetism—and far more profound in his physical investigations than his cotemporaries were aware of—when he passes from the description of the habits of animals to that of the landscape, shews in his artificially-constructed periods, more rhetorical pomp than individual truth to nature; rather disposing the mind generally to the reception of exalted impressions, than taking hold of it by such visible paintings of the actual life of nature, as should render her actually present to the imagination. In perusing even his most justly celebrated efforts in this department, we are made to feel that he has never quitted middle Europe, and never actually beheld the tropical world which he engages to describe. What, however, we particularly miss in the works of this great writer, is the harmonious connection of the representation of nature with the expression of awakened emotion; we miss in him almost all that flows from the mysterious analogy between the movements of the mind and the phænomena perceived by the senses.

Greater depth of feeling, and a fresher spirit of life, breathe in Jean Jacques Rousseau, in Bernardin de St.-Pierre, and in Chateaubriand. If in the first-named writer (whose principal works were twenty years earlier than Buffon's fanciful Epoques de la Nature) ([101]) I allude to his fascinating eloquence, and to the picturesque descriptions of Clarens and La Meillerie on Lake Leman, it is because, in the most celebrated works of this ardent but little informed plant-collector, poetical inspiration shews itself principally in the inmost peculiarities of the language, breaking forth no less overflowingly, in his prose, than in Klopstock's, Schiller's, Goethe's, and Byron's imperishable verse. Even where an author has no purpose in view immediately connected with the study of nature, our love for that study may still be enhanced by the magic charm of a poetic representation of the life of nature, although in regions of the earth already familiar to us.

In referring to modern prose writers, I dwell with peculiar complacency on that small production of the creative imagination to which Bernardin de St.-Pierre owes the fairest portion of his literary fame—I mean Paul and Virginia: a work such as scarcely any other literature can shew. It is the simple but living picture of an island in the midst of the tropic seas, in which, sometimes smiled on by serene and favouring skies, sometimes threatened by the violent conflict of the elements, two young and graceful forms stand out picturesquely from the wild luxuriance of the vegetation of the forest, as from a flowery tapestry. Here, and in the Chaumière Indienne, and even in the Etudes de la Nature, (which are unhappily disfigured by extravagant theories and erroneous physical views), the aspect of the sea, the grouping of the clouds, the rustling of the breeze in the bushes of the bamboo, and the waving of the lofty palms, are painted with inimitable truth. Bernardin de St.-Pierre's master-work, Paul and Virginia, accompanied me into the zone to which it owes its origin. It was read there for many years by my dear companion and friend Bonpland and myself, and there—(let this appeal to personal feelings be forgiven)—under the silent brightness of the tropical sky, or when, in the rainy season on the shores of the Orinoco, the thunder crashed and the flashing lightning illuminated the forest, we were deeply impressed and penetrated with the wonderful truth with which this little work paints the power of nature in the tropical zone in all its peculiarity of character. A similar firm grasp of special features, without impairing the general impression or depriving the external materials of the free and animating breath of poetic imagination, characterises in an even higher degree the ingenious and tender author of Atala, René, the Martyrs, and the Journey to Greece and Palestine. The contrasted landscapes of the most varied portions of the earth's surface are brought together and made to pass before the mind's eye with wonderful distinctness of vision: the serious grandeur of historic remembrances could alone have given so much of depth and repose to the impressions of a rapid journey.

In our German fatherland, the love of external nature showed itself but too long, as in Italian and Spanish literature, under the forms of the idyl, the pastoral romance, and didactic poems: this was the course followed by the Persian traveller Paul Flemming, Brockes, Ewald von Kleist, in whom we recognise a mind full of feeling, Hagedorn, Solomon Gessner, and by one of the greatest naturalists of all times, Haller, whose local descriptions present, however, better defined outlines and more objective truth of colour.

At that time the elegiac idyllic element predominated in a heavy style of landscape poetry, in which, even in Voss, the noble and profound classical student of antiquity, the poverty of the materials could not be veiled by happy and elevated, as well as highly finished diction. It was not until the study of the earth's surface gained depth and variety, and natural science, no longer limited to tabular enumerations of extraordinary occurrences and productions, rose to the great views of comparative geography, that this finish of language could become available in aiding to impart life and freshness to the pictures of distant zones.

The older travellers of the middle ages, such as John Mandeville (1353), Hans Schiltberger of Munich (1425), and Bernhard von Breytenbach (1486), still delight us by an amiable naïveté, by the freedom with which they write, and the apparent feeling of security with which they come before a public who, being wholly unprepared, listen with the greater curiosity and readiness of belief, because they have not yet learnt to feel ashamed of being amused or even astonished. The interest of books of travels was at that period almost wholly dramatic; and the indispensable mixture of the marvellous which they so easily and naturally acquired, gave them also somewhat of an epic colouring. The manners of the inhabitants of the different countries are not so much described, as shewn incidentally in the contact between the travellers and the natives. The vegetation is unnamed and unheeded, excepting where a fruit of particularly pleasant flavour or curious form, or a stem or leaves of extraordinary dimensions, induce a special notice. Amongst animals, the kinds which they are most fond of remarking are, first, those which shew some resemblance to the human form, and next those which are most wild and most formidable to man. The cotemporaries of these travellers gave the fullest credence to dangers which few among them had shared; the slowness of navigation, and the absence of means of communication, caused the Indies, as all tropical countries were then called, to appear at an immeasurable distance. Columbus was as yet scarcely justified in saying, as he did in his letter to Queen Isabella, "the earth is not very large: it is much less than people imagine" (102).

In respect to composition, these almost-forgotten books of travels of the middle ages had, notwithstanding the poverty of their materials, great advantages over most of our modern voyages. They had the unity which every work of art requires: everything was connected with an action, *i. e.* subordinated to the journey itself. The interest arose from the simple, animated, and usually implicitly believed narrative of difficulties overcome. Christian travellers, unacquainted with the previous travels of Arabs, Spanish Jews, and proselytizing Buddhists, always supposed themselves to be the first to see and describe everything. The remoteness and even the dimensions of objects were magnified by the obscurity which seemed to veil the East and the interior of Asia. This attractive unity of composition is necessarily wanting in the greater part of modern travels, and especially in those undertaken for scientific purposes; in these, what is done yields precedence to what is observed; the action almost disappears under the multitude of observations. A true dramatic interest can now only be looked for, in arduous, though perhaps little instructive ascents of mountains, and above all adventurous navigations of untraversed seas in voyages of discovery properly so called, and in the awful solitudes of the Polar regions, where the surrounding desolation and the lonely situation of the mariners, cut off from all human aid, isolate the picture, and cause it to act more stirringly on the imagination of the reader. If the above considerations render it undeniably evident that in modern books of travels the active element necessarily falls into the background, affording for the most part merely a connecting thread whereby the successive observations of nature or of manners are linked together, yet ample compensation may be derived from the treasures of observation, from grand views of the universe, and from the laudable endeavour in each writer to avail himself of the peculiar advantages which his native language may possess for clear and animated description. The benefits for which we are indebted to modern cultivation are the constantly advancing enlargement of our field of view, the increasing wealth in ideas and feelings, and their active mutual influence. Without leaving our native soil, we may now not only be informed what is the character and form of the earth's crust in the most distant zones, and what are the plants and animals which enliven its surface, but we may also expect to be presented with such pictures as may produce in ourselves a vivid participation in a portion at least of those impressions which in each zone man receives from external nature. To satisfy these demands,—this requirement of a species of intellectual delight unknown to the ancient world,—is one of the efforts of modern times; the effort prospers, and the work advances, both because it is the common work of all cultivated nations, and because the increasing improvement of the means of transport, both by sea and land, renders the whole earth more accessible, and brings into comparison its remotest portions.

I have here attempted to indicate, however vaguely, the manner in which the traveller's power of presenting the result of his opportunities of observation, the infusion of a fresh life into the descriptive element of literature, and the variety of the views which are continually opening before us on the vast theatre of the producing and destroying forces, may all tend to enlarge the scientific study of nature and to incite to its pursuit. The writer who, in our German literature, has, according to my feelings, opened the path in this direction with the greatest degree of vigour and success, was my distinguished teacher and friend George Forster. Through him has been commenced a new era of scientific travelling, having for its object the comparative knowledge of nations and of nature in different parts of the earth's surface. Gifted with refined æsthetic feeling, and retaining the fresh and living pictures with which Tahiti and the other fortunate islands of the Pacific had filled his imagination (as in later years that of Charles Darwin) ([103]), George Forster was the first gracefully and pleasingly to depict the different gradations of vegetation, the relations of climate, and the various articles of food, in their bearing on the habits and manners of different tribes according to their differences of race and of previous habitation. All that can give truth, individuality, and graphic distinctness to the representation of an exotic nature, is united in his writings: not only his excellent account of the second voyage of Captain Cook, but still more his smaller works, contain the germ of much which, at a later period, has been brought to maturity ([104]). But, for this noble, sensitive, and ever-hopeful spirit, a fortunate and happy life was not reserved.

If a disparaging sense has sometimes been attached to the terms "descriptive and landscape poetry," as applied to the numerous descriptions of natural scenes and objects which in the most modern times have more especially enriched German, French, English, and North American literatures, yet such censure is only properly applicable to the abuse of the supposed enlargement of the field of art. Versified descriptions of natural objects, such as at the close of a long and distinguished literary career were given by Delille, cannot be regarded, notwithstanding the refinements of language and of metre expended on them, as the poetry of external nature in the higher sense of the term: they lack poetic inspiration, and are therefore strangers on true poetic ground; they are cold and meagre, as is all that glitters with mere outward ornament. But if what has been called (as a distinct and independent form) "descriptive poetry," be justly blamed, such disapprobation cannot assuredly apply to an earnest endeavour, by the force of language,—by the power of significant words,—to bring the richer contents of our modern knowledge of nature before the contemplation of the imagination as well as of the intellect. Should means be left unemployed whereby we may have brought home to us not only the vivid picture of distant zones over which others have wandered, but also a portion even of the enjoyment afforded by the immediate contact with nature? The Arabs say figuratively but truly that the best description is that in which the ear is transformed into an eye (105). It is one of the evils of the present time that an unfortunate predilection for an empty species of poetic prose, and a tendency to indulge in sentimental effusions, has seized simultaneously in different countries on authors otherwise possessed of merit as travellers, and as writers on subjects of natural history. This mixture is still more unpleasing, when the style, from the absence of literary cultivation, and especially of all true inward spring of emotion, degenerates into rhetorical inflation and spurious sentimentality. Descriptions of nature, I would here repeat, may be sharply defined and scientifically correct, without being deprived thereby of the vivifying breath of imagination. The poetic element must be derived from a recognition of the links which unite the sensuous with the intellectual; from a feeling of the universal extension, the reciprocal limitation, and the unity of the forces which constitute the life of Nature. The more sublime the objects, the more carefully must all outward adornment of language be avoided. The true and proper effect of a picture of nature depends upon its composition, and the impression produced by it can only be disturbed and marred by the intrusions of elaborate appeals on the part of its presenter. He who, familiar with the great works of antiquity, and in secure possession of the riches of his native tongue, knows how to render with simplicity and characteristic truth that which he has received by his own contemplation, will not fail in the impression which he desires to convey; and the risk of failure will be less, as in depicting external nature, and not his own frame of mind, he leaves unfettered the freedom of feeling in others.

But it is not alone the animated description of those richly adorned lands of the equinoctial zone, in which intensity of light and of humid warmth accelerates and heightens the development of all organic germs, which has furnished in our days a powerful incentive to the general study of nature: the secret charm excited by a deep insight into organic life is not limited to the tropical world; every region of the earth offers the wonders of progressive formation and development, and the varied connection of recurring or slightly deviating types. Everywhere diffused is the awful domain of those powerful forces, which in the dark storm clouds that veil the sky, as well as in the delicate tissues of organic substances, resolve the ancient discord of the elements into harmonious union. Therefore, wherever spring unfolds a bud, from the equator to the frigid zone, our minds may receive and may rejoice in the inspiration of nature pervading every part of the wide range of creation. Well may our German fatherland cherish such belief; where is the more southern nation who would not envy us the great master of our poetry, through all whose works there breathes a profound feeling of external nature, seen alike in the Sorrows of Werter, in the Reminiscences of Italy, in the Metamorphoses of Plants, and in his Miscellaneous Poems. Who has more eloquently excited his cotemporaries to "solve the sacred enigma of the universe" ("des Weltalls heilige Räthsel zu lösen"); and to renew the ancient alliance which in the youth of human kind united philosophy, physical science, and poetry in a common bond? Who has pointed with more powerful charm to that land, his intellectual home, where

> Ein sanfter Wind vom blauem Himmel weht,
> Die Myrte still, und hoch der Lorbeer steht?

INCITEMENTS TO THE STUDY OF NATURE.

II.—Landscape painting—Graphical representation of the physiognomy of plants—Characteristic form and aspect of vegetation in different zones.

As fresh and vivid descriptions of natural scenes and objects are suited to enhance a love for the study of nature, so also is landscape painting. Both shew to us the external world in all its rich variety of forms, and both are capable, in various degrees, according as they are more or less happily conceived, of linking together the outward and the inward world. It is the tendency to form such links which marks the last and highest aim of representative art; but the scientific object to which these pages are devoted, restricts them to a different point of view; and landscape painting can be here considered only as it brings before us the characteristic physiognomy of different positions of the earth's surface, as it increases the longing desire for distant voyages, and as, in a manner equally instructive and agreeable, it incites to fuller intercourse with nature in her freedom.

In classical antiquity, from the peculiar direction of the Greek and Roman mind, landscape painting, like the poetic description of scenery, could scarcely become an independent object of art: both were used only as auxiliaries. Employed in complete subordination to other objects, landscape painting long served merely as a background to historical composition, or as an accidental ornament in the decoration of painted walls. The epic poet, in a similar manner, sometimes marked the locality of particular events by a picturesque description of the landscape, or, as I might again term it, of the background, in front of which the acting personages were moving. The history of art teaches how the subordinate auxiliary gradually became itself a principal object, until landscape painting, separated from true historical painting, took its place as a distinct form. Whilst this separation was being gradually effected, the human figures were sometimes inserted as merely secondary features in a mountainous or woodland scene, a marine or a garden view. It has been justly remarked, in reference to the ancients, that not only did painting remain subordinate to sculpture, but more especially, that the feeling for picturesque beauty of landscape reproduced by the pencil was not entertained by them at all, but is wholly of modern growth.

Graphical indications of the peculiar features of a district must, however, have existed in the earliest Greek paintings, if (to cite particular instances) Mandrocles of Samos, as Herodotus tells us (106), had a painting made for the great Persian king of the passage of the army across the Bosphorus; or if Polygnotus (107) painted the destruction of Troy in the Lesche at Delphi. Among the pictures described by the elder Philostratus mention is even made of a landscape, in which smoke was seen to issue from the summit of a volcano, and the stream of lava to pour itself into the sea. In the very complicated composition of a view of seven islands, the most recent commentators think that they recognise the representation of a real district; viz. the small volcanic group of the Æolian or Lipari islands, north of Sicily (108).

Perspective scene painting, which was made to contribute to the theatrical representation of the master-works of Æschylus and Sophocles, gradually extended this department of art (109), by increasing a demand for the illusive imitation of inanimate objects, such as buildings, trees, and rocks. In consequence of the improvement which followed this extension, landscape painting passed with the Greeks and Romans from the theatre into halls adorned with columns, where long surfaces of wall were covered, at first with more restricted scenes (110), but afterwards with extensive views of cities, sea-shores, and wide pastures with grazing herds of cattle (111). These pleasing decorations were not, indeed, invented by the Roman painter, Ludius, in the Augustan age, but were rendered generally popular (112) by him, and enlivened by the introduction of small figures (113). Almost at the same period, and even half a century earlier, amongst the Indians, in the brilliant epoch of Vikramaditya, we find landscape painting referred to as a much practised art. In the charming drama of "Sacontala," the king, Dushmanta, has the picture of his beloved shewn him; but not satisfied with her portrait only, he desires that "the paintress should draw the places which Sacontala most loved:—the Malini river, with a sandbank on which the red flamingoes are standing; a chain of hills, which rest against the Himalaya, and gazelles reposing on the hills." These are no small requisitions: they indicate a belief, at least, in the possibility of executing complicated representations.

In Rome, from the time of the Cæsars, landscape painting became a separate branch of art, but so far as we can judge by what the excavations at Herculaneum, Pompeiï, and Stabia, have shewn us, the pictures were often mere bird's-eye views, resembling maps, and aimed rather at the representation of seaport towns, villas, and artificial gardens, than of nature in her freedom. That which the Greeks and the Romans regarded as attractive in a landscape, seems to have been almost exclusively the agreeably habitable, and not what we call the wild and romantic. In their pictures, the imitation might possess as great a degree of exactness as could consist with frequent inaccuracy in regard to perspective, and with a disposition to conventional arrangement; their compositions of the nature of arabesques, to the use of which the severe Vitruvius was averse, contained rhythmically recurring and tastefully arranged forms of plants and animals; but, to avail myself of an expression of Otfried Müller's, "the soul of the landscape did not appear to the ancients an object for imitative art: their sketches were conceived sportively, rather than with earnestness and feeling."

The specimens of ancient landscape-painting in the manner of Ludius, which have been brought to light by the excavations at Pompeii (now happily continued), belong most probably to a single and very limited epoch (115), viz. from Nero to Titus; for the town had been entirely destroyed by earthquake sixteen years before the catastrophe caused by the celebrated eruption of Vesuvius.

From Constantine the Great to the beginning of the middle ages, painting, though connected with Christian subjects, preserved a close affinity to its earlier character.

An entire treasury of old memorials is found both in the miniatures (116) adorning superb manuscripts still in good condition, and in the scarcer mosaics of the same period. Rumohr mentions a manuscript Psalter, in the Barberina at Rome, containing a miniature in which "David is seen playing on the harp, seated in a pleasant grove from amongst the branches of which nymphs look forth and listen: this personification marks the antique character of the whole picture." From the middle of the sixth century, when Italy was impoverished and in a state of utter political confusion, it was Byzantine art in the eastern empire which did most to preserve the lingering echoes and types of a more flourishing period. Memorials, such as we have spoken of, form a kind of transition to the more beautiful creations of the later middle ages: the fondness for ornamented manuscripts spread from Greece in the east to the countries of the west and the north,—into the Frankish monarchy, among the Anglo-Saxons, and into the Netherlands. It is therefore a fact of no little importance in respect to the history of modern art, "that the celebrated brothers, Hubert and John van Eyck, belonged essentially to a school of miniature painters, which, since the second half of the fourteenth century, had reached a high degree of perfection in Flanders" (117).

It is in the historical paintings of the brothers Van Eyck that we first meet with a careful elaboration of the landscape portion of the picture. Italy was never seen by either of them; but the younger brother, John, had enjoyed an opportunity of beholding a south European vegetation, having, in 1428, accompanied the embassy which Philip the Good, Duke of Burgundy, sent to Lisbon, to prefer his suit to the daughter of King John I. of Portugal. We possess, in the Berlin Museum, the volets of the magnificent painting which these artists, the true founders of the great Netherlands school of painting, executed for the cathedral at Ghent. On the sides which present the holy hermits and pilgrims, John van Eyck has adorned the landscape with orange trees, date palms, and cypresses, which are marked by an extreme fidelity to nature, and cast over the other dark masses a shade which imparts to them a grave and elevated character. In viewing this picture, we feel that the painter had himself received the impression of a vegetation fanned by soft and warm breezes.

The master-works of the brothers Van Eyck belong to the first half of the fifteenth century, when oil painting, though it had only just begun to supersede fresco, had already attained high technical perfection. The desire to produce an animated representation of natural forms was now awakened; and if we would trace the gradual extension and heightening of the feelings connected therewith, we should recal how Antonello of Messina, a scholar of the brothers Van Eyck, transplanted to Venice a fondness for landscape; and how, even in Florence, the pictures of the Van Eyck school exerted a similar influence over Domenico Ghirlandaio, and other masters ([118]). At this period, the efforts of the painters were, for the most part, directed to a careful, but almost painfully solicitous and minute imitation of natural forms. The representation of nature first appears conceived with freedom and with grandeur in the master-works of Titian, to whom, in this respect also, Giorgione had served as an example. I had the opportunity, during many years, of admiring, at Paris, Titian's painting of the death of Peter Martyr (119), attacked in a forest by an Albigense in the presence of another Dominican monk. The form of the forest trees, their foliage, the blue mountainous distance, the management of the light and the subdued tone of colouring, produce an impression of grandeur, solemnity, and depth of feeling, pervading the whole composition of the landscape, which is of exceeding simplicity. Titian's feeling of nature was so lively, that not only in paintings of beautiful women, as in the background of the Venus in the Dresden Gallery, but also in those of a severer class, as in the portrait of the poet Pietro Aretino, he gives to the landscape or to the sky a character corresponding to that of the subject of the picture. In the Bolognese school, Annibal Caracci and Domenichino remained faithful to this elevation of style and character. If, however, the sixteenth century was the greatest epoch of historic painting, the seventeenth is that of landscape. As the riches of nature became better known and more carefully studied, artistic feeling could extend itself over a wider and more varied range of subjects; and, at the same time, the technical means of representation had also attained a higher degree of perfection. Meanwhile, the landscape painter's art becoming more often and more intimately connected and associated with inward tone and feeling, the tender and mild expression of the beautiful in nature was enhanced thereby, as well as the belief in the power of the emotions which the external world can awaken within us. When, conformably to the elevated aim of all art, this awakening power transforms the actual into the ideal, the enjoyment produced is accompanied by emotion; the heart is touched whenever we look into the depths either of nature or of humanity (120).

We find assembled, in the same century, Claude Lorraine, the idyllic painter of light and of aerial distance; Ruysdael's dark forest masses and threatening clouds; Gaspar and Nicholas Poussin's heroic forms of trees; and the faithful and simply natural representations of Everdingen, Hobbima, and Cuyp (121). This flourishing period in the development of art comprised happy imitations of the vegetation of the north of Europe, of southern Italy, and of the Iberian peninsula: the painters adorned their landscapes with oranges and laurels, with pines and date trees. The date (the only member of the magnificent family of Palms which the artists had themselves seen, except the small native European species, the Chamærops maritima) was usually represented conventionally, with scaly and serpentlike trunks (122), and long served as the representative of tropical vegetation generally,—much as Pinus pinea (the stone pine) is, by a still widely prevailing idea, regarded as exclusively characteristic of Italian vegetation. The outlines of lofty mountains were yet but little studied: and naturalists and landscape painters still regarded the snowy summits, which rise above the green pastures of the lower Alps, as inaccessible. The particular characters of masses of rock were rarely made objects of careful imitation, except where associated with the foaming waterfall. We may here remark another instance of the comprehensiveness with which the varied forms of nature are seized by a free and artistic spirit. Rubens, who in his great hunting pieces has depicted with inimitable truth and animation the wild movements of the beasts of the forest, has also apprehended, with peculiar felicity, the characteristics of the inanimate surface of the earth, in the arid desert and rocky plateau on which the Escurial is built (123).

The department of art to which we are now referring might be expected to advance in variety and exactness as the geographical horizon became enlarged, and as voyages to distant climates facilitated the perception of the relative beauty of different vegetable forms, and their connection in groups of natural families. The discoveries of Columbus, Vasco de Gama, and Alvarez Cabral in Central America, Southern Asia, and Brazil, the extensive commerce in spices and drugs carried on by the Spaniards, Portuguese, Italians, Dutch, and Flemings, and the establishment, between 1544 and 1568, of botanic gardens (not yet however furnished with regular hothouses), at Pisa, Padua, and Bologna, did indeed afford to painters the opportunity of becoming acquainted with many remarkable exotic productions even of the tropical world; and single fruits, flowers, and branches, were represented with the utmost fidelity and grace by John Breughel, whose celebrity had commenced before the close of the sixteenth century; but until near the middle of the seventeenth century there were no landscapes which reproduced the peculiar aspect of the torrid zone from actual impressions received by the artist himself on the spot. The first merit of such representation probably belongs (as I learn from Waagen), to a painter of the Netherlands, Franz Post of Haarlem, who accompanied Prince Maurice of Nassau to Brazil, where that prince, who took great interest in tropical productions, was the Stattholder for Holland in the conquered Portuguese possessions from 1637 to 1644. Post made many studies from nature near Cape St. Augustine, in the bay of All Saints, on the shores of the Rio San Francisco, and on those of the lower part of the river of the Amazons ([124]). Some of these were afterwards executed by himself as pictures, and others were etched with much spirit. There are preserved in Denmark, (in a gallery of the fine castle at Frederiksborg), some large oil paintings of great merit belonging to the same epoch by the painter Eckhout, who, in 1641, was also in Brazil with Prince Maurice. In these pictures, palms, papaws (Carica papaya), bananas, and heliconias, are most characteristically pourtrayed, as are likewise the native inhabitants, birds of many-coloured plumage, and small quadrupeds.

These examples were followed by few artists of merit until Cook's second voyage of circumnavigation: what Hodge did for the western islands of the Pacific, and our distinguished countryman, Ferdinand Bauer, for New Holland and Van Diemen Island, has been since done in very recent times in a much grander style, and with a more masterly hand, for tropical America, by Moritz, Rugendas, Count Clarac, Ferdinand Bellermann, and Edward Hildebrandt; and for many other parts of the earth by Heinrich von Kittlitz, who accompanied the Russian admiral, Lutke, on his voyage of circumnavigation ([125]).

He who with feelings alive to the beauties of nature in mountain, river, or forest scenery, has himself wandered in the torrid zone, and beheld the variety and luxuriance of the vegetation, not merely on the well-cultivated coasts, but also on the declivities of the snow-crowned Andes the Himalaya or the Neilgherries of Mysore, or in the virgin forests watered by the network of rivers between the Orinoco and the Amazons, can feel,—and he alone can feel,—how almost infinite is the field which still remains to be opened to landscape painting in the tropical portions of either continent, and in the islands of Sumatra, Borneo, and the Philippines; and how all that this department of art has yet produced, is not to be compared to the magnitude of the treasures, of which at some future day it may become possessed. Why may we not be justified in hoping that landscape painting may hereafter bloom with new and yet unknown beauty, when highly-gifted artists shall oftener pass the narrow bounds of the Mediterranean, and shall seize, with the first freshness of a pure youthful mind, the living image of the manifold beauty and grandeur of nature in the humid mountain valleys of the tropical world?

Those glorious regions have been hitherto visited chiefly by travellers to whom the want of previous artistic training, and a variety of scientific occupations, allowed but little opportunity of attaining perfection in landscape painting. But few among them were able, in addition to the botanical interest excited by individual forms of flowers and leaves, to seize the general characteristic impression of the tropical zone. The artists who accompanied great expeditions supported at the expense of the states which sent them forth, were too often chosen as it were by accident, and were thus found to be less prepared than the occasion demanded; and perhaps the end of the voyage was approaching, when even the most talented among them, after a long enjoyment of the spectacle of the great scenes of nature, and many attempts at imitation, were just beginning to master a certain degree of technical skill. Moreover, in voyages of circumnavigation, artists are seldom conducted into the true forest regions, to the upper portions of the course of great rivers, or to the summits of the mountain chains of the interior. It is only by coloured sketches taken on the spot, that the artist, inspired by the contemplation of these distant scenes, can hope to reproduce their character in paintings executed after his return. He will be able to do so the more perfectly, if he has also accumulated a large number of separate studies of tops of trees, of branches clothed with leaves, adorned with blossoms, or laden with fruit, of fallen trunks of trees overgrown with pothos and orchideæ, of portions of rocks and river banks, as well as of the surface of the ground in the forest, all drawn or painted directly from nature. An abundance of studies of this kind, in which the outlines are well and sharply marked, will furnish him with materials enabling him, on his return, to dispense with the misleading assistance afforded by plants grown in the confinement of hot-houses, or by what are called botanical drawings.

Great events in the world's history, the independence of the Spanish and Portuguese Americas, and the spread and increase of intellectual cultivation in India, New Holland, the Sandwich Islands, and the southern colonies of Africa, cannot fail to procure, not only for meteorology and other branches of natural knowledge, but also for landscape painting, a new and grander development which might not have been attainable without these local circumstances. In South America populous cities are situated 13,000 feet above the level of the sea. In descending from them to the plains, all climatic gradations of the forms of plants are offered to the eye. What may we not expect from the picturesque study of nature in such scenes, if after the termination of civil discord and the establishment of free institutions, artistic feeling shall at length awaken in those elevated highlands!

All that belongs to the expression of human emotion and to the beauty of the human form, has attained perhaps its highest perfection in the northern temperate zone, under the skies of Greece and Italy. By the combined exercise of imitative art and of creative imagination, the artist has derived the types of historical painting, at once from the depths of his own mind, and from the contemplation of other beings of his own race. Landscape painting, though no merely imitative art, has, it may be said, a more material substratum and a more terrestrial domain: it requires a greater mass and variety of direct impressions, which the mind must receive within itself, fertilize by its own powers, and reproduce visibly as a free work of art. Heroic landscape painting must be a result at once of a deep and comprehensive reception of the visible spectacle of external nature, and of this inward process of the mind.

Nature, in every region of the earth, is indeed a reflex of the whole; the forms of organised being are repeated everywhere in fresh combinations; even in the icy north, herbs covering the earth, large alpine blossoms, and a serene azure sky, cheer a portion of the year. Hitherto, landscape painting has pursued amongst us her pleasing task, familiar only with the simpler forms of our native floras, but not therefore without depth of feeling or without the treasures of creative imagination. Even in this narrower field, highly-gifted painters, the Caracci, Gaspar Poussin, Claude Lorraine, and Ruysdael, have, with magic power, by the selection of forms of trees and by effects of light, found scope wherein to call forth some of the most varied and beautiful productions of creative art. The fame of these master works can never be impaired by those which I venture to hope for hereafter, and to which I could not but point, in order to recal the ancient and deeply-seated bond which unites natural knowledge with poetry and with artistic feeling; for we must ever distinguish, in landscape painting as in every other branch of art, between productions derived from direct observation, and those which spring from the depths of inward feeling and from the power of the idealising mind. The great and beautiful works which owe their origin to this creative power of the mind applied to landscape-painting, belong to the poetry of nature, and like man himself and the imagination with which he is gifted, are not rivetted to the soil or confined to any single region. I allude here more particularly to the gradation in the forms of trees from Ruysdael and Everdingen, through Claude Lorraine to Poussin and Annibal Caracci. In the great masters of the art we perceive no trace of local limitation; but an enlargement of the visible horizon, and an increased acquaintance with the nobler and grander forms of nature, and with the luxuriant fulness of life in the tropical world, offer the advantage not only of enriching the material substratum of landscape painting, but also of affording a more lively stimulus to less gifted artists, and of thus heightening their power of production.

I would here be permitted to recal some considerations which I communicated to the public nearly half a century ago, and which have an intimate connection with the subject which is at present under notice; they were contained in a memoir which has been but little read, entitled "Ideen zu einer Physiognomik der Gewächse" (226) (Ideas towards a physiognomy of plants). When rising from local phenomena we embrace all nature in one view, we perceive the increase of warmth from the poles to the equator accompanied by the gradual advance of organic vigour and luxuriance. From Northern Europe to the beautiful coasts of the Mediterranean this advance is even less than from the Iberian Peninsula, Southern Italy and Greece, to the tropic zone. The carpet of flowers and of verdure spread over our bare and naked earth is unequally woven; thicker where the sun rises high in a sky either of a deep azure purity or veiled with light semi-transparent clouds; and thinner towards the gloomy north, where returning frosts are often fatal to the opening buds of spring, or destroy the ripening fruits of autumn. If in the frigid zone the bark of trees is covered with lichens or with mosses, in the zone of palms and finely-feathered arborescent ferns, the trunks of Anacardias and of gigantic species of Ficus are enlivened by Cymbidium and the fragrant vanilla. The fresh green of the Dracontias, and the deep-cut leaves of the Pothos, contrast with the many-coloured flowers of the Orchideæ. Climbing Bauhinias, Passifloras, and yellow flowering Banisterias, entwining the stems of the forest trees, spread far and wide, and rise high in air; delicate flowers unfold themselves from the roots of the Theobromas, and from the thick and rough bark of the Crescentias and the Gustavia. In the midst of this abundance of leaves and blossoms, this luxuriant growth and profusion of climbing plants, the naturalist often finds it difficult to discover to which stem different flowers and leaves belong; nay, a single tree adorned with Paullinias, Bignonias, and Dendrobium, presents a mass of vegetation and a variety of plants which, if detached from each other, would cover a considerable space of ground.

But to each zone of the earth are allotted peculiar beauties; to the tropics, variety and grandeur in the forms of vegetation; to the north, the aspect of its meadows and green pastures, and the periodic long-desired reawakening of nature at the first breath of the mild air of spring. As in the Musaceæ we have the greatest expansion, so in the Casuarinæ and needle trees we have the greatest contraction of the leafy vessels. Firs, Thuias, and Cypresses, constitute a northern form which is extremely rare in the low grounds of the tropics. Their ever-fresh verdure cheers the winter landscape; and tells to the inhabitants of the north, that when snow and ice cover the earth, the inward life of plants, like the Promethean fire, is never extinct upon our planet.

Each zone of vegetation, besides its peculiar beauties, has also a distinct character, calling forth in us a different order of impressions. To recal here only forms of our native climates, who does not feel himself differently affected in the dark shade of the beech or on hills crowned with scattered firs, and on the open pasture where the wind rustles in the trembling foliage of the birch? As in different organic beings we recognise a distinct physiognomy, and as descriptive botany and zoology, in the more restricted sense of the terms, imply an analysis of peculiarities in the forms of plants and animals, so is there also a certain natural physiognomy belonging exclusively to each region of the earth. The idea which the artist indicates by the expressions "Swiss nature," "Italian sky," &c. rests on a partial perception of local character. The azure of the sky, the form of the clouds, the haze resting on the distance, the succulency of the herbage, the brightness of the foliage, the outline of the mountains, are elements which determine the general impression. It is the province of landscape painting to apprehend these, and to reproduce them visibly. The artist is permitted to analyse the groups, and the enchantment of nature is resolved under his hands, like the written works of men (if I may venture on the figurative expression), into a few simple characters.

Even in the present imperfect state of our pictorial representations of landscape, the engravings which accompany, and too often only disfigure, our books of travels, have yet contributed not a little to our knowledge of the aspect of distant zones, to the predilection for extensive voyages, and to the more active study of nature. The improvement in landscape painting on a scale of large dimensions (as in decorative or scene painting, in panoramas, dioramas, and neoramas), has of late years increased both the generality and the strength of these impressions. The class of representations which Vitruvius and the Egyptian Julius Pollux satirically described as "rustic adornments of the stage," which, in the middle of the sixteenth century, were, by Serlio's arrangement of coulisses, made to increase theatrical illusion, may now, in Barker's panoramas, by the aid of Prevost and Daguerre, be converted into a kind of substitute for wanderings in various climates. More may be effected in this way than by any kind of scene painting; and this partly because in a panorama, the spectator, enclosed as in a magic circle and withdrawn from all disturbing realities, may the more readily imagine himself surrounded on all sides by nature in another clime. Impressions are thus produced which in some cases mingle years afterwards by a wonderful illusion with the remembrances of natural scenes actually beheld. Hitherto, panoramas, which are only effective when they are of large diameter, have been applied chiefly to views of cities and of inhabited districts, rather than to scenes in which nature appears decked with her own wild luxuriance and beauty. Enchanting effects might be obtained by means of characteristic studies sketched on the rugged mountain declivities of the Himalaya and the Cordilleras, or in the recesses of the river country of India and South America; and still more so if these sketches were aided by photographs, which cannot indeed render the leafy canopy, but would give the most perfect representation possible of the form of the giant trunks, and of the mode of ramification characteristic of the different kinds of trees. All the methods to which I have here alluded are fitted to enhance the love of the study of nature; it appears, indeed, to me, that if large panoramic buildings, containing a succession of such landscapes, belonging to different geographical latitudes and different zones of elevation, were erected in our cities, and, like our museums and galleries of paintings, thrown freely open to the people, it would be a powerful means of rendering the sublime grandeur of the creation more widely known and felt. The comprehension of a natural whole, the feeling of the unity and harmony of the Cosmos, will become at once more vivid and more generally diffused, with the multiplication of all modes of bringing the phænomena of nature generally before the contemplation of the eye and of the mind.

INCITEMENTS TO THE STUDY OF NATURE.

III.—Cultivation of tropical plants—Assemblage of contrasted forms—Impression of the general characteristic physiognomy of the vegetation produced by such means.

THE effect of landscape painting, notwithstanding the multiplication of its productions by engravings and by the modern improvements of lithography, is still both more limited and less vivid, than the stimulus which results from the impression produced on minds alive to natural beauty by the direct view of groups of exotic plants in hot-houses or in the open air. I have already appealed on this subject to my own youthful experience, when the sight of a colossal dragon tree and of a fan palm in an old tower of the botanic garden at Berlin, implanted in my breast the first germ of an irrepressible longing for distant travel. Those who are able to reascend in memory to that which may have given the first impulse to their entire course of life, will recognise this powerful influence of impressions received through the senses.

I would here distinguish between those plantations which are best suited to afford us the picturesque impression of the forms of plants, and those in which they are arranged as auxiliaries to botanical studies; between groups distinguished for their grandeur and mass, as clumps of Bananas and Heliconias alternating with Corypha Palms, Araucarias and Mimosas, and moss-covered trunks from which shoot Dracontias, Ferns with their delicate foliage, and Orchideæ rich in varied and beautiful flowers, on the one hand; and on the other, a number of separate low-growing plants classed and arranged in rows for the purpose of conveying instruction in descriptive and systematic botany. In the first case, our consideration is drawn rather to the luxuriant development of vegetation in Cecropias, Carolinias, and light-feathered Bamboos; to the picturesque apposition of grand and noble forms, such as adorn the banks of the upper Orinoco and the forest shores of the Amazons, and of the Huallaga described with such truth to nature by Martius and Edward Poppig; to impressions which fill the mind with longing for those lands where the current of life flows in a richer stream, and of whose glorious beauty a faint but still pleasing image is now presented to us in our hot-houses, which formerly were mere hospitals for languishing unhealthy plants.

Landscape painting is, indeed, able to present a richer and more complete picture of nature than can be obtained by the most skilful grouping of cultivated plants. Almost unlimited in regard to space, it can pursue the margin of the forest until it becomes indistinct from the effect of aerial perspective; it can pour the mountain torrent from crag to crag, and spread the deep azure of the tropic sky above the light tops of the palms, or the undulating savannah which bounds the horizon. The illumination and colouring, which between the tropics are shed over all terrestrial objects by the light of the thinly veiled or perfectly pure heaven, give to landscape painting, when the pencil succeeds in imitating this mild effect of light, a peculiar and mysterious power. A deep perception of the essence of the

Greek tragedy led my brother to compare its *chorus* with the *sky in the landscape* (127).

The multiplied means which painting can command for stimulating the fancy, and concentrating in a small space the grandest phænomena of sea and land, are indeed denied to our plantations in gardens or in hot-houses; but the inferiority in general impression is compensated by the mastery which the reality every where exerts over the senses. When in the palm house of Loddiges, or in that of the Pfauen-insel near Potsdam (a monument of the simple feeling for nature of our noble departed monarch), we look down from the high gallery, during a bright noonday sunshine, upon the abundance of reed-like and arborescent palms, a complete illusion in respect to the locality in which we are placed is momentarily produced; we seem to be actually in the climate of the tropics, looking down from the summit of a hill upon a small thicket of palms. The aspect of the deep blue sky, and the impression of a greater intensity of light, are indeed wanting, but still the illusion is greater, and the imagination more vividly active, than from the most perfect painting: we associate with each vegetable form the wonders of a distant land; we hear the rustling of the fan-like leaves, and see the changing play of light, as, gently moved by slight currents of air, the waving tops of the palms come into contact with each other. So great is the charm which reality can give. The recollection of the needful degree of artificial care bestowed no doubt returns to disturb the impression; for a perfectly flourishing condition, and a state of freedom, are inseparable in the realm of nature as elsewhere; and in the eyes of the earnest and travelled botanist, the dried specimen in an herbarium, if actually gathered on the Cordilleras of South America, or the plains of India, often has a greater value than the living plant in an European hot-house: cultivation effaces somewhat of the original natural character; the constraint which it produces disturbs the free organic development of the separate parts.

The physiognomic character of plants, and their assemblage in happily contrasted groups, is not only an incitement to the study of nature, and itself one of the objects of that study, but attention to the physiognomy of plants is also of great importance in landscape gardening—in the art of composing a garden landscape. I will resist the temptation to expatiate in this closely adjoining field of disquisition, and content myself with bringing to the recollection of my readers that, as in the earlier portion of the present volume, I found occasion to notice the more frequent manifestation of a deep feeling for nature among the Semitic, Indian, and Iraunian nations, so also the earliest ornamental parks mentioned in history belonged to middle and southern Asia. The gardens of Semiramis, at the foot of the Bagistanos mountains ([128]), are described by Diodorus, and the fame of them induced Alexander to turn aside from the direct road, in order to visit them during his march from Chelone to the Nysäic horse pastures. The parks of the Persian kings were adorned with cypresses, of which the form, resembling obelisks, recalled the shape of flames of fire, and which, after the appearance of Zerdusht (Zoroaster), were first planted by Gushtasp around the sanctuary of the fire temple. It was, perhaps, thus that the form of the tree led to the fiction of the Paradisaical origin of cypresses ([129]). The Asiatic terrestrial paradises (παραδεισοι), were early celebrated in more western countries ([130]); and the worship of trees even goes back among the Iraunians to the rules of Hom, called, in the Zend-Avesta, the promulgator of the old law. We know from Herodotus the delight which Xerxes took in the great plane tree in Lydia, on which he bestowed golden ornaments, and appointed for it a sentinel in the person of one of the "immortal ten thousand" ([231]). The early veneration of trees was associated, by the moist and refreshing canopy of foliage, with that of sacred fountains. In similar connection with the early worship of nature, were, amongst the Hellenic nations, the fame of the great palm tree of Delos, and of an aged plane tree in Arcadia. The Buddhists of Ceylon venerate the colossal Indian fig tree (the Banyan) of Anurahdepura, supposed to have sprung from the branches of the original tree under which Buddha, while inhabiting the ancient Magadha, was absorbed in beatification, or "self-extinction" (nirwana) ([132]). As single trees thus became objects of veneration from the beauty of their form, so did also groups of trees, under the name of "groves of the gods." Pausanias is full of the praise of a grove belonging to the temple of Apollo, at Grynion, in Æolis ([133]); and the grove of Colonè is celebrated in the renowned chorus of Sophocles.

The love of nature which showed itself in the selection and care of these venerated objects of the vegetable kingdom, manifested itself with yet greater vivacity, and in a more varied manner, in the horticultural arrangements of the early civilised nations of Eastern Asia. In the most distant part of the old continent, the Chinese gardens appear to have approached most nearly to what we now call English parks. Under the victorious dynasty of Han, gardens of this class were extended over circuits of so many miles that agriculture was affected, [134] and the people were excited to revolt. "What is it," says an ancient Chinese writer, Lieu-tscheu, that we seek in the pleasures of a garden? It has always been agreed that these plantations should make men amends for living at a distance from what would be their more congenial and agreeable dwelling-place, in the midst of nature, free, and unconstrained. The art of laying out gardens consists, therefore, in combining cheerfulness of prospect, luxuriance of growth, shade, retirement, and repose, so that the rural aspect may produce an illusion. Variety, which is a chief merit in the natural landscape, must be sought by the choice of ground with alternation of hill and dale, flowing streams, and lakes covered with aquatic plants. Symmetry is wearisome; and a garden where every thing betrays constraint and art becomes tedious and distasteful." [135] A description which Sir George Staunton has given us of the great imperial garden of Zhe-hol, [136] north of the Chinese wall, corresponds with these precepts of Lieu-tscheu—precepts to which our ingenious contemporary, who formed the beautiful park of Moscow, [137] would not refuse his approbation.

The great descriptive poem, composed in the middle of the last century by the Emperor Kien-long to celebrate the former Mantchou imperial residence, Moukden, and the graves of his ancestors, is also expressive of the most thorough love of nature sparingly embellished by art. The royal poet knows how to blend the cheerful images of fresh and rich meadows, wood-crowned hills, and peaceful dwellings of men, all described in a very graphic manner, with the graver image of the tombs of his forefathers. The offerings which he brings to his deceased ancestors, according to the rites prescribed by Confucius, and the pious remembrance of departed monarchs and warriors, are the more special objects of this remarkable poem. A long enumeration of the wild plants, and of the animals which enliven the district, is tedious, as didactic poetry always is; but the weaving together the impression received from the visible landscape (which appears only as the background of the picture,) with the more elevated objects taken from the world of ideas, with the fulfilment of religious rites, and with allusions to great historical events, gives a peculiar character to the whole composition. The consecration of mountains, so deeply rooted among the Chinese, leads the author to introduce careful descriptions of the aspect of inanimate nature, to which the Greeks and the Romans shewed themselves so little alive. The forms of the several trees, their mode of growth, the direction of the branches, and the shape of the leaves, are dwelt on with marked predilection. (138)

As I do not participate in that distaste to Chinese literature which is too slowly disappearing amongst us, and as I have dwelt, perhaps, at too much length on the work of a cotemporary of Frederic the Great, it is the more incumbent on me to go back to a period seven centuries and a half earlier, for the purpose of recalling the poem of "The Garden," by See-ma-kuang, a celebrated statesman. It is true that the pleasure grounds described in this poem are, in part, overcrowded with numerous buildings, as was the case in the ancient villas of Italy; but the minister also describes a hermitage, situated between rocks, and surrounded by lofty fir trees. He praises the extensive prospect over the wide river Kiang, with its many vessels: "here he can receive his friends, listen to their verses, and recite to them his own." (139) See-ma-kuang wrote in the year 1086, when, in Germany, poetry, in the hands of a rude clergy, did not even speak the language of the country. At that period, and, perhaps, five centuries earlier, the inhabitants of China, Transgangetic India, and Japan, were already acquainted with a great variety of forms of plants. The intimate connection maintained between the Buddhistic monasteries was not without influence in this respect. Temples, cloisters, and burying-places were surrounded with gardens, adorned with exotic trees, and with a carpet of flowers of many forms and colours. The plants of India were early conveyed to China, Corea, and Nipon. Siebold, whose writings afford a comprehensive view of all that relates to Japan, was the first to call attention to the cause of the intermixture of the floras of widely-separated Buddhistic countries. (140)

The rich and increasing variety of characteristic vegetable forms which, in the present age, are offered both to scientific observation and to landscape painting, cannot but afford a lively incentive to trace out the sources which have prepared for us this more extended knowledge and this increased enjoyment. The enumeration of these sources is reserved for the succeeding section of my work, *i. e.* the history of the contemplation of the universe. In the section which I am now closing, I have sought to depict those incentives, due to the influence exerted on the intellectual activity and the feelings of men by the reflected image of the external world, which, in the progress of modern civilisation, have tended so materially to encourage and vivify the study of nature. Notwithstanding a certain degree of arbitrary freedom in the development of the several parts, primary and deep-seated laws of organic life bind all animal and vegetable forms to firmly established and ever recurring types, and determine in each zone the particular character impressed on it, or *the physiognomy of nature.* I regard it as one of the fairest fruits of general European civilisation, that it is now almost every where possible for men to obtain,—by the cultivation of exotic plants, by the charm of landscape painting, and by the power of the inspiration of language,—some part, at least, of that enjoyment of nature, which, when pursued by long and dangerous journeys through the interior of continents, is afforded by her immediate contemplation.

HISTORY OF THE PHYSICAL CONTEMPLATION OF THE UNIVERSE.

Principal epochs of the progressive development and extension of the idea of the Cosmos as an organic whole.

THE history of the physical contemplation of the universe is the history of the recognition of nature as a whole; it is the recital of the endeavours of man to conceive and comprehend the concurrent action of natural forces on the earth and in the regions of space: it accordingly marks the epochs of progress in the generalisation of physical views. It is that part of the history of our world of thought which relates to objects perceived by the senses, to the form of conglomerated matter, and to the forces by which it is pervaded.

In the first portion of this work, in the section on the limitation and scientific treatment of a physical description of the universe, I have endeavoured to point out the true relation which the separate branches of natural knowledge bear to that description, and to shew that the science of the Cosmos derives from those separate studies only the materials for its scientific foundation. ([141]) The history of the recognition or knowledge of the universe as a whole,—of which history I now propose to present the leading ideas, and which, for the sake of brevity, I here term sometimes the "history of the Cosmos," and sometimes the "history of the physical contemplation of the universe,"—must not, therefore, be confounded with the "history of the natural sciences," as it is given in several of our best elementary books of physics, or in those of the morphology of plants and animals. In order to afford some preliminary notion of the import and bearing of what is to be here contemplated as historic periods or epochs, it may be useful to give instances, shewing on the one hand what is to be treated of, and on the other hand what is to be excluded. The discoveries of the compound microscope, of the telescope, and of colored polarisation, belong to the history of the science of the Cosmos,—because they have supplied the means of discovering what is common to all organic bodies, of penetrating into the most distant regions of space, and of distinguishing borrowed or reflected light from that of self-luminous bodies, *i. e.* of determining whether the light of the sun proceeds from a solid mass, or from a gaseous envelope; whilst, on the other hand, the relation of the experiments which, from the time of Huygens, have gradually led to Arago's discovery of colored polarisation, is reserved for the history of optics. In like manner the development of the principles according to which the varied mass of vegetable forms may be arranged in families is left to the history of phytognosy or botany; whilst what relates to the geography of plants, or to the insight into the local and climatic distribution of vegetation over the whole globe, on the dry land and in the algæferous basin of the sea, constitutes an important section in the history of the physical contemplation of the universe.

The thoughtful consideration of that which has conducted men to their present degree of insight into nature as a whole, is assuredly far from embracing the entire history of human cultivation. Even were we to regard the insight into the connection of the animating forces of the material universe as the noblest fruit of that cultivation, as tending towards the loftiest pinnacle which the intelligence of man can attain, yet that which we here propose to indicate would still be but one portion of a history, of which the scope should comprehend all that marks the progress of different nations in all directions in which moral, social, or mental improvement can be attained. Restricted to physical associations, we necessarily study but one part of the history of human knowledge; we fix our eyes especially on the relation which progressive attainment has borne to the whole which nature presents to us; we dwell less on the extension of the separate branches of knowledge, than on what different ages have furnished either of results capable of general application, or of powerful material aids contributing to the more exact observation of nature.

We must first of all distinguish carefully and accurately between early presage and actual knowledge. With increasing cultivation much passes from the former into the latter by a transition which obscures the history of discoveries. Presage or conjecture is often unconsciously guided by a meditative combination of what previous investigation has made known, and is raised by it as by an inspiring power. Among the Indians, the Greeks, and in the middle ages, much was enunciated concerning the connection of natural phænomena, which, at first unproved, and mingled with the most unfounded speculations, has at a later period been confirmed by sure experience, and has since become matter of scientific knowledge. The presentient imagination, the all-animating activity of spirit, which lived in Platò, in Columbus, and in Kepler, must not be reproached as if it had effected nothing in the domain of science, or as if it tended necessarily to withdraw the mind from the investigation of the actual.

Since we have defined the subject before us as the history of nature as a whole, or of unity in the phænomena and concurrence in the action of the forces of the universe, our method of proceeding must be to select for our notice those subjects by which the idea of the unity of phænomena has been gradually developed. We distinguish in this respect, 1°, the efforts of reason to attain the knowledge of natural laws by a thoughtful consideration of natural phænomena; 2°, events in the world's history which have suddenly enlarged the horizon of observation; 3°, the discovery of new means of perception through the senses, whereby observations are varied, multiplied, and rendered more accurate, and men are brought into closer communication both with terrestrial objects and with the most distant regions of space. This threefold view must be our guide in determining the principal epochs of the history of the science of the Cosmos. For the sake of illustrating what has been said, we will again adduce particular instances, characteristic of the different means by which men have gradually arrived at the intellectual possession of a large part of the material universe. I take, therefore, examples of "the enlarged knowledge of nature,"—of "great events,"—and of the "invention or discovery of new organs."

The "knowledge of nature" in the oldest Greek physics, was derived more from inward contemplation and from the depths of the mind, than from the observation of phænomena. The natural philosophy of the Ionic physiologists was directed to the primary principle of origin or production, or to the changes of form, of a single elementary substance. In the mathematical symbolism of the Pythagoreans, in their considerations on number and form, there is disclosed, on the other hand, a philosophy of measure and of harmony. This Doric Italic school, in seeking every where for numerical elements, from a certain predilection for the relations of number which it recognized in space and in time, may be said to have laid the foundation, in this direction, of the future progress of our modern experimental sciences. The history of the contemplation of the universe, in my view, records not so much the often recurring fluctuations between truth and error, as the principal epochs of the gradual approximation towards a just view of terrestrial forces and of the planetary system. It shews that the Pythagoreans, according to the report of Philolaus of Croton, taught the progressive movement of the non-rotating earth, or its revolution around the hearth or focus of the universe (the central fire, Hestia); whereas Plato and Aristotle imagined the earth to have neither a rotatory nor a progressive movement, but to rest immoveably in the center. Hicetas of Syracuse (who is at least more ancient than Theophrastus), Heraclides Ponticus, and Ecphantus, were acquainted with the rotation of the earth around its axis; but Aristarchus of Samos, and especially Seleucus the Babylonian who lived a century and a half after Alexander, were the first who knew that the earth not only rotates, but also at the same time revolves around the sun as the center of the whole planetary system. And if, in the middle ages, fanaticism, and the still prevailing influence of the Ptolemaic system, combined to bring back a belief in the immobility of the earth, and if, in the view of the Alexandrian Cosmas Indicopleustes, its form even became again that of the disk of Thales,—on the other hand it should be remembered that a German Cardinal, Nicolaus de Cuss, almost a century before Copernicus, had the mental freedom and the courage to reascribe to our planet both a rotation round its axis, and a progressive movement round the sun. After Copernicus, Tycho Brahe's doctrine was a step backwards; but the retrogression was of short duration. When once a considerable mass of exact observations had been assembled, to which Tycho himself largely contributed, the true view of the structure of the universe could not be long repressed. We have here shewn how the period of fluctuations is especially one of presentiment and speculation.

Next to the "enlarged knowledge of nature," resulting at once from observation and from ideal combinations, I have proposed to notice "great events," by which the horizon of the contemplation of the universe has been extended. To this class belong the migration of nations, remarkable voyages, and military expeditions; these have been instrumental in making known the natural features of the earth's surface, such as the form of continents, the direction of mountain chains, the relative elevation of high plateaus, and sometimes by the wide range over which they extended, have even provided materials for the establishment of general laws of nature. In these historical considerations, it will not be necessary to present a connected tissue of events; it will be sufficient to notice those occurrences which, at each period, have exerted a decisive influence on the intellectual efforts of man, and on a more enlarged and extended view of the universe. Such have been, to the nations settled round the basin of the Mediterranean, the navigation of Colæus of Samos beyond the Pillars of Hercules; the expedition of Alexander to Western India; the empire of the world obtained by the Romans; the spread of Arabian cultivation; and the discovery of the new Continent. I propose not so much to dwell on the narration of occurrences, as to indicate the influence which events,—such as voyages of discovery, the predominance and extension of a highly polished language possessing a rich literature, or the suddenly acquired knowledge of the Indo-African monsoons,—have exerted in developing the idea of the Cosmos.

Having among these heterogeneous examples alluded thus early to the influence of languages, I would here call attention generally to their immeasurable importance in two very different ways. Single languages widely extended operate as means of communication between distant nations; — a plurality of languages, by their intercomparison, and by the insight obtained into their internal organisation and their degrees of relationship, operate on the deeper study of the history of the human race. The Greek language, and the national life of the Greeks so intimately connected with their language, have exercised a powerful influence on all the nations with whom they have been brought in contact. (142) The Greek tongue appears in the interior of Asia, through the influence of the Bactrian empire, as the conveyer of knowledge which more than a thousand years afterwards the Arabs brought back to the extreme west of Europe, mingled with additions from Indian sources. The ancient Indian and Malayan languages promoted trade and national intercourse in the south-eastern Asiatic islands, and in Madagascar; and it is even probable that through intelligence from the Indian trading stations of the Banians, they had a large share in occasioning the bold enterprise of Vasco de Gama. The wide predominance of particular languages, though unfortunately it prepared the early destruction of the displaced idioms, has contributed beneficially to bring mankind together; resembling in this, one of the effects which have followed the extension of Christianity, and which has also been produced by the spread of Buddhism.

Languages, compared with each other, and considered as objects of the natural history of the human mind, being divided into families according to the analogy of their internal structure, have become, (and it is one of the most brilliant results of modern studies in the last sixty or seventy years), a rich source of historical knowledge. Products of the mental power, they lead us back, by the fundamental characters of their organisation, to an obscure and otherwise unknown distance. The comparative study of languages shews how races of nations, now separated by wide regions, are related to each other, and have proceeded from a common seat; it discloses the direction and the path of ancient migrations; in tracing out epochs of development, it recognises in the more or less altered characters of the language, in the permanency of certain forms, or in the already advanced departure from them, which portion of the race has preserved a language nearest to that of their former common dwelling-place. The long chain of the Indo-Germanic languages, from the Ganges to the Iberian extremity of Europe, and from Sicily to the North Cape, furnishes a large field for investigations of this nature into the first or most ancient conditions of language. The same historical comparison of languages leads us to trace the native country of certain productions which, since the earliest times, have been important objects of trade and barter. We find that the Sanscrit names of true Indian productions,—rice, cotton, nard, and sugar,—have passed into the Greek, and partly even into the Semitic languages. (143)

The considerations here indicated, and illustrated by examples, lead us to regard the comparative study of languages as an important means towards arriving, through scientific and true philologic investigations, at a generalisation of views in regard to the relationships of different portions of the human race, which, it has been conjectured, have extended themselves by lines radiating from several points.

We see from what has been said, that the intellectual aids to the gradual development of the science of the Cosmos are of very various kinds; they include, for example, the examination of the structure of language, the decipherment of ancient inscriptions and historical monuments in hieroglyphics and arrow-headed characters, and the increased perfection of mathematics, and especially of that powerful analytical calculus, which brings within our intellectual grasp the figure of the earth, the tides of the ocean, and the regions of space. To these aids we must add, lastly, the material inventions, which have made for us, as it were, new organs, heightening the power of the senses, and bringing men into closer communication with terrestrial forces, and with distant worlds. Noticing here only those instruments which mark great epochs in the history of the knowledge of nature, we may name the telescope, and its too long delayed combination with instruments for angular determinations;—the compound microscope, which affords us the means of following the processes of development in organisation (the formative activity, the origin of being or production, as Aristotle finely says); the compass, with the different mechanical contrivances for investigating the earth's magnetism; the pendulum, employed as a measure of time; the barometer; the thermometer; hygrometric and electrometric apparatus; and the polariscope, in its application to the phænomena of colored polarisation of light, either of the heavenly bodies or of the illumined atmosphere.

The history of the physical contemplation of the universe, based, as we have seen, on the thoughtful consideration of natural phænomena, on the occurrence of influential events and on discoveries which have enlarged our sphere of perception, is, however, to be here presented only in its leading features, and in a fragmentary and general manner. I flatter myself with the hope, that brevity in the treatment may enable the reader more easily to apprehend the spirit in which an image, so difficult to be defined, should, at some future day, be traced. Here, as in the "picture of nature" contained in the first volume of Cosmos, I aim not at completeness in the enumeration of separate parts, but at a clear development of leading ideas, seeking, in the present case, to indicate some of the paths which may be traversed by the physical inquirer in historical investigations. I assume on the part of the reader such a knowledge of the different events, and of their connection and causal relations, as may render it sufficient to name them, and to shew the influence which they have exerted on the gradually increasing knowledge and recognition of nature as a whole. Completeness, I think it necessary to repeat, is neither attainable, nor is it to be regarded as the object of such an undertaking. In making this announcement, for the sake of preserving to my work on the Cosmos the peculiar character which can alone render its execution possible, I doubtless expose myself anew to the strictures of those who dwell less on that which a book contains, than on that which, according to their individual views, ought to be found in it. I have purposely entered far more into detail in the earlier than in the later portions of history. Where the sources from whence the materials are to be drawn are less abundant, combination is less easy, and the opinions propounded may require a fuller reference to authorities less generally known. I have also freely permitted myself to treat my materials at unequal length, where the narration of particulars could impart a more lively interest.

As the recognition of the Cosmos began with intuitive presentiments, and with only a few actual observations made on detached portions of the great realm of nature, so it appears to me, that the historical representation of the contemplation of the universe may fitly proceed first from a limited portion of the earth's surface. I select for this purpose the basin of the Mediterranean, around which dwelt those nations from whose knowledge our western cultivation (the only one of which the progress has been almost uninterrupted), is immediately derived. We may indicate the principal streams through which have flowed the elements of the civilisation, and of the enlarged views of nature, of western Europe; but we cannot trace back these streams to one common primitive fountain. A deep insight into the forces and a recognition of the unity of nature, does not belong to an original and so-called primitive people, notwithstanding that such an insight has been attributed at different periods, and according to different historical views, at one time to a Semitic race in Northern Chaldea, (Arpaxad [144], the Arrapachitis of Ptolemy), and at another, to the race of the Indians and Iraunians in the ancient land of the Zend [145], near the sources of the Oxus and the Jaxartes. History, as founded on testimony, recognises no such primitive people occupying a primary seat of civilisation, and possessing a primitive physical science or knowledge of nature, the light of which was subsequently darkened by the vicious barbarism of later ages. The student of history has to pierce through many superimposed strata of mist, composed of symbolical myths, in order to arrive at the firm ground beneath, on which appear the first germs of human civilisation unfolding according to natural laws. In the early twilight of history, we perceive several shining points already established as centers of civilisation, radiating simultaneously towards each other. Such was Egypt at least five thousand years before our Era; [146] such also were Babylon, Niniveh, Kashmeer, and Iran; such too was China, after the first colony had migrated from the north-eastern declivity of the Kuen-lun into the lower valley of the Hoang-ho. These central points remind us involuntarily of the larger among the sparkling fixed stars, those suns of the regions of space, of which we know, indeed, the brightness, but, with few exceptions, [147] we are not yet acquainted with their relative distances from our planet.

A supposed primitive physical knowledge made known to the first race of men—a wisdom or science of nature possessed by savage nations, and subsequently obscured by civilisation—can find no place in the history of which we treat. We meet with such a belief deeply rooted in the earliest Indian doctrine of Krishna. [148] "Truth was originally deposited with men, but gradually slumbered and was forgotten; the knowledge of it returns like a recollection." We willingly leave it undecided whether the nations which we now call savage are all in a condition of original natural rudeness, or whether, as the structure of their languages often leads us to conjecture, many of them are not rather to be regarded as tribes having lapsed into a savage state,—fragments remaining from the wreck of a civilisation which was early lost. Closer communication with these so-called children of nature discloses nothing of that superior knowledge of terrestrial forces, which the love of the marvellous has sometimes chosen to ascribe to rude nations. There rises, indeed, in the bosom of the savage a vague and awful feeling of the unity of natural forces; but such a feeling has nothing in common with the endeavours to embrace intellectually the connection of phenomena. True cosmical views are the results of observation and ideal combination; they are the fruit of long-continued contact between the mind of man and the external world. Nor are they the work of a single people; in their formation, mutual communication is required, and great if not general intercourse between various nations.

As in the considerations on the reflex action of the external world on the imaginative faculties, which formed the first portion of the present volume, I gathered, from the general history of literature, that which relates to the expression of a vivid feeling of nature, so in the "history of the contemplation of the universe," I select, from the history of general intellectual cultivation, that which marks progress in the recognition of a natural whole. Both these portions, not detached arbitrarily, but according to determinate principles, bear to each other the same relations as do the subjects of study from which they are taken. The history of the intellectual cultivation of mankind includes the history of the elementary powers of the human mind, and therefore, also, of the works in which these powers have manifested themselves in the domains of literature and art. In a similar manner we recognise in the depth and vividness of the feeling for nature, which has been described as differently manifested at different epochs and among different nations, influential incitements to a more sedulous regard to phænomena, and to a grave and earnest investigation of their cosmical connection.

The very variety of the streams by which the elements of the enlarged knowledge of nature have been conveyed, and spread unequally in the course of time over the earth's surface, renders it advisable, as I have already remarked, to begin the history of cosmical contemplation with a single group of nations, viz. with that from which our present western scientific culture is derived. The mental cultivation of the Greeks and Romans is, indeed, of very recent origin compared with that of the Egyptians, the Chinese, and the Indians: but that which the Greeks and Romans received from without, from the east and from the south, associated with that which they themselves originated or carried onwards towards perfection, has been handed down on European ground without interruption, notwithstanding the constant changes of events, and the admixture of foreign elements by the arrival of fresh immigrating races.

The countries, on the other hand, in which many departments of knowledge were cultivated at a much earlier period, have either lapsed into a state of barbarism, whereby this knowledge has been lost, or, whilst preserving their ancient civilisation and firmly established complex civil institutions, as is the case with China, they have made extremely little progress in science and in the industrial arts, and have been still more deficient in participation in that intercourse with the rest of the world, without which general views cannot be formed. The cultivated nations of Europe, and their descendants transplanted to other continents, have, by the gigantic extension of their maritime enterprises, made themselves, as it were, at home simultaneously on almost every coast; and those shores which they do not yet possess they threaten. In their almost uninterruptedly inherited knowledge, and in their far-descended scientific nomenclature, we may discover land-marks in the history of mankind, recalling the various paths or channels by which important discoveries or inventions, or at least their germs, have been conveyed to the nations of Europe. Thus from Eastern Asia has been handed down the knowledge of the directive force and declination of a freely-suspended magnetic bar; from Phœnicia and Egypt, the knowledge of chemical preparations (as glass, animal and vegetable colouring substances, and metallic oxides); and from India, the general use of *position* in determining the greater or less value of a few numerical signs.

Since civilisation has left its early seats in the tropical or sub-tropical zone, it has fixed itself permanently in that part of the world, of which the most northern portions are less cold than the same latitudes in Asia and America. I have already shewn how the continent of Europe is indebted for the mildness of its climate, so favourable to general civilisation, to its character as a western peninsula of Asia; to the broken and varied configuration of its coast line, extolled by Strabo; to its position relatively to Africa, a broad expanse of land within the torrid zone; and to the circumstance that the prevailing winds from the west are warm winds in winter, owing to their passing over a wide extent of ocean. [149] The physical constitution of the surface of Europe has moreover offered fewer impediments to the spread of civilisation, than have the long-extended parallel chains of mountains, the lofty plateaus, and the sandy wastes, which, in Asia and Africa, form barriers between different nations over which it is difficult to pass.

In the enumeration of the leading epochs in the history of the physical contemplation of the universe, I propose, therefore, to dwell first on a small portion of the earth's surface where intercourse between nations, and the enlargement of cosmical views which results from such intercourse, have been most favoured by geographical relations.

PRINCIPAL EPOCHS IN THE HISTORY OF THE PHYSICAL CONTEMPLATION OF THE UNIVERSE.

I.

The Mediterranean taken as the point of departure for the representation of the relations which led to the gradual extension of the idea of the Cosmos.—Connection with the earliest Greek cultivation.—Attempts at distant navigation towards the north-east (the Argonauts); towards the south (Ophir); and towards the west (Colæus of Samos).

PLATO describes the narrow limits of the Mediterranean in a manner quite appropriate to enlarged cosmographical views. He says, in the Phædo, (150) "we who dwell from the Phasis to the Pillars of Hercules, inhabit only a small portion of the earth, in which we have settled round the (interior) sea, like ants or frogs around a marsh." It is from this narrow basin, on the margin of which Egyptian, Phœnician, and Hellenic nations flourished and attained a brilliant civilisation, that the colonisation of great territories in Asia and Africa has proceeded; and that those nautical enterprises have gone forth, which have lifted the veil from the whole western hemisphere of the globe.

The present form of the Mediterranean shews traces of a former subdivision into three smaller closed basins. (151) The Ægean portion is bounded to the south by a curved line, which, commencing at the coast of Caria in Asia Minor, is formed by the islands of Rhodes, Crete, and Cerigo, joining the Peloponnesus not far from Cape Malea. More to the west we have the Ionian Sea, or the Syrtic basin, in which Malta is situated: the western point of Sicily approaches to within forty-eight geographical miles of the African shore; and we might almost regard the sudden but transient elevation of the burning island of Ferdinandea (1831), to the southwest of the limestone rocks of Sciacca, as an effort of nature to reclose the Syrtic basin, by connecting together Cape Grantola, the Adventure bank (examined by Captain Smith), the island of Pantellaria, and the African Cape Bon,—and thus to divide it from the third, the westernmost, or Tyrrhenian basin. This last receives the influx from the western ocean through the passage opened between the Pillars of Hercules, and contains Sardinia and Corsica, the Balearic Islands, and the small volcanic group of the Spanish Columbratæ.

The peculiar form of the Mediterranean was very influential on the early limitation and later extension of Phœnician and Grecian voyages of discovery, of which the latter were long restricted to the Ægean and Syrtic basins. In the Homeric times, continental Italy was still an "unknown land." The Phocæans first opened the Tyrrhenian basin west of Sicily, and navigators to Tartessus reached the Pillars of Hercules. It should not be forgotten that Carthage was founded near the limits of the Tyrrhenian and Syrtic basins. The march of events, the direction of nautical undertakings, and changes in the possession of the empire of the sea, reacting on the enlargement of the sphere of ideas, have all been influenced by the physical configuration of coasts.

A more richly varied and broken outline gives to the northern shore of the Mediterranean an advantage over the southern or Lybian shore, which, according to Strabo, was remarked by Eratosthenes. The three great peninsulas, (153) the Iberian, the Italian, and the Hellenic, with their sinuous and deeply indented shores, form, in combination with the neighbouring islands and opposite coasts, many straits and isthmuses. The configuration of the continent and of the islands, the latter either severed from the main or volcanically elevated in lines, as if over long fissures, early led to geognostical views respecting eruptions, terrestrial revolutions, and overpourings of the swollen higher seas into those which were lower. The Euxine, the Dardanelles, the Straits of Gades, and the Mediterranean with its many islands, were well fitted to give rise to the view of such a system of sluices. The Orphic Argonaut, who probably wrote in Christian times, wove antique legends into his song; he describes the breaking up of the ancient Lyktonia into several islands, when "the dark-haired Poseidon, being wroth with Father Kronion, smote Lyktonia with the golden trident." Similar phantasies, which, indeed, may often have arisen from imperfect knowledge of geographical circumstances, proceeded from the Alexandrian school, where erudition abounded, and a strong predilection was felt for antique legends. It is not necessary to determine here whether the myth of the Atlantis broken into fragments, should be regarded as a distant and western reflex of that of Lyktonia (as I think I have elsewhere shewn to be probable), or whether, as Otfried Müller considers, "the destruction of Lyktonia (Leuconia) refers to the Samothracian tradition of a great flood, which had changed the form of that district." (154)

But, as has already been often remarked, the circumstances which have most of all rendered the geographical position of the Mediterranean so beneficently favourable to the intercourse of nations, and the progressive extension of the knowledge of the world, are the neighbourhood of the peninsula of Asia Minor, projecting from the eastern continent; the numerous islands of the Ægean which have formed a bridge for the passage of civilisation; and the fissure between Arabia, Egypt, and Abyssinia, by which the great Indian ocean, under the name of the Arabian Gulf or Red Sea, advances so as to be only divided by a narrow isthmus from the Delta of the Nile, and from the south-eastern coast of the Mediterranean. By means of these geographical relations, the influence of the sea, as the "uniting element," shewed itself in the increasing power of the Phœnicians, and subsequently also in that of the Hellenic nations, and in the rapid enlargement of the circle of ideas. Civilisation in its earlier seats, in Egypt, on the Euphrates and the Tigris, in the Indian Pentapotamia, and in China, had been confined to the rich alluvial lands watered by wide rivers; but it was otherwise in Phœnicia and in Hellas. The early impulse to maritime undertakings, which shewed itself in the lively and mobile minds of the Greeks and especially of the Ionic branch, found a rich and varied field in the remarkable forms of the Mediterranean, and in its position relatively to the oceans to the south and west.

The Red Sea, formed by the entrance of the Indian Ocean through the Straits of Bab el Mandeb, belongs to a class of great physical phænomena which modern geology has made known to us. The European continent has its principal axis in a north-east and south-west line; but, almost at right angles to this direction, there exists a system of fissures, which have given occasion, in some cases, to the entrance of the water of the sea, and in others, to the elevation of parallel ridges of mountains. We may trace this transverse strike in a south-east and north-west direction, from the Indian Ocean to the mouth of the Elbe in northern Germany; it shews itself in the Red Sea which, in its southern portion, is bordered on both sides by volcanic rocks;—in the Persian Gulf, with the lowlands of the double river Euphrates and Tigris;—in the Zagros mountain chain in Louristan;—in the mountain chains of Greece and the neighbouring islands of the Archipelago; in the Adriatic Sea;—and in the Dalmatian limestone Alps. This intersection (156) of two systems of geodesic lines, N.E.—S.W. and S.E.—N.W. (concerning which I believe the S.E.—N.W. to be the more recent, and that both result from the direction of deep-seated earthquake movements in the interior of the globe), has had an important influence on the destinies of men, and in facilitating the intercourse between nations. The relative positions of Eastern Africa, Arabia, and the peninsula of Hindostan, and their very unequal heating by the sun's rays at different seasons of the year, produce a regular alternation of currents of air (Monsoons), (157) favouring navigation to the Myrrhifera Regio of the Adramites in Southern Arabia, and to the Persian Gulf, India, and Ceylon. During the season of north winds in the Red Sea (April and May to October), the south-west Monsoon prevails from the eastern shore of Africa to the coast of Malabar; whilst from October to April, the north-east Monsoon, which is favourable to the return, coincides with the period of southerly winds between the Straits of Bab-el-Mandeb and the Isthmus of Suez.

Having thus described the theatre on which the Greeks

might receive from different quarters foreign elements of mental cultivation and the knowledge of other countries, I will next notice other nations dwelling near the Mediterranean, who enjoyed an early and high degree of civilisation—the Egyptians, the Phœnicians with their north and west African colonies, and the Etruscans. Immigration and commercial intercourse were powerful agents: the more our historical horizon has been extended in the most recent times, as by the discovery of monuments and inscriptions, and by philosophical investigations into languages, the greater we find to have been the influence which, in the earliest times, the Greeks experienced even from the Euphrates, from Lycia, and through the Phrygians allied to the Thracian tribes.

Concerning the valley of the Nile, which plays so large a part in history, I follow the latest investigations of Lepsius, ([158]) and the results of his important expedition which throws light on the whole of antiquity, in saying that "there exist well-assured cartouches of kings belonging to the commencement of the fourth dynasty of Manetho, which includes the builders of the great pyramids of Gizeh (Chephren or Schafra, Cheops-Chufu, and Menkera or Mencheres). This dynasty commenced thirty-four centuries before our Christian era, and twenty-three centuries before the Doric immigration of the Heraclides into the Peloponnesus. ([159]) The great stone pyramids of Daschur, a little to the south of Gizeh and Sakara, are considered by Lepsius to have been the work of the third dynasty: there are sculptural inscriptions on the blocks of which they are composed, but as yet no kings' names have been discovered. The latest dynasty of the "old kingdom," which terminated at the invasion of the

Hyksos, 1200 years before Homer, was the twelfth of Manetho, to which belonged Amenemha III. who made the original labyrinth, and formed Lake Moeris artificially by excavation and by large dykes of earth to the north and west. After the expulsion of the Hyksos, the "new kingdom" begins with the eighteenth dynasty (1600 B.C.) The great Ramses Miamoun (Ramses II.) was the second monarch of the nineteenth dynasty. The representations on stone which perpetuated the record of his victories were explained to Germanicus by the priests of Thebes. ([160]) He was known to Herodotus under the name of Sesostris, probably from a confusion with the almost equally warlike and powerful conqueror Seti (Setos), who was the father of Ramses II."

I have thought it right to notice these few chronological points, in order that, where we have solid historical ground, we may determine approximately the relative antiquity of great events in Egypt, Phœnicia, and Greece. As I before described in a few words the Mediterranean and its geographical relations, so I have thought it necessary here to indicate the centuries by which the civilisation of the Valley of the Nile preceded that of Greece. Without this double reference to place and time, we cannot, from the very nature of our mental constitution, form to ourselves any clear and satisfactory picture of history.

Civilisation, early awakened and arbitrarily modelled in Egypt by the mental requirements of the people, by the peculiar physical constitution of their country, and by their hierarchical and political institutions, produced there, as everywhere else on the globe, a tendency to intercourse with foreign nations, and to distant military expeditions and settlements. But the records preserved to us by history and by monumental remains indicate only transitory conquests by land, and but little extensive navigation by the Egyptians themselves. This civilised nation, so ancient and so powerful, appears to have done less to produce a permanent influence beyond its own borders, than other races less numerous but more active and mobile. The national cultivation, favourable rather to the masses than to individuals, was, as it were, geographically insulated, and remained, therefore, probably unfruitful as respects the extension of cosmical views. Ramses Miamoun (from 1388 to 1322 B.C., 600 years, therefore, before the first Olympiad of Corœbus) undertook, according to Herodotus, extensive military expeditions into Ethiopia (where Lepsius considers that his most southern works are to be found near Mount Barkal); through Palestinian Syria; and passing from Asia Minor into Europe, to the Scythians, Thracians, and finally to Colchis and the Phasis, on the banks of which, part of his army, weary of their wanderings, finally settled. Ramses was also the first—so said the priests—who, with long ships, subjected to his dominion the dwellers on the coast of the Erythrean, until at length, sailing onwards, he arrived at a sea so shallow as to be no longer navigable. (161) Diodorus says expressly, that Sesoosis (the great Ramses) advanced in India beyond the Ganges, and that he also brought back captives from Babylon. "The only well-assured fact in relation to the nautical pursuits of the native ancient Egyptians is, that from the earliest times they navigated not only the Nile, but also the Arabian Gulf. The famous copper mines near Wadi Magara, on the peninsula of Sinai, were worked as early as in the time of the fourth dynasty, under Cheops-

Chufu. The inscriptions of Hamamat on the Cosseir road, which connected the Valley of the Nile with the western coast of the Red Sea, reach back as far as the sixth dynasty. The canal from Suez was attempted under Ramses the Great, (162) the immediate motive being probably the intercourse with the Arabian copper district." Greater maritime enterprises, such even as the often-contested, but I think, not improbable, circumnavigation of Africa under Nechos II. (611—595 B.C.), were entrusted to Phœnician vessels. Nearly at the same period, but rather earlier, under Nechos's father, Psammetichus (Psemetek), and also somewhat later, after the close of the civil war under Amasis (Aahmes), hired Greek troops, by their settlement at Naucratis, laid the foundation of a permanent foreign commerce, of the introduction of foreign ideas, and of the gradual penetration of Hellenism into Lower Egypt. Thus was deposited a germ of mental freedom,—of a greater independence of local influences,—which developed itself with rapidity and vigour in the new order of things which followed the Macedonian conquest. The opening of the Egyptian ports under Psammetichus marks an epoch so much the more important, since until that period, Egypt, or at least her northern coast, had been as completely closed against all foreigners as Japan now is. (164)

Amongst the cultivated nations, not Hellenic, who dwelt around the Mediterranean in the ancient seats where our modern knowledge originated, we must place the Phœnicians next after the Egyptians. They must be regarded as the most active intermediaries and agents in the connection of nations from the Indian ocean to the west and north of Europe. Limited in many spheres of intellectual development, and addicted rather to the mechanical than to the fine arts, with little of the grand and creative genius of the more thoughtful inhabitants of the Valley of the Nile, the Phœnicians, as an adventurous and far ranging commercial people, and by the formation of colonies, one of which far surpassed the parent city in political power, did nevertheless, earlier than all the other nations surrounding the Mediterranean, influence the course and extension of ideas, and promote richer and more varied views of the physical universe. The Phœnicians had Babylonian weights and measures, (165) and, at least after the Persian dominion, employed for monetary purposes a stamped metallic currency, which, singularly enough, was not possessed by the Egyptians, notwithstanding their advanced political institutions and skill in the arts. But that by which the Phœnicians contributed most to the intellectual advancement of the nations with whom they came in contact, was by the communication of alphabetical writing, of which they had themselves long made use. Although the whole legendary history of a particular colony, founded in Bœotia by Cadmus, may remain wrapped in mythological obscurity, yet it is not the less certain, that it was through the commercial intercourse of the Ionians with the Phœnicians that the Greeks received the characters of their alphabetical writing, which were long termed Phœnician signs. (166) According to the views which, since Champollion's great discovery, have prevailed more and more respecting the early conditions of the development of alphabetical writing, the Phœnician, and all the Semitic written characters, though they may have been originally formed from pictorial writing, are to be regarded as a *phonetic alphabet; i. e.* as an alphabet in which the ideal signification of the pictured signs is wholly disregarded, and these signs or characters are treated exclusively as signs of sound. Such a phonetic alphabet, being in its nature and fundamental form a syllabic alphabet, was suited to satisfy all the requirements of a graphical representation of the phonetic system of a language. "When the Semitic writing," says Lepsius, in his treatise on the alphabet, "passed into Europe to Indo-Germanic nations, who all shew a much stronger tendency to a strict separation between vowels and consonants (a separation to which they could not but be led by the much more significant import of vowels in their languages), this syllabic alphabet underwent very important and influential changes." (167) Amongst the Greeks, the tendency to do away with the syllabic character proceeded to its full accomplishment. Thus not only did the communication of the Phœnician signs to almost all the coasts of the Mediterranean, and even to the north-west coast of Africa, facilitate commercial intercourse and form a common bond between several civilised nations, but this system of written characters, generalised by its graphic flexibility, had a yet higher destination. It became the depository of the noblest results attained by the Hellenic race in the two great spheres of the intellect and the feelings, by investigating thought and by creative imagination; and the medium of transmission through which this imperishable benefit has been bequeathed to the latest posterity.

Nor is it solely as intermediaries, and by conveying an impulse to others, that the Phœnicians have enlarged the elements of cosmical contemplation. They also independently, and by their own discoveries, extended the sphere of knowledge in several directions. Industrial prosperity, founded on extensive maritime commerce, and on the products of labour and skill in the manufactures of Sidon in white and coloured glass, in tissues, and in purple dyes, led, as every where else, to advances in mathematical and chemical knowledge, and especially in the technical arts. "The Sidonians," says Strabo, "are described as active investigators in astronomy as well as in the science of numbers, having been conducted thereto by arithmetical skill and by the practice of nocturnal navigation, both of which are indispensable to trade and to maritime intercourse." In order to indicate the extent of the earth's surface first opened by Phœnician navigation and the Phœnician caravan trade, we must name the settlements on the Bythinian coast (Pronectus and Bythinium), which were probably of very early formation; the Cyclades and several islands of the Ægean visited in the Homeric times; the south of Spain, from whence silver was obtained (Tartessus and Gades); the north of Africa, west of the lesser Syrtis (Utica, Hadrumetum, and Carthage); the countries in the north of Europe, from whence tin (169) and amber were derived; and two trading factories (170) in the Persian gulf, the Baharein islands Tylos and Aradus.

The amber trade, which was probably first directed to the west Cimbrian coasts, (171) and only subsequently to the Baltic and the country of the Esthonians, owes its first origin to the boldness and perseverance of Phœnician coast navigators. In its subsequent extension it offers, in the point of view of which we are treating, a remarkable instance of the influence which may be exerted by a predilection for even a single foreign production, in opening an inland trade between nations, and in making known large tracts of country. In the same way that the Phocæan Massilians brought the British tin across France to the Rhone, the amber was conveyed from people to people through Germany, and by the Celts on either declivity of the Alps to the Padus, and through Pannonia to the Borysthenes. It was this inland traffic which first brought the coasts of the northern ocean into connection with the Euxine and the Adriatic.

Phœnicians from Carthage, and probably from the settlements of Tartessus and Gades which were founded two centuries earlier, visited an important part of the northwest coast of Africa, extending much beyond Cape Bojador; although the Chretes of Hanno is neither the Chremetes of Aristotle's Meteorology, nor yet our Gambia. ([172]) This was the locality of the many towns of Tyrians (according to Strabo even as many as 300,) which were destroyed by Pharusians and Nigritians. ([173]) Among them, Cerne (Dicuil's Gaulea, according to Letronne) was the principal naval station and chief staple for the settlements on the coast. In the west the Canary islands and the Azores (which latter the son of Columbus, Don Fernando, considered to be the first Cassiterides discovered by the Carthaginians), and in the north the Orkneys, the Faroe islands, and Iceland, became the intermediary stations of transit to the New Continent. They indicate the two paths by which the European portion of mankind became acquainted with Central and North America. This consideration gives to the question of the period when Porto Santo, Madeira, and the Canaries were first known to the Phœnicians, either of the mother country or of the cities planted in Iberia and Africa, a great, I might almost say a universal, importance in the history of the world. In a long protracted chain of events we love to trace the first links. It is probable that, from the foundations of Tartessus and Utica by the Phœnicians, fully 2000 years elapsed before the discovery of America by the northern route, *i. e.* before Eric Rauda crossed the ocean to Greenland (an event which was soon followed by voyages to North Carolina), and 2500 years before its discovery by the south western route taken by Columbus from a point of departure near the ancient Phœnician Gadeira.

In following out that generalisation of ideas which belongs to the object of this work, I have here regarded the discovery of a group of islands situated only 168 geographical miles from the coast of Africa, as the first link in a long series of efforts tending in the same direction, and have not connected it with the poetic fiction, sprung from the inmost depths of the mind, of the Elysium, the Islands of the Blest, placed in the far ocean at earth's extremest bounds, and warmed by the near presence of the disk of the setting sun. In this remotest distance was placed the seat of all the charms of life, and of the most precious productions of the earth; ([174]) but as the Greeks' knowledge of the Mediterranean extended, the ideal land, the geographical mythus of the Elysium, was moved farther and farther to the west, beyond the Pillars of Hercules. True geographical knowledge, the discoveries of the Phœnicians,—of the epoch of which we have no certain information,—did not probably first originate the mythus of the Fortunate Islands; but the application was made afterwards, and the geographical discovery did but embody the picture which the imagination had formed, and of which it became, as it were, the substratum.

Later writers, such as the unknown compiler of the

"Collection of Wonderful Narrations," which was ascribed to Aristotle, and of which Timæus made use, and such as the still more circumstantial Diodorus Siculus, when speaking of lovely islands, which may be supposed to be the Canaries, allude to the storms which may have occasioned their accidental discovery. Phœnician and Carthaginian ships, it is said, sailing to the settlements already existing on the Coast of Lybia, were driven out to sea; the event is placed at the early period of the Tyrrhenian naval power, during the strife between the Tyrrhenian Pelasgians and the Phœnicians. Statius Sebosus and the Numidian King Juba first gave names to the different islands, but unfortunately not Punic names, although certainly according to notices drawn from Punic books. Plutarch having said that Sertorius, when driven out of Spain, and after the loss of his fleet, thought of taking refuge "in a group, consisting of only two islands, situated in the Atlantic, ten thousand stadia to the west of the mouth of the Betis," he has been supposed to refer to the two islands of Porto Santo and Madeira, (175) indicated not obscurely by Pliny as Purpurariæ. The strong current which, beyond the Pillars of Hercules, sets from north west to south east, may long have prevented the coast navigators from discovering the islands most distant from the continent, of which only the smaller (Porto Santo) was found inhabited in the fifteenth century. The curvature of the earth would prevent the summit of the great volcano of Teneriffe from being seen, even with a strong refraction, by the Phœnician ships sailing along the coast of the continent; but it appears from my researches (176) that it might have been discovered from the heights near Cape

Bojador under favourable circumstances, and especially during eruptions, and by the aid of reflection from an elevated cloud above the volcano. It has even been asserted that eruptions of Etna have been seen in recent times from Mount Taygetos. (177)

In noticing the elements of a more extended knowledge of the earth, which early flowed in to the Greeks from other parts of the Mediterranean, we have hitherto followed the Phœnicians and Carthaginians in their intercourse with the northern countries from whence tin and amber were derived, and in their settlements near the tropics on the west coast of Africa. We have now to speak of a southern navigation of the same people to far within the torrid zone, four thousand geographical miles east of Cerne and Hanno's western horn, in the Prasodic and Indian Seas. Whatever doubts may remain as to the particular locality of the distant "gold lands" Ophir and Supara, — whether these gold lands were on the west coast of the Indian peninsula, or on the east coast of Africa,—it is not the less certain that this active Semitic race, early acquainted with written characters, roving extensively over the surface of the earth, and bringing its various inhabitants into relation with each other, came into contact with the productions of the most varied climates, ranging from the Cassiterides to south of the Straits of Bab-el-Mandeb, and far within the region of the tropics. The Tyrian flag waved at the same time in Britain and in the Indian ocean. The Phœnicians had formed trading settlements in the most northern part of the Arabian Gulf, in the harbours of Elath and Ezion Geber, as well as in the Persian Gulf at Aradus and Tylos, where, according to Strabo, there were temples similar in their style of architecture to those of the Mediterranean. ([178]) The caravan trade which the Phœnicians carried on, in order to procure spices and incense, was directed by Palmyra to Arabia Felix, and to the Chaldean or Nabathæic Gerrha, on the western or Arabian shore of the Persian Gulf.

The expeditions of Hiram and Solomon, conjoint undertakings of the Tyrians and Israelites, sailed from Ezion Geber through the Straits of Bab-el-Mandeb to Ophir (Opheir, Sophir, Sophara, the Sanscrit Supara ([179]) of Ptolemy). Solomon, who loved magnificence, caused a fleet to be built in the Red Sea, and Hiram supplied him with Phœnician mariners well acquainted with navigation, and also Tyrian vessels, "ships of Tarshish." ([180]) The articles of merchandise which were brought back from Ophir were gold, silver, sandal wood (algummim), precious stones, ivory, apes (kophim), and peacocks (thukkiim). The names by which these articles are designated are not Hebrew but Indian. ([181]) The researches of Gesenius, Benfey, and Lassen, have made it extremely probable that the western shores of the Indian peninsula were visited by the Phœnicians, who, by their colonies in the Persian Gulf, and by their intercourse with the Gerrhans, were early acquainted with the periodically blowing monsoons. Columbus was even persuaded that Ophir (the El Dorado of Solomon), and the mountain Sopora, were a part of Eastern Asia—of the Chersonesus Aurea of Ptolemy. ([182]) If it seem difficult to view Western India as a country productive in gold, it will be sufficient, without referring to the "gold-seeking ants," or to Ctesias's unmistakable description of a foundry, (in which, however, according to his account, gold and iron were melted together), ([183]) to remember the vicinity of several places notable in this respect. Such are the Southern part of Arabia, the Island of Dioscorides (Diu Zokotora of the moderns, a corruption of the Sanscrit Dvipa Sukhatara), cultivated by Indian settlers,—and the auriferous East African coast of Sofala. Arabia, and the island just mentioned to the south east of the Straits of Bab-el-Mandeb, formed for the combined Phœnician and Hebrew commerce intermediate and uniting links between the Indian peninsula and the East Coast of Africa. Indians had settled on the latter from the earliest times as on a shore opposite to their own, and the traders to Ophir might find in the basin of the Erythrean and Indian Seas other sources of gold than India itself.

Less influential than the Phœnicians in connecting different nations and in extending the geographical horizon, and early subjected to the Greek influence of Pelasgic Tyrrhenians arriving from the sea, we have next to consider the austere and gloomy nation of the Etruscans. A not inconsiderable inland trade with the remote amber countries was carried on by them, passing through Northern Italy, and across the Alps, where a "via sacra" (184) was protected by all the neighbouring tribes. It seems to have been almost by the same route that the primitive Tuscan people, the Rasenæ, came from Rhætia to the Padus, and even still farther southward. That which is most important to notice, according to the point of view which we have selected, and in which we seek always to seize what is most general and permanent, is the influence exerted by the commonwealth of Etruria on the earliest Roman civil institutions, and thus upon the whole of Roman life. The reflex action of this influence, in its remotely derived consequences, may be said to be still politically operative even at the present day, in as far as through Rome it has for centuries promoted, or at least has given a peculiar character to the civilisation of a large portion of the human race. (185)

A peculiar characteristic of the Tuscans, which is especially deserving of notice in the present work, was the disposition to cultivate intimate relations with certain natural phænomena. Divination, which was the occupation of the caste of equestrian and warrior priests, occasioned the daily observation of the meteorological processes of the atmosphere. The "Fulguratores" occupied themselves with the examination of the direction of lightnings, with "drawing them down," and "turning them aside." (186) They distinguished carefully between lightnings from the elevated region of clouds, and lightnings sent from below by Saturn (an earth god), (187) and called Saturnian lightnings: a distinction which modern physical science has considered deserving of particular attention. Thus there arose official records of the occurrence of thunderstorms. (188) The "Aquælicium" practised by the Etruscans, the supposed art of finding water and drawing forth hidden springs, implied in the Aquileges an attentive examination of the natural indications of the stratification of rocks, and of the inequalities of the ground. Diodorus praises their habits of investigating nature; it may be remarked in addition, that the high-born and powerful sacerdotal caste of the Tarquinii offered the rare example of favouring physical knowledge.

Before proceeding to the Greeks,—to that highly gifted race in whose intellectual culture our own is most deeply rooted, and through whom has been transmitted to us an important part of all the earlier views of nature, and knowledge of countries and of nations,—we have named the more ancient seats of civilisation in Egypt, Phœnicia, and Etruria; and have considered the basin of the Mediterranean, in its peculiarities of form and of geographical position relatively to other portions of the earth's surface, and in regard to the influence which these have exerted on commercial intercourse with the West Coast of Africa, with the North of Europe, and with the Arabian and Indian Seas. No portion of the earth has been the theatre of more frequent changes in the possession of power, or of more active and varied movement under mental influences. The progressive movement propagated itself widely and enduringly through the Greeks and the Romans, and especially after the latter had broken the Phœnicio-Carthaginian power. That which we call the beginning of history, is but the record of later generations. It is a privilege of the period at which we live, that by brilliant advances in the general and comparative study of languages, by the more careful search for monuments, and by their more certain interpretation, the historical investigator finds that his scope of vision enlarges daily; and penetrating through successive strata, a higher antiquity begins to reveal itself to his eyes. Besides the different cultivated nations of the Mediterranean which we have named, there are also others shewing traces of ancient civilisation,—as in Western Asia the Phrygians and Lycians,—and in the extreme west the Turduli and Turdetani. (189) Strabo says of the latter, "they are the most civilised of all the Iberians; they have the art of writing, and possess written books of old memorials, and also poems and laws in metrical verse, to which they ascribe an age of six thousand years." I have referred to these particular instances as indicating how much of ancient cultivation, even in European nations, has disappeared without leaving traces which we can follow; and for the sake of shewing that the history of early cosmical views, or of the physical contemplation of which we treat, is necessarily confined within restricted limits.

Beyond the 48th degree of latitude, north of the sea of Azof and of the Caspian, between the Don, the Volga, and the Jaik, where the latter flows from the southern and auriferous portion of the Ural, Europe and Asia melt as it were into each other in wide plains or steppes. Herodotus, and before him Pherecydes of Syros, considered the whole of Northern Scythian Asia (Siberia), as belonging to Sarmatic Europe, ([190]) and even as forming a part of Europe itself. Towards the south, Europe and Asia are distinctly separated; but the far projecting peninsula of Asia Minor, and the varied shores and islands of the Ægean Sea, forming, as it were, a bridge between the two continents, have afforded an easy transit to races, languages, manners, and civilisation. Western Asia has been from the earliest times the great highway of nations migrating from the East, as was the north west of Hellas for the Illyrian races. The archipelago of the Ægean, divided under Phœnician, Persian, and Greek dominion, formed the intermediate link between the Greek world and the far East.

When the Phrygian was incorporated with the Lydian and the latter with the Persian empire, the circle of ideas of the Asiatic and European Greeks was enlarged by the contact. The Persian sway was extended by the warlike enterprises of Cambyses and Darius Hystaspes, from Cyrene and the Nile to the fruitful lands on the Euphrates and the Indus. A Greek, Scylax of Karyanda, was employed to examine the course of the Indus, from the then kingdom of Kashmeer (Kaspapyrus), (191) to the mouth of the river. The Greeks had carried on an active intercourse with Egypt (with Naucratis and the Pelusiac arm of the Nile) under Psammetichus and Amasis, (192) before the Persian conquest. In these various ways many Greeks were withdrawn from their native land, not only in the plantation of distant colonies which we shall have occasion to refer to in the sequel, but also as hired soldiers, forming the nucleus of foreign armies, in Carthage, (193) Egypt, Babylon, Persia, and the Bactrian country round the Oxus.

A deeper consideration of the individual character and popular temperament of the different Greek races has shewn, that if a grave and exclusive reserve in respect to all beyond their own boundaries prevailed amongst the Dorians, and partially among the Æolians, the gayer Ionic race, on the other hand, were distinguished by a vividness of life, incessantly stimulated by energetic love of action, and by eager desire of investigation, to expand towards the world without as well as to expatiate in inward contemplation. Directed by the objective tendency of their mode of thought, and embellished by the richest imagination in poetry and art, Ionic life, when transplanted in the colonised cities to other shores, scattered every where the beneficent germs of progressive cultivation.

As the Grecian landscape possesses in a high degree the peculiar charm of the intimate blending of land and sea, (194) so likewise was the broken configuration of the coast line, which produced this blending, well fitted to invite to early navigation, active commercial intercourse, and contact with strangers. The dominion of the sea by the Cretans and

Rhodians was followed by the expeditions of the Samians, Phocæans, Taphians, and Thesprotians, which, it must be admitted, were at first directed to carrying off captives and to plunder. Hesiod's aversion to a maritime life may probably be regarded as an individual sentiment, though it may also indicate that at an early stage of civilisation inexperience and timidity arising from want of knowledge of nautical affairs prevailed on the mainland of Greece. On the other hand, the most ancient legendary stories and myths relate to extensive wanderings, as if the youthful fancy of mankind delighted in the contrast between these ideal creations and the restricted reality. Examples of these are seen in the journeyings of Dionysus and of the Tyrian Hercules (Melkart, in the temple at Gadeira), the wanderings of Io, (195) and those of the often resuscitated Aristeas, of the marvellous Hyperborean Abaris, in whose guiding arrow (196) some have thought that they recognised the compass. We see in these journeyings the reciprocal reflection of occurrences and of ancient views of the world, and we can even trace the reaction of the progressive advance in the latter on the mixed mythical and historical narrations. In the wanderings of the heroes returning from Troy, Aristonichus makes Menelaus circumnavigate Africa, (197) and sail from Gadeira to India five hundred years before Nechos.

In the period of which we are now treating, *i. e.* in the history of the Greek world previous to the Macedonian expeditions to Asia, three classes of events especially influenced the Hellenic view of the universe; these were the attempts made to penetrate beyond the basin of the Mediterranean towards the East, the attempts towards the West, and the foundation of numerous colonies from the Straits of Hercules to the

North Eastern part of the Euxine. These Greek colonies were far more varied in their political constitution, and far more favourable to the progress of intellectual cultivation, than those of the Phœnicians and Carthaginians in the Ægean Sea, in Sicily, Iberia, and on the North and West Coasts of Africa.

The pressing forwards towards the East about twelve centuries before our era and a century and a half after Ramses Miamoun (Sesostris), when regarded as an historical event, is called the "expedition of the Argonauts to Colchis." The actual reality which, in this narration, is clothed in a mythical garb, or mingled with ideal features to which the minds of the narrators gave birth, was the fulfilment of a national desire to open the inhospitable Euxine. The legend of Prometheus, and the unbinding the chains of the fire-bringing Titan on the Caucasus by Hercules in journeying eastward,—the ascent of Io from the valley of the Hybrites (198) towards the Caucasus,—and the mythus of Phryxus and Helle,—all point to the same path on which Phœnician navigators had earlier adventured.

Before the Doric and Æolic migration, the Bœotian Orchomenus, near the north end of the Lake of Copais, was a rich commercial city of the Minyans. The Argonautic expedition, however, began at Iolchus, the chief seat of the Thessalian Minyans on the Pagasæan Gulf. The locality of the legend, which, as respects the aim and supposed termination of the enterprise, has at different times undergone various modifications, (199) became attached to the mouth of the Phasis (Rion), and to Colchis, a seat of more ancient civilisation, instead of to the undefined distant land of Æa. The voyages of the Milesians, and the numerous towns planted by them on the Euxine, procured a more exact knowledge of the north and east boundaries of that sea, thus giving to the geographical portion of the mythus more definite outlines. An important series of new views began at the same time to open; the west coast of the neighbouring Caspian had long been the only one known, and Hecatæus still regarded this western shore (200) as that of the encircling eastern ocean; it was the venerable father of history who first taught the fact, which after him was again contested for six centuries until the time of Ptolemy, that the Caspian Sea is a closed basin, surrounded by land on every side.

In the north east corner of the Black Sea an extensive field was also opened to ethnology. Men were astonished at the multiplicity of languages which they encountered; (201) and the want of skilful interpreters (the first aids and rough instruments of the comparative study of languages) was strongly felt. The exchange of commodities led traders beyond the Mæotic Gulf (which was supposed to be of far larger dimensions than it really is), through the steppe where the horde of the central Kirghis now pasture their herds,—and through a chain of Scythian-Scolotic tribes of the Argippæans and Issedones (who I take to be of Indo-Germanic (202) origin), to the Arimaspes (203) dwelling on the northern declivity of the Altai, and possessing much gold. Here is the ancient "kingdom of the Griffin," the site of the meteorological mythus of the Hyperboreans, (204) which has wandered with Hercules far to the westward.

It may be conjectured that the part of Northern Asia above alluded to (which has again been rendered celebrated in our own days by the Siberian gold washings), as well as the large quantity of gold which, in the time of Herodotus, had been accumulated among the Massagetae (a tribe of Gothic descent), became, by means of the intercourse opened with the Euxine, an important source of wealth and luxury to the Greeks. I place the locality of this source between the 53d and 55th degrees of latitude. The region of auriferous sand, of which the Daradas (Darders or Derders, mentioned in the Mahabharata, and in the fragments of Megasthenes,) gave intelligence to the travellers, and with which the often repeated fable of the gigantic ants became connected, owing to the accidental double meaning of a name, (205) belongs to a more southern latitude 35° or 37°. It would fall (according to which of two combinations was preferred), either in the Thibetian high land east of the Bolor chain, between the Himalaya and Kuen-lün, west of Iskardo; or north of those mountains, towards the desert of Gobi, which is also described as being rich in gold by the Chinese traveller and accurate observer Hiuen-thsang, in the beginning of the seventh century of our era. How much more accessible to the trade of the Milesian colonies on the north east of the Euxine, must have been the gold of the Arimaspes and the Massagetæ! It has appeared to me suitable to the subject of the present portion of my work, to allude thus generally to all that belongs to an important and still recently operating result of the opening of the Euxine, and of the first advances of the Greeks towards the East.

The great event, so productive of change, of the Doric migration and the return of the Heraclides to the Peloponnesus, falls about a century and a half after the semi-mythical expedition of the Argonauts, *i. e.* after the opening of the Euxine to Greek navigation and commerce. This migration, together with the foundation of new states and new institutions, first gave rise to the *systematic* establishment of colonial cities, which marks an important epoch in the history of Greece, and which became most influential on intellectual cultivation based on enlarged views of the natural world. The more intimate connection of Europe and Asia was especially dependent on the establishment of colonies; they formed a chain from Sinope, Dioscurias, and the Tauric Panticapæum, to Saguntum and Cyrene; the latter founded from the rainless Thera.

By no ancient nation were more numerous, or for the most part more powerful, colonial cities established; but it should also be remarked, that four or five centuries elapsed from the foundation of the oldest Æolian colonies, among which Mytilene and Smyrna were chiefly distinguished, to the foundation of Syracuse, Croton, and Cyrene. The Indians and the Malays only attempted the formation of feeble settlements on the East Coast of Africa, in Socotora (Dioscorides), and in the South Asiatic Archipelago. The Phœnicians had, it is true, a highly advanced colonial system, extending over a still larger space than the Grecian, stretching (although with wide interruptions between the stations) from the Persian Gulf to Cerne on the West Coast of Africa. No mother country has ever founded a colony which became at once so powerful in conquest and in commerce as Carthage. But Carthage, notwithstanding her greatness, was far inferior to the Greek colonial cities in all that belongs to intellectual culture, and to the most noble and beautiful creations of art.

Let us not forget that there flourished at the same time

many populous Greek cities in Asia Minor, on the shores of the Ægean Sea, in Lower Italy, and in Sicily; that Miletus and Massilia became, like Carthage, the founders of fresh colonies; that Syracuse, at the summit of its power, fought against Athens, and against the armies of Hannibal and of Hamilcar; and that Miletus was for a long time the first commercial city in the world after Tyre and Carthage.

Whilst a life so rich in intellectual movement and animation was thus developed externally by the activity of a people whose internal state was so often violently agitated, and whilst the native cultivation, transplanted to other shores, propagated itself afresh, and prosperity increased, new germs of mental national development were every where elicited. Community of language and of worship bound together the most distant members, and through them the mother country took part in the wide circle of the life of other nations. Foreign elements were received into the Greek world without detracting anything from the greatness of its own independent character. No doubt the influence of contact with the East, and with Egypt before it had become Persian, more than a hundred years before the invasion of Cambyses,—must have been more permanent in its nature, than the influence of the settlements of Cecrops from Sais, of Cadmus from Phœnicia, and of Danaus from Chemmis, the reality of which has been much contested, and is at least wrapped in obscurity.

The peculiar characteristics which, pervading the whole organisation of the Greek colonies, distinguished them from all others, and especially from the Phœnician, arose from the distinctness and original diversity of the races into which the parent nation was divided. In the Hellenic colonies, as in all that belonged to ancient Greece, there existed a mixture of uniting and dissevering forces, which by their opposition imparted variety of tone, form, and character, not only to ideas and feelings, but also to poetic and artistic conceptions, and gave to all that rich luxuriance and fulness of life, in which apparently hostile forces are resolved, according to a higher universal order, into combining harmony.

If Miletus, Ephesus, and Colophon were Ionic, Cos, Rhodes, and Halicarnassus Doric, and Croton and Sybaris Achaian, yet in the midst of all this diversity, and even where, as in lower Italy, towns founded by different races stood side by side, the power of the Homeric songs exercised over all alike its uniting spell. Notwithstanding the deeply rooted contrasts of manners and of political institutions, and notwithstanding the fluctuations of the latter, still Greek nationality remained unbroken and undivided, and the wide range of ideas and of types of art, achieved by the several races, was regarded as the common property of the entire united nation.

There still remains to notice, in the present section, the third point to which I before referred, as having been, concurrently with the opening of the Euxine, and the establishment of colonies along the margin of the Mediterranean, influential on the enlargement of physical views. The foundation of Tartessus and Gades, where a temple was dedicated to the wandering divinity Melkart (a son of Baal), and the colony of Utica, more ancient than Carthage, remind us that Phœnician ships had sailed in the open ocean for several centuries, when the straits, which Pindar termed the "Gadeirian Gate" (206), were still closed to the Greeks. As the Milesians in the East, by opening the

Euxine (207), laid the groundwork of communications which led to an active overland commerce with the north of Europe and Asia, and in much later times with the Oxus and the Indus, so the Samians (208) and the Phocæans (209) were the first among the Greeks who sought to penetrate to the west beyond the limits of the Mediterranean.

Colæus of Samos sailed for Egypt, where at that time an intercourse with the Greeks (which perhaps was only the renewal of former communications) had begun to take place under Psammetichus; he was driven by easterly winds and tempests to the island of Platea, and thence, Herodotus significantly adds "not without divine direction," through the Straits into the ocean. It was not merely the magnitude of the unexpected gain of a commerce opened with the Iberian Tartessus, but still more the discovery in space, the entrance into a world before unknown or thought of only in mythical conjectures, which gave to this event grandeur and celebrity throughout the Mediterranean, wherever the Greek tongue was understood. Here, beyond the Pillars of Hercules (earlier called the Pillars of Briareus, of Ægæon, and of Cronos), at the western margin of the Earth, on the way to the Elysian regions and to the Hesperides, the Greeks first saw the primeval waters of the all-encircling ocean (ωρανος) (210), the origin, as they believed, of all rivers.

On arriving at the Phasis, the explorers of the Euxine had found that sea terminated by a shore, beyond which a fabled "Sun lake" was supposed to exist; but the Greeks who reached the Atlantic, on looking southward from Gadeira and Tartessus, gazed onward into a boundless region. It was this which, for fifteen hundred years, gave to the "gate of the interior sea" a peculiar importance, Ever stretching forwards towards that which lay beyond, one maritime people after another, Phœnicians, Greeks, Arabians, Catalans, Majorcans, Frenchmen from Dieppe and La Rochelle, Genoese, Venetians, Portuguese, and Spaniards, made successive efforts to penetrate onwards in the Atlantic Ocean, which was long regarded as a miry, shallow, misty sea of darkness (mare tenebrosum); until, as it were station by station, by the Canaries and the Azores, they at last arrived at the New Continent, which, however, Northmen had already reached at an earlier period and by another route.

When the expeditions of Alexander were making known to the Greeks the regions of the East, considerations on the form of the Earth were leading the great Stagyrite ([211]) to the idea of the nearness of India to the Pillars of Hercules; Strabo even formed the conjecture, that in the northern hemisphere—perhaps in the parallel which passes through the Pillars, through the island of Rhodes, and through Thinæ—"there might exist intermediately between the shores of western Europe and eastern Asia *several other habitable lands*" ([212]). The assignment of the locality of such lands in the continuation of the length of the Mediterranean was connected with a grand geographical view put forward by Eratosthenes and extensively entertained in antiquity, according to which the whole of the old continent, in its widest extent from west to east, nearly in the parallel of 36°, would form an almost continuous line of elevation ([213]).

But the expedition of Colæus of Samos not only marked an epoch which offered to the Greek races, and to the nations which inherited their civilisation, new prospects and a new outlet for maritime enterprises,—it was also the means of making known a fact by which the range of physical ideas was more immediately enlarged. A great natural phenomenon which, by the periodical upraising of the level of the sea, renders visible the relations which connect the Earth with the Moon and the Sun, now first permanently arrested attention. When seen in the Syrtes of Africa, this phenomenon had appeared to the Greeks accidental and irregular, and had been sometimes even an occasion of danger. Posidonius now observed the ebb and flood at Ilipa and Gadeira, and compared his observations with what the experienced Phœnicians were able to tell him respecting the influence of the Moon. ([214])

EPOCHS IN THE HISTORY OF THE CONTEMPLATION OF THE UNIVERSE. CONQUESTS OF ALEXANDER.

II.

Military Expeditions of the Macedonians under Alexander the Great. —Change in the mutual relations of different parts of the World.—Fusion of the West with the East, by the promotion, through Greek influence, of a union between different nations from the Nile to the Euphrates, the Jaxartes and the Indus.— The knowledge of nature possessed by the Greeks suddenly enlarged, both by direct observation, and by intercourse with nations addicted to industry and commerce, and possessing an ancient civilization.

THE Macedonian Expeditions under Alexander the Great, the downfal of the Persian Empire, the beginning of intercourse with Western India, and the influence of the 116 years' duration of the Greco-Bactrian kingdom, mark one of the most important epochs of General History; or of that part of the progressive development of the History of the Human Race, which treats of the more intimate communication and union of the European countries of the West with South-Western Asia, the Valley of the Nile and Lybia. The sphere of the development of community of life, or of the common action and mutual influence of different nations, was not only immensely enlarged in material space, but it was also powerfully strengthened, and its moral grandeur increased, by the constant tendency of the unceasing efforts of the conqueror towards a blending of all the different races, and the formation of a general unity, under the animating influences of the Grecian spirit (215). The foundation of so many new cities at points the selection of which indicates higher and more general aims, the formation and arrangement of an independent community for the government of those cities, the tenderness of treatment towards national usages and native worship, all testify that the plan for a great organic whole was laid. At a later period, as is always the case, much which may not have been originally comprehended in the plan, developed itself from the nature of the relations established. If we remember that only 52 Olympiads elapsed from the battle of the Granicus to the destructive irruption of the Sacæ and Tochari into Bactria, we shall look with admiration on the permanent influence, and the wonderfully uniting and combining power of the Greek cultivation thus introduced from the West; which mingled with Arabian, and with later Persian and Indian knowledge, exerted its action until far into the middle ages, so as to render it often doubtful what to ascribe to Grecian influence, and what to the original spirit of invention or discovery of those Asiatic nations.

All the civil institutions and measures of this daring conqueror shew that the principle of union and unity, or rather a sense of the useful political influence of this principle, was deeply seated in his mind. Even as applied to Greece, it had been early impressed upon him by his great teacher. In the Politics of Aristotle (216) we read:— "The Asiatic nations are not wanting in activity of mind and skill in art; yet they live listlessly in subjection and servitude, while the Greeks, vigorous and susceptible, living in freedom and therefore well governed, *might, if they were united in one state, subdue and rule over all barbarians.*" Thus the Stagirite wrote during his second stay at Athens (217), before Alexander had yet passed the Granicus. These maxims, however, the Stagirite might elsewhere have spoken of an unlimited dominion (πανβασιλεία) as unnatural, doubtless made a more powerful impression on the mind of the conqueror, than the imaginative accounts of India given by Ctesias, to which August Wilhelm von Schlegel, and before him Ste. Croix, attributed so much importance (218).

The preceding section was devoted to a brief description of the influence of the sea as the combining and uniting element; we have shewn how this influence was extended by the navigation of the Phœnicians, Carthaginians, Tyrrhenians, and Tuscans; and how the Greeks, having their naval power strengthened by numerous colonies, advanced from the Basin of the Mediterranean towards the east and the west, by the Argonauts from Iolchos and by the Samian Colæus; and how towards the south the expeditions of Solomon and Hiram passing through the Red Sea, visited the distant Gold lands in voyages to Ophir. The present section will conduct us principally into the interior of a great continent, on paths opened by land traffic and by river navigation. In the short interval of twelve years there followed successively, the expeditions into Western Asia and Syria, with the battles of the Granicus and of the passes of the Issus; the siege and taking of Tyre; the easy possession of Egypt; the Babylonian and Persian campaign, in which at Arbela (in the plain of Gaugamela) the world-dominion of the Achæmenides was annihilated; the expedition to Bactria and Sogdiana, between the Hindoo Coosh and the Jaxartes (Syr); and, lastly, the daring advance into the country of the five rivers (Pentapotamia) of Western India. Alexander planted Greek settlements almost every where, and diffused Grecian manners over the immense region extending from the temple of Ammon in the Lybian Oasis, and from Alexandria on the western Delta of the Nile, to the Northern Alexandria on the Jaxartes, the present Kodjend in Fergana.

The extension of the new field opened to consideration —and this is the point of view from which we must regard the enterprises of the Macedonian conqueror and the continuance of the Bactrian Empire,—proceeded from the large geographical space made known, and the diversity of climates, from Cyropolis on the Jaxartes in the latitude of Tiflis and Rome, to the eastern Delta of the Indus, near Tira, under the tropic of Cancer. Let us add the wonderful variety in the character and elevation of the ground, including rich and fruitful lands, desert wastes, and snowy mountains; the novelty and gigantic size of the productions of the animal and vegetable kingdoms; the aspect and geographical distribution of races of men differing in colour; the living contact with the nations of the East, highly gifted in some respects, enjoying a civilization of high antiquity, with their religious myths, their systems of philosophy, their astronomical knowledge, and their astrological phantasies. At no other epoch (with the exception, eighteen centuries and a half later, of the discovery and opening of tropical America), was there offered, at one time and to one part of the human race, a greater influx of new views of nature, and more abundant materials for the foundation of physical geography and comparative ethnological studies. The vividness of the impression produced thereby is testified by the whole of western literature; it is testified even by the doubts (always attendant on what speaks to our imagination in the description of scenes of nature), which the accounts of Megasthenes, Nearchus, Aristobulus, and other followers of Alexander, raised in the minds of Greek and subsequently of Roman writers. Those narrators, subject to the colouring and influence of the period in which they lived, and constantly mixing up facts and individual opinions or conjectures, have experienced the changeful fate of all travellers, from bitter blame at first to subsequent milder criticism and justification. The latter has especially prevailed in our days, when a deep study of Sanscrit, a more general knowledge of native geographical names, Bactrian coins discovered in Topes, and above all the immediate view of the country itself and of its organic productions, have furnished to critics elements which were wanting to the partial knowledge of Eratosthenes so frequent in censure, of Strabo, and of Pliny ([219]).

If we compare in difference of longitude the length of the Mediterranean with the distance from west to east which divides Asia Minor from the shores of the Hyphasis (Beas), and from the "Altars of Return," we perceive that the geography of the Greeks was doubled in the course of a few years. In order to indicate more particularly the character of that which I have termed the rich increase of materials for physical geography and natural knowledge obtained by the expeditions of Alexander, I would refer first to the remarkable diversity presented by the earth's surface. In the countries which the army traversed, low lands,—deserts devoid of vegetation or salt steppes, (as on the north of the Asferah chain which is a continuation of the Tian-schan),—and the four large, cultivated, and rich alluvial districts of the Euphrates, the Indus, the Oxus, and the Jaxartes, — contrasted with snowy mountains of nearly 20,000 feet of elevation. The Hindoo Coosh, or Indian Caucasus of the Macedonians, is a continuation of the Kuen-lun of North Thibet, and in its further extension towards Herat, on the west of the transverse north and south chain of Bolor, it divides into two great chains bounding Kafiristan ([220]), the southern of which is the loftiest and most important. Alexander passed over the plateau of Bamian, which has more than 8000 feet of elevation, and in which the cave of Prometheus was supposed to be seen ([221]), gained the crest of Kohibaba, and passed over Kabura, and along the course of the Choes to cross the Indus above the present Attock. The Hindoo Coosh, crowned with eternal snow, which, according to Burnes, begins near Bamian at an elevation of 12,200 French feet, must, when compared with the humbler height of the Taurus to which the Greeks were accustomed, have given to them occasion to recognise on a more colossal scale the superposition of different zones of climate and vegetation. That which elementary nature unfolds thus visibly, when presented to the senses of men produces in susceptible minds a deep and lasting effect. Strabo gives a highly graphic description of the passage over the mountainous land of the Paropanisadæ, where the army opened for itself with toil a passage through the snow, and where all arborescent vegetation ceases (222).

The dwellers in the west received through the Macedonian settlements accurate accounts of Indian pro ductions of nature and of art, of which little more than the names were previously known by reports derived either through more ancient commercial connections, or through Ctesias of Cnidos who had lived for seventeen years at the Persian court as the physician of Artaxerxes Mnemon. Such were the watered rice fields, of the cultivation of which Aristobulus gave a particular account; the cotton shrub and the fine tissues and paper (223) for which it furnished the materials; spices and opium; wine made from rice, and from the juice of palms the Sanscrit name of which, tala, has been preserved by Arrian (224); sugar from the sugar-cane (225), which, indeed, is often confounded by the Greek and Roman writers with the Tabaschir of bamboo stems; wool from the great Bombax trees (226); shawls from the wool of the Thibetian goat; silken (Seric) tissues (227); oil of white sesamum (Sanscrit, tila); oil of roses and other perfumes; lac (Sanscrit, lakscha, and in the vulgar tongue, lakkha) (228); and, lastly, the hardened Indian wootz steel.

Besides the knowledge of these productions, which soon became the objects of an extensive commerce, and of which several were transplanted into Arabia by the Seleucidæ (229), the aspect of nature in these richly adorned subtropical regions procured for the Greeks enjoyments of a different kind. Gigantic forms of plants and animals never before seen filled the imagination with exciting imagery. Writers from whose severe and scientific style any degree of inspiration is elsewhere entirely absent, become poetical when describing the habits of the elephant,— the height of the trees, "to the summit of which an arrow cannot reach, and whose leaves are broader than the shields of infantry,"— the bamboo, a light, feathery, arborescent grass, "of which single joints (internodia) served as four-oared boats,—and the Indian fig-tree, whose pendant branches take root around the parent stem, which attains a diameter of 28 feet, "forming," as Onesicritus expresses himself with great truth to nature, "a leafy canopy similar to a many-pillared tent." The tree-fern, which according to my feelings is the greatest ornament of the tropics, is never mentioned by Alexander's companions ([230]); but they speak of the magnificent fan-like umbrella palm, and of the delicate and ever fresh green of the cultivated banana ([231]).

Now for the first time the knowledge of a large part of the earth's surface was truly opened. The world of objects came forward with preponderating power to meet that of subjective creation; and while the Grecian language and literature, and their fertilising influence on the human mind, were widely diffused through the medium of Alexander's conquests, at the same time scientific observation and the systematic availment of the knowledge obtained, were brought into clear light by the teaching and example of Aristotle ([232]). We touch here on the happy coincidence by which, at the very same epoch when there was suddenly offered so immense a supply of new materials of human knowledge, their co-ordination and intellectual availment were facilitated and multiplied, through the new direction given by the Stagirite to the empirical research of facts in the domain of nature, to the workings of the mind when plunging into the depths of speculation, and to the formation of a *scientific language*, by which everything may be accurately defined. Thus Aristotle remains, for thousands of years to come, according to Dante's fine expression, "il maestro di color che sanno" ([233]).

The belief in an immediate enrichment of Aristotle's zoological knowledge by the campaigns of Alexander has been rendered very uncertain, if not entirely dissipated by recent and very careful researches. The miserable compilation of a life of the Stagirite, which was long ascribed to Ammonius the son of Hermias, has given rise, among many other historical errors ([234]), to that of the philosopher having accompanied his pupil at least as far as the banks of the Nile ([235]). The great work on animals appears to have been of very little later date than the Meteorologica, and the latter is shewn by internal evidence ([236]) to belong either to the 106th or at the utmost to the 111th Olympiad; therefore, either fourteen years before Aristotle came to the court of Philip, or, at the latest, three years before the passage of the Granicus. Some particular notices contained in the nine books of the history of animals, have indeed been brought forward in opposition to the view here taken of their early completion: particularly the exact knowledge which Aristotle appears to have had of the elephant, of the bearded horse-stag (hippelaphos), of the Bactrian camel with two humps, of the hippardion supposed to be the hunting tiger (Guépard), and of the Indian buffalo which was first brought to Europe at the time of the Crusades. It should be remarked, however, that the native place of the remarkable large stag with the horse's mane, which Diard and Duvaucel sent from Eastern India to Cuvier, (and to which Cuvier gave the name of Cervus aristotelis) is, according to the Stagirite's own notice, not the Indian Pentapotamia traversed by Alexander, but Arachosia, a country west of Candahar, which together with Gedrosia formed an ancient Persian Satrapy ([237]). May not the notices, mostly so brief, on the forms and habits of the above named animals, have been derived by Aristotle from information obtained by him, quite independently of the Macedonian expeditions, from Persia and from Babylon, the centre of such widely extended trading intercourse? It should be remembered that when preparations by means of alcohol ([338]) were wholly unknown, it was only skins and bones, and not the soft parts susceptible of dissection, which under any circumstances could be sent from remote parts of Asia to Greece. Probable as it is that Aristotle received both from Philip and Alexander the most liberal support in the prosecution of his studies in physics and in natural history,—in procuring immense zoological materials from the whole of Greece and from the Grecian seas, and even in laying the grounds of a collection of books unique for the period, and which passed afterwards to Theophrastus and subsequently to Neleus of Scepsis,—yet we must regard the stories of presents of eight hundred talents, and the "maintenance of many thousands of collectors, overseers of fish-ponds, and bird-keepers" as exaggerations of a later period ([339]), or as traditions misunderstood by Pliny, Athenæus, and Ælian.

The Macedonian expedition, which opened so large and fair a portion of the earth's surface to a single nation of such high intellect and cultivation, may therefore be regarded in the strictest sense of the term as a scientific expedition; and, indeed, as the first in which a conqueror surrounded himself with learned men of all departments of knowledge—naturalists, historians, philosophers, and artists. We should attribute to Aristotle not only that which he himself produced;—he acted also through the intelligent men of his school who accompanied the army. Amongst these shone pre-eminently the near relation of the Stagirite, Callisthenes of Olynthus, who, even previous to the Asiatic campaigns, had been the author of botanical works, and of a delicate anatomical examination of the eye. The grave severity of his manners, and the unmeasured freedom of his language, rendered him hateful both to the flatterers, and to the monarch himself already fallen from his higher thoughts and nobler dispositions. Callisthenes unshrinkingly preferred liberty to life; and when in Bactra he was implicated, though guiltless, in the conspiracy of Hermolaus and the pages, he became the unhappy occasion of Alexander's exasperation against his former teacher. Theophrastus, the genuine friend and fellow disciple of Callisthenes, uprightly and worthily undertook his defence after his fall. From Aristotle we only know that before Callisthenes' departure, the Stagirite recommended to him prudence; and apparently well versed in the knowledge of courts by his long sojourn at that of Philip of Macedon, advised him to "speak with the king as little as possible, and if it must be, always in agreement with him" (240).

Callisthenes, as a philosopher familiar with the study of nature before leaving Greece, and supported by chosen men of the school of Aristotle, directed to higher views the researches of his companions in the new and wider sphere of investigation now opened to them. It was not only the grander forms of the animal kingdom, the luxuriance of vegetation, the variations of the surface, and the periodical swelling of the great rivers which arrested his attention;—man and his varieties, with their many gradations of form and colour, could not but appear in accordance with Aristotle's own saying (241), as "the centre and object of the whole creation, the conscious possessor of thought derived from the divine source of thought." From the little that remains to us of the accounts of Onesicritus (much censured by the ancients), we see that in the Macedonian expedition great surprise was felt when in advancing far towards the east, the Indian races spoken of by Herodotus, "dark coloured and resembling Ethiopians," were indeed met with; but the African negro with curly hair, was not found (242). The influence of the atmosphere on colour, and the different effects of dry and humid warmth, were carefully noticed. In the early Homeric times, and for a long subsequent period, the dependence of the temperature of the air on latitude was completely overlooked. Eastern and Western relations determined the whole thermic meteorology of the Greeks. The parts of the earth towards the sun-rising were regarded as near to the sun, or "sun lands." "The God in his course colours the skin of man with a dark sooty lustre, and parches and curls his hair" (243).

The campaigns of Alexander first afforded an opportunity of comparing on a large scale the African races, assembled in Egypt especially, with Arian races beyond the Tigris, and with the very dark coloured, but not woolly haired, Indian aborigines. The subdivision of mankind into varieties, their distribution over the earth's surface, (the result rather of historical events than of a long continuance of climatic influences, when the types have been once firmly established,) and the apparent contradiction between colour and situation, must have awakened the liveliest interest in thoughtful observers. We still find in the interior of India an extensive territory peopled by very dark coloured, almost black, aboriginal inhabitants, quite distinct from the lighter coloured and later-immigrating Arian races. To these belong among the Vindhya nations, the Gondas, the Bhillas (Bheels) in the forest-covered mountains of Malwa and Guzerat, and the Kolas of Orissa. The acute Lassen considers it probable that in the time of Herodotus, the black Asiatic race, —the "Ethiopians of the sun-rising," resembling the Lybian Ethiopians in the colour of the skin but not in the quality of the hair,—extended much farther towards the north-west than at present ([244]). Thus also in the ancient Egyptian kingdom, the habitations of the true woolly-haired, often-conquered Negro races extended far into northern Nubia ([245]).

The enlargement of the sphere of ideas which arose from the aspect of many new physical phænomena, as well as from contact with different races of men and with their civilisation and the contrasts which it presented, was unfortunately not accompanied by the fruits of an ethnological comparison of languages, either philosophical, regarding the fundamental relations of ideas ([246]),—or simply historical. What we call classical antiquity was wholly a stranger to this class of investigations. On the other hand, the expeditions of Alexander offered to the Greeks scientific materials taken from the long accumulated treasures of more anciently cultivated nations. What I would more especially refer to is the fact that, with an increased knowledge of the earth and its productions, we find by recent and careful investigations that the Greeks obtained from Babylon an important augmentation of their knowledge of the heavens. The conquest of Cyrus had indeed already caused the downfal of the glories of the Astronomical College of priests in the capital of the eastern world: the terraced pyramid of Belus, (at once a temple, a tomb, and an astronomical observatory whence the nocturnal hours were proclaimed), had been given over to destruction by Xerxes, and already lay in ruins when the Macedonians came. But the very fact of the close sacerdotal caste being dissolved, and of many astronomical schools having formed themselves (247), rendered it possible for Callisthenes to send to Greece, (by the advice of Aristotle according to Simplicius), observations of stars for a very long period; Porphyry says for a period of 1903 years before Alexander's entry into Babylon, Ol. 112, 2. The oldest Chaldean observations referred to in the Almagest, (probably the oldest which Ptolemy found suitable for his objects) go back indeed only to 721 years before our era, or to the first Messenian War. It is certain that "the Chaldeans knew the mean motions of the moon with an exactness which caused the Greek astronomers to employ them for the foundation of the theory of the moon (248)." Their planetary observations, to which they were stimulated by the old love of astrology, appear also to have been used for the construction of astronomical tables.

This is not the place to examine how much of the earliest Pythagorean views of the true fabric of the heavens, of the course of the planets, and of that of comets which according to Apollonius Myndius ([249]) return in long regulated paths, belonged to the Chaldeans: Strabo calls the "mathematician Seleucus" a Babylonian, and distinguishes him from the Erythrean who measured the tide of the sea. ([250]) It is sufficient to remark as highly probable that the Greek Zodiac is borrowed from the Dodecatemoria of the Chaldeans, and according to Letronne's important investigations does not go back farther than to the beginning of the sixth century before our era ([251]).

The immediate results of the contact of the Greeks with the nations of Indian origin, at the period of the Macedonian campaigns, are wrapped in much obscurity. In science, little was probably gained; as after traversing the kingdom of Porus, between the cedar fringed ([252]) Hydaspes (Jelum), and the Acesines (Tschịnab), Alexander only advanced in the Pentapotamia (the Pantschanada), as far as the Hyphasis,—below the junction, however, of that river, with its tributary the Satadru, the Hesidrus of Pliny. Distrust of his soldiers, and uneasiness respecting a dreaded general insurrection in the Persian and Syrian provinces, forced the warrior king, who would fain have advanced to the Ganges, to the great catastrophe of his return. The countries passed through by the Macedonians were inhabited by very imperfectly civilised races. In the space between the Satadru and the Yamuna (the region of the Indus and the Ganges), the sacred Sarasvati, an inconsiderable stream, forms a classic boundary of the highest antiquity between the "pure, worthy, pious" worshippers of Brahma on the East, and the "impure, kingless" tribes, not divided into castes, on the West ([253]). Alexander, therefore, did not reach the proper seat of the higher Indian civilisation. Seleucus Nicator, the founder of the great empire of the Seleucidæ, was the first who advanced from Babylon towards the Ganges, and by the repeated missions of Megasthenes to Pataliputra ([254]) connected himself by political relations with the powerful Sandracottus (Chandragupta).

Thus first arose an animated and lasting contact with the civilised parts of the Madhya-desa ("the central land"). There were indeed in the Pendschab (Punjaub, or Pentapotamia) learned Brahmins living as hermits. We do not know, however, whether those Brahmins and Gymnosophists were acquainted with the fine Indian system of numbers, in which a few characters receive their value merely by "position;" nor are we even certain whether at that period the method of assigning value by position was known even in the most cultivated parts of India, although it is highly probable that such was the case. What a revolution would have been effected in the more rapid development of mathematical knowledge, and in the facilities of its application, if the Brahmin Sphines (called by the Greeks Calanos) who accompanied Alexander's army;—or at a later period, in the time of Augustus, the Brahmin Bargosa,—before they voluntarily ascended the funeral pile at Susa and at Athens, had been able to communicate the knowledge of the Indian system of numbers to the Greeks, so that it might have been brought into general use! The acute and comprehensive researches of Chasles have indeed shewn, that what is called the method of the Pythagorean Abacus or Algorismus, as we find it described in Boethius' Geometry, is almost identical with the position-value of the

Indian system: but that method, long unfruitful with the Greeks and Romans, first obtained general extension in the middle ages, especially after the zero sign had superseded the vacant space. The most beneficial discoveries often require centuries for their recognition and completion.

EPOCHS IN THE HISTORY OF THE CONTEMPLATION OF THE UNIVERSE. EPOCH OF THE PTOLEMIES.

III.

Progress of the contemplation of the Universe under the Ptolemies—Museum at Serapeum.—Peculiar character of the scientific direction of the period.—Encyclopedic learning.—Generalisation of the views of nature regarding both the earth and the regions of space.

After the dissolution of the great Macedonian Empire comprising territories in the three Continents, the germs which the uniting and combining system of the government of Alexander had deposited in a fruitful soil, began to develop themselves every where, although with much diversity of form. In proportion as the national exclusiveness of the Hellenic character of thought vanished, and its creative inspiring power was less strikingly characterised by depth and intensity, increasing progress was made in the knowledge of the connection of phenomena, by a more animated and more extensive intercourse between nations, as well as by a generalisation of the views of Nature based on argumentative considerations. In the Syrian kingdom, by the Attalidæ of Pergamos, and under the Seleucidæ and the Ptolemies, this progress was favoured and promoted every where and almost at the same time by distinguished sovereigns. Grecian Egypt enjoyed the advantage of political unity, as well as that of geographical position; the influx of the Red Sea through the Straits of Bab-el-Mandeb to Suez and Akaba, (occupying one of the SSE.-NNW. fissures, of which I have elsewhere spoken), (255), bringing the traffic and intercourse of the Indian Ocean within a few miles of the coasts of the Mediterranean.

The kingdom of the Seleucidæ did not enjoy the advantages of sea traffic, which the distribution of land and water, and the configuration of the coast line, offered to that of the Lagidæ; and its stability was endangered by the divisions produced by the diversity of the nations of which the different Satrapies were composed. The intercourse and traffic enjoyed by the kingdom of the Seleucidæ was mostly an inland one, confined either to the course of rivers, or to caravan tracks, which braved every natural obstacle,—snowy mountain chains, lofty plateaus, and deserts. The great caravan conveying merchandise, of which silk was the most valuable article, travelled from the interior of Asia, from the high plain of the Seres north of Uttara-kuru, by the "stone tower" (256) (probably a fortified Caravanserai) south of the sources of the Jaxartes, to the valley of the Oxus, and to the Caspian and Black Seas. In the kingdom of the Lagidæ, on the other hand, animated as was the river navigation of the Nile, and the communication between its banks and the artificial roads along the shores of the Red Sea, the principal traffic was, nevertheless, in the strictest sense of the word, a sea traffic. In the grand views formed by Alexander, the newly founded Egyptian Alexandria in the West, and the very ancient City of Babylon in the East, were designed to be the two metropolitan cities of the Macedonian universal empire; Babylon, however, never in later times fulfilled these expectations; and the flourishing prosperity of Seleucia, founded by Seleucus Nicator on the lower Tigris, and united with the Euphrates by means of canals (257), contributed to its complete decline.

Three great rulers, the three first Ptolemies, whose reigns occupied a whole century, by their love of the sciences, by their brilliant institutions for the promotion of intellectual cultivation, and by their uninterrupted endeavours to promote and extend commerce, caused the knowledge of Nature and of distant countries to receive a greater and more rapid increase than had yet been achieved by any single nation. This treasure of true scientific cultivation passed from the Greeks settled in Egypt to the Romans. Even under Ptolemy Philadelphus, hardly half a century after the death of Alexander, and before the first Punic war had shaken the aristocratic republic of Carthage, Alexandria was the port of greatest commerce in the world. The nearest and most commodious route from the basin of the Mediterranean to South Eastern Africa, Arabia and India, was by Alexandria. The Lagidæ availed themselves with unexampled success of the road which Nature had as it were marked out for the commerce of the world by the direction of the Red Sea or Arabian Gulf (258);—a route which will never be fully appreciated until the wildness of Eastern life, and the jealousies of the Western powers, shall both diminish. Even when Egypt became a Roman province, it continued to be the seat of almost boundless riches; the increasing luxury of Rome under the Cæsars reacted on the land of the Nile, and sought the means of its satisfaction principally in the universal commerce of Alexandria.

The important extension of the knowledge of Nature, and

of different countries under the Lagidæ, was derived from the caravan traffic in the interior of Africa by Cyrene and the Oases; from the conquests in Ethiopia and Arabia Felix under Ptolemy Euergetes; and from commerce by sea with the whole Western Peninsula of India, from the Gulf of Barygaza (Guzerat and Cambay), along the coasts of Canara and Malabar (Malaya-vara, territory of Malaya), to the Brahminical Sanctuaries of Cape Comorin (Kumari), (259) and to the great Island of Ceylon, (Lanka in the Ramayana, and called by Alexander's cotemporaries Taprobane by the mutilation of a native name). (260) An important advance in nautical knowledge had previously been obtained, by the laborious five months' voyage of Nearchus along the coasts of Gedrosia and Caramania, between Pattala at the mouth of the Indus and the mouths of the Euphrates.

Alexander's companions were not ignorant of the existence of the periodical winds or monsoons, which favour so materially the navigation between the East coast of Africa and the North and West coasts of India. At the end of ten months, spent by the Macedonians in navigating and examining the Indus, between Nicea on the Hydaspes and Pattala, with the view of opening that river to the commerce of the world, Nearchus hastened at the beginning of October (Ol. 113, 3) to sail away from the mouth of the Indus at Stura, because he knew that his voyage to the Persian Gulf along a coast running on a parallel of latitude, would be favoured by the North East and East monsoon. The farther knowledge acquired by experience of this remarkable local law of the direction of the wind, subsequently emboldened navigators sailing from Ocelis in the Straits of Bab-el-Mandeb, to hold a direct course across the open sea to Muziris, the great mart on the Malabar coast (south of Mangalor), to which internal traffic brought articles of commerce from the Eastern coast of the Indian Peninsula, and even gold from the remote Chrysa (Borneo ?). The honour of being the first to apply this new system of Indian navigation is ascribed to an otherwise unknown mariner, Hippalus ; and even the precise period at which he lived is doubtful. ([261])

Whatever brings nations together, and by rendering large portions of the Earth more accessible, enlarges the sphere of men's knowledge, belongs to the history of the contemplation of the Universe. The opening of a water communication between the Red Sea and the Mediterranean by means of the Nile, holds an important place in this respect. At the part where a slender line of junction barely unites the two continents, and which offers the deepest maritime inlets, the excavation of a canal had been commenced, not indeed by the great Sesostris (Ramses Miamoun) to whom Aristotle and Strabo ascribe it, but by Nechos (Neku), who, however, was deterred by oracles given by the priests from prosecuting the undertaking. Herodotus saw and described a finished canal which entered the Nile somewhere above Bubastis, and was the work of the Achæmenian, Darius Hystaspes. Ptolemy Philadelphus restored this canal which had fallen into decay, in so complete a manner, that although (notwithstanding a skilful arrangement of locks and sluices it was not navigable at all seasons of the year), it long aided and greatly promoted traffic with Ethiopia, Arabia, and India, continuing to do so under the Roman sway as late as the reign of Marcus Aurelius, and perhaps even as late as that of Septimius Severus, a period of four centuries and a half. With a similar purpose of encouraging intercourse by means of the Red Sea, harbour works were sedulously carried on at Myos Hormos and Berenice, and were connected with Coptos by the formation of an excellent artificial road. (262) All these different enterprises of the Lagidæ, commercial as well as scientific, were based on the idea of connection and union, on a ceaseless tendency to embrace a wider whole, remoter distances, larger masses, more extensive and varied relations, and greater and more numerous objects of contemplation. This direction of the Hellenic mind, so fruitful in results, had been long preparing in silence, and became manifested on a great scale in the expeditions of Alexander, in his endeavours to blend the Western and Eastern worlds. In its continued extension under the Lagidæ it characterised the epoch which I here desire to pourtray, and must be regarded as having effected an important advance in the progressive recognition and knowledge of the Universe as a whole.

So far as an abundant supply of objects of direct contemplation is required for this increasing and advancing knowledge, the frequent intercourse of Egypt with distant countries, scientific exploring journies into Ethiopia at the cost of the Government, (263) distant ostrich and elephant hunts, (264) and menageries of wild and rare beasts in the "kings' houses of Bruchium," might act as incitements to the study of natural history, (265) and contribute data to empirical knowledge; but the peculiar character of the Ptolemaic epoch, as well as of the whole "Alexandrian School," which, indeed, preserved the same direction until the third and fourth centuries, manifested itself in a different path; it occupied itself less with the immediate observation of particulars, than with the laborious assemblage of all that was already obtained, and in the arrangement, comparison, and intellectual fructification of that which had long been collected. During the long period of many centuries, and until the powerful genius of Aristotle appeared, natural phenomena, not regarded as objects of accurate observation, were subjected in their interpretation to the exclusive sovereignty of ideas, and even given over to the sway of vague presentiments and unstable hypotheses. There was now, however, manifested a higher appreciation of empirical knowledge. Men examined and sifted what they possessed. Natural philosophy becoming less bold in her speculations and less fanciful in her images, at length approached nearer to a searching empirical investigation in treading the sure path of induction. A laborious tendency to accumulate materials had enforced the acquisition of a corresponding amount of technical information; and although in the works of distinguished and thoughtful men, an extensive and varied knowledge presented valuable results, yet in the decline of the creative power of the Greek mind this knowledge appeared too often to want an animating spirit, and wore the character of mere erudition. The absence of due care in respect to composition, as well as want of animation and grace of style, have also contributed to expose Alexandrian learning to the severe censure of posterity.

It particularly belongs to these pages to bring forward that which the epoch of the Ptolemies contributed towards the contemplation of the physical Universe, whether by the concurrent action of external relations, by the foundation and suitable endowment of two great establishments (the Alexandrian Institution, and the libraries of Bruchium (266) and Rhakotis), or by the collegiate assemblage of so many learned men of active and practical minds. An encyclopædic knowledge was favourable to the comparison of the results of observation, and thus tended to facilitate generalisations in the view of Nature. The great scientific Institution which owed its origin to the two first Ptolemies, long maintained amongst other privileges that of its members being free to labour in wholly different directions ([267]); and thus, although settled in a foreign country, and surrounded by men of many different races and nations, they preserved the peculiar Hellenic character of thought, and the acute Hellenic ingenuity.

In accordance with the spirit and form of the present historic representation, a few examples may suffice to shew the manner in which, under the protecting influence of the Ptolemies, observation and experiment assumed their appropriate places, as the true sources of knowledge respecting the heavens and the earth; and how, in the Alexandrian period, in combination with a diligent accumulation of the mere materials of knowledge, a happy tendency to generalisation was also at all times manifested. Although the different Greek schools of philosophy transplanted to Lower Egypt did not escape a certain degree of Oriental degeneracy, and gave occasion to many mythical interpretations of Nature and of physical phenomena, yet in the Alexandrian school the Platonic doctrines ([268]) still remained as the most secure support of mathematical knowledge. The progressive advances made in this knowledge embraced almost at the same time pure mathematics, mechanics and astronomy. In Plato's high esteem for mathematical development of thought, as well as in Aristotle's morphological views embracing all organic beings, were contained the germs of all later advances in natural science; they became the guiding stars which conducted the human intellect securely through the mazes of fanaticism in the dark ages; and did not suffer healthy scientific intellectual power to perish.

The mathematician and astronomer Eratosthenes of Cyrene, the most celebrated of the Alexandrian librarians, availed himself of the treasures at his command by working them up into a systematic "universal geography." He freed geographical description from mythical legends, and, although himself occupied with chronology and history, even separated from it the historical admixtures by which it had been previously not ungracefully enlivened. Their absence was satisfactorily supplied by mathematical considerations on the more or less articulated form of continents, and on their extent; and by geological conjectures on the connection of chains of mountains, the action of currents, and the former presence of an aqueous covering over the surface of lands still bearing traces of having been once the bottom of the sea. Regarding with favour the oceanic sluice theory of Strato of Lampsacus, the Alexandrian librarian was led by the belief of the former swollen state of the Euxine, the disruption of the Dardanelles, and the consequent opening of the pillars of Hercules, to the important investigation of the problem of the equality of level of the "outward sea encompassing all continents." (269) A farther instance of happy generalisation on the part of Eratosthenes is his assertion that the whole continent of Asia is traversed in the parallel of Rhodes, (the diaphragm of Dicearchus), by a connected chain of mountains running East and West. (270)

A lively desire for generalisation, the result of the intellectual movement of the period, also led Eratosthenes to set on foot the first (Hellenic) measurement of an arc of the

Meridian, having its extremities at Alexandria and Syene, and for its object the approximate determination of the earth's circumference. It is not the result that he obtained, based as it was upon imperfect data, furnished by pedestrians, which awakens our interest; it is the endeavour of the philosopher to rise from the narrow limits of a single country to the knowledge of the magnitude of the entire globe.

A similar tendency towards generalization of view is manifested in the brilliant advances made in the epoch of the Ptolemies towards a scientific knowledge of the heavens: I allude here to the determination of the places of the fixed stars by the earliest Alexandrian astronomers, Aristyllus and Timocharis;—to Aristarchus of Samos, the cotemporary of Cleanthes, who, familiar with the old Pythagorean views, adventured an inquiry into the relations in space of the whole fabric of the Universe, and who first recognised the immeasurable distance of the heaven of the fixed stars from our little planetary system, and even conjectured the twofold movement of the earth, *i. e.* her rotation round her axis, and her progressive movement around the sun;—to Seleucus of Erythrea, or of Babylon, (271) who, a century later, sought to support the views of the Samian philosopher (views which we may term Copernican, and which at that period found little acceptance);—and to Hipparchus, the creator of scientific astronomy, and the greatest of observing astronomers in all antiquity. Among the Greeks, Hipparchus was the true and proper author of astronomical tables, (272) and the discoverer of the precession of the equinoxes. His own observations of fixed stars (made at Rhodes, not at Alexandria), when compared with those of Timocharis and Aristyllus, led him (probably without the sudden apparition of a new star ([273]))to this great discovery; to which the long-continued observation of the heliacal rising of Sirius ought indeed to have conducted the earlier Egyptians. ([274])

Another peculiar feature in the proceedings of Hipparchus, was his endeavouring to avail himself of celestial phenomena for determinations of geographical position. Such a combination of the study of the heavens and the earth, the knowledge of the one becoming reflected on the other, served by its uniting tendency to give a lively impulse to the great idea of the Cosmos. In a new map of the world, constructed by Hipparchus, and founded on that of Eratosthenes, wherever the application of astronomical observations was possible, the geographical positions were assigned by longitudes and latitudes, obtained, the former from lunar eclipses, and the latter from lengths of the solar shadow measured by the gnomon. The hydraulic clock of Ctesibius, an improvement upon the ancient Clepsydra, might afford the means of making more exact measurements of time; whilst, for determinations in space, gradually improved means of angular measurement were offered to the Alexandrian astronomers, from the old gnomon and scaphe to the invention of astrolabes, solstitial armills, and dioptras. Thus men arrived by successive steps, as if by the acquisition of new organs, to a more exact knowledge of the movements of the planetary system. It was only the knowledge of the absolute magnitudes, forms, masses, and physical constitution of the heavenly bodies, which made no progress for many centuries.

Not only were several practical astronomers of the Alexandrian school themselves distinguished geometricians, but the epoch of the Ptolemies was moreover the most brilliant epoch of the cultivation of mathematical knowledge. There flourished in the same century Euclid the creator of mathematics as a science, Apollonius of Perga, and Archimedes, who visited Egypt and was connected through Conon with the Alexandrian school. The long path of time which leads from what is called the geometric analysis of Plato, and the three conic sections of Menæchmes, ([275]) to the age of Kepler and Tycho, Euler and Clairaut, d'Alembert and Laplace, is marked by a series of mathematical discoveries, without which the laws of the motions of the heavenly bodies, and their mutual relations in space, would never have been disclosed to mankind. The telescope pierces space, and brings distant worlds near through our sense of vision. Mathematical knowledge forms a no less powerful instrument of another class: ever leading us onward through the connection of ideas, it conducts us to those distant regions of space, of part of which it has taken secure possession. In our own times so favoured in the extension of knowledge, by the application of all the resources afforded by modern astronomy, a heavenly body has even been seen by the intellectual eye, and its place, its path, and its mass pointed out, before a single telescope had been directed towards it. ([276])

EPOCHS IN THE HISTORY OF THE CONTEMPLATION OF THE UNIVERSE. ROMAN EMPIRE.

IV.

Roman Universal Empire—Influence on Cosmical Views of a great Political Union of Countries—Progress of Geography through Commerce by Land—Strabo and Ptolemy—Commencement of Mathematical Optics and Chemistry—Pliny's Attempt at a Physical Description of the Universe—The Rise of Christianity produces and favours the Feeling of the Unity of Mankind.

In tracing the intellectual progress of mankind and the gradual extension of cosmical views, the period of the Roman universal Empire presents itself as one of the most important epochs. We now for the first time find all those fertile regions of the globe which surround the basin of the Mediterranean connected in a bond of close political union, which also comprehended extensive countries to the eastward. I may here appropriately notice, (277) that this political union gives to the picture which I endeavour to trace, (that of the history of the contemplation of the universe), an objective unity of presentation. Our civilization, *i. e.* the intellectual development of all the nations of the European Continent, may be regarded as based on that of the dwellers around the Mediterranean, and more immediately on that of the Greeks and the Romans. That which we term, perhaps too exclusively, classical literature, has received this denomination through men's recognition of the source from whence our earliest know ledge has largely flowed, and which gave the first impulse to a class of ideas and feelings most intimately connected with the civilization and intellectual elevation of a nation or a race. (278) We do not by any means regard as unimportant the elements of knowledge, which, flowing through the great current of Greek and Roman cultivation, were yet derived in a variety of ways from other sources—from the valley of the Nile, Phœnicia, the banks of the Euphrates, and India; but even for these we are indebted, in the first instance, to the Greeks, and to Romans surrounded by Etruscans and Greeks. At how late a period have the great monuments of more anciently civilized nations been directly examined, interpreted, and arranged according to their relative antiquity! It is only within a very recent period that hieroglyphics and cuneiform inscriptions have been read, after having been for thousands of years passed by armies and caravans, who divined nothing of their import.

From the shores of the Mediterranean, and especially from its Italic and Hellenic peninsulas, have indeed proceeded the intellectual character and political institutions of those nations who now possess the daily increasing treasures of scientific knowledge and creative artistic activity, which we would fain regard as imperishable; nations which spread civilization, and with it, first servitude, and then, involuntarily, liberty, over another hemisphere. Yet in modern Europe too, as it were by a favour of destiny, unity and diversity are still happily associated. The elements received have been various, and no less various have been their appropriation and transformation, according to the sharply contrasted peculiarities, and individual tone of mind and disposition, of the different races by which Europe has been peopled. Her civilization has been carried beyond the ocean to another hemisphere, where the reflex of these contrasts is still preserved in colonies and settlements, some of which have formed, and others it may be hoped may yet form, powerful free states.

The Roman state, as a monarchy under the Cæsars, when considered only in regard to superficial extent, (279) was inferior in absolute magnitude to the Chinese empire under the dynasty of Thsin and the eastern Han (from 30 years before to 116 years after the commencement of the Christian era); it was inferior in extent to the empire of Ghengis Khan, and to the present area of the Russian dominions in Europe and Asia; but with the single exception of the Spanish monarchy at the period when it extended over the New World, never has there been combined under one sceptre a greater mass of countries so favoured in climate, fertility, and geographical position, as the Roman empire from Augustus to Constantine.

This empire, stretching from the western extremity of Europe to the Euphrates, from Britain and part of Caledonia to Getulia and the limits of the Lybian Desert, not only offered the greatest variety of form of ground, organic productions, and physical phenomena, but also presented mankind in every gradation from cultivation to barbarism, and from the possession of ancient knowledge and long practised arts, to the first twilight of intellectual awakening. Distant expeditions to the North and to the South, to the Amber Coasts, and, (under Ælius Gallus and Balbus) to Arabia and the Garamantes, were carried out with unequal success. Measurements of the whole empire were begun even under Augustus, by Greek geometers, Zenodorus and Polycletus; and itineraries and special topographies were prepared (as had indeed been done some centuries earlier in the Chinese empire), for distribution amongst the several governors of provinces. (280) These were the first statistical works which Europe produced. Many extensive prefectures were traversed by Roman roads, divided into miles; and Hadrian even visited the different parts of his empire, though not without interruption, in an eleven years' journey, from the Iberian peninsula to Judea, Egypt, and Mauritania. Thus a large portion of the globe, subject to the Roman dominion, was opened and made traversable; "pervius orbis," as the chorus in Seneca's Medea less justly prophesies of the whole earth. (281)

We might, perhaps, have expected that during the enjoyment of long-continued peace, and the union under a single monarchy of such extensive countries and different climates, the facility and frequency with which the provinces were traversed by civil and military functionaries, often accompanied by a numerous train of educated men possessed of varied information, would have been productive of extraordinary advances, not only in geography, but also in the knowledge of nature generally, and in the formation of higher views concerning the connection of phenomena. Such high expectations were not, however, realised. In the long period of the undivided Roman empire, occupying almost four centuries, there arose as observers of nature only Dioscorides the Cilician, and Galen of Pergamos. The first of these, who augmented considerably the number of described species of plants, is far inferior to the philosophically combining Theophrastus;—whereas Galen, who extended his observations to many genera of animals, by the fineness of his distinctions, and the comprehensiveness of his physiological discoveries, "may be placed very near to Aristotle, and in most respects even above him." It is Cuvier who has pronounced this judgment. ([282])

By the side of Dioscorides and Galen shines a third and great name—that of Ptolemy. I do not cite him here as the author of an astronomical system, or as a geographer; but as an experimental physical philosopher, who measured refractions, and, therefore, as the first founder of an important part of optical science. His incontestable rights in this respect were not recognised until very lately. ([283]) Important as were the advances made in the department of organic life, and in the general views of comparative zootomy, physical experiments on the passage of rays of light, at a period five centuries anterior to that of the Arabians, must arrest our attention yet more forcibly; they form, as it were, the first step in a newly-opened course,—in the vast career of mathematical physics.

The distinguished men whom we have named as shedding scientific lustre on the period of the Roman empire, were all of Grecian origin; (the profound arithmetical algebraist Diophantus, ([284]) who, however, was still without the use of symbols, belonging to a later time.) In the two chief divisions in respect to intellectual cultivation which the Roman empire presents to us, the palm was still with the Hellenes, the older and more happily organised nation; but the gradual decline of the Egyptian Alexandrian school was followed by the dispersion of the still remaining, but weakened, points of light in scientific knowledge and rational investigation; and it was only at a later period that they reappeared in Greece and Asia Minor. As in all unlimited monarchies of enormous extent, and composed of heterogeneous elements, the efforts of the government were principally directed to avert by military force, and by the internal rivalries of a divided administration, impending dismemberment and dissolution—to conceal family discords in the house of the Cæsars by alternate mildness and severity,—and, under a few nobler rulers, to give to the nations beneath their sway the repose which unresisted despotism can at times afford.

The attainment of the Roman universal empire was itself a fruit of the greatness of the Roman character, of a long preserved severity of manners, and of an exclusive love of country, united with high individual feeling; but after this universal empire was attained, these noble qualities became gradually weakened, and were perverted even by the inevitable influences which new circumstances called forth. As the national spirit became extinct, the same deadening effect extended to individual life; publicity and individuality,—the two chief supports of free institutions,—disappeared at the same time. The eternal city had become the centre of too great a circle; the spirit which could permanently animate a body so vast, and composed of so many members, was wanting. Christianity became the religion of the state when the empire was already shaken to its foundations; and the mildness of the new doctrine, and its beneficent influences, were soon disturbed by the dogmatic strife of parties. Then also began the "unfortunate contest between knowledge and faith," which, under various forms, all tending to impede investigation, has been continued through succeeding centuries.

Although, however, the vastness of the Roman empire, and the institutions which that vastness rendered necessary, were strongly contrasted with the independent life of the small Hellenic republics, and tended rather to deaden than to cherish creative intellectual power among its citizens, yet there resulted from the same cause some peculiar advantages, which should be noticed here. A rich accession of ideas was the fruit of experience and varied observation; the world of objects was considerably augmented, and the ground was thus laid for a thoughtful contemplation of natural phenomena at a later epoch. The Roman empire gave animation to the intercourse between nations, and extended the Roman language over the whole of the West, and over a portion of Northern Africa. In the East, Greek influence survived, as if naturalised, long after the Bactrian empire had been destroyed under Mithridates I. (thirteen years before the attack of the Sacæ, or Scythians.)

In point of geographical extent, the Roman language gained upon the Greek, even before the seat of empire was transferred to Byzantium. The interpenetration of two highly-gifted idioms, rich in literary monuments, became a means of farther blending and uniting different nations and races, and of increasing civilization and susceptibility to mental culture; it tended, as Pliny says, ([285]) "to humanize men, and to give them a common country." However much the language of the barbarians (the dumb, αγλωσσοι, as Pollux calls them) may have been contemned, yet there were instances in which the translation of a literary work from the Punic to the Roman language was desired by the public authorities: Mago's Treatise on Agriculture is known to have been translated by the command of the Roman

Senate. The Lagidæ had previously given examples of a similar kind.

Whilst the Roman empire extended westward to the extremity of the old Continent (at least on the northern side of the Mediterranean), its eastern limit, under Trajan, who navigated the Tigris, reached only to the meridian of the Persian Gulf. It was in this direction that, at the period we are describing, the greatest intercourse between different nations took place in a shape very conducive to the progress of geography, viz. that of commerce by land. After the fall of the Greco-Bactrian empire, the rising and flourishing power of the Arsacides favoured intercourse with the Seres; but to the Romans this communication was only an indirect one, their immediate contact with the interior of Asia being impeded by the active carrying trade of the Parthians. Movements which proceeded from the most distant parts of China produced sudden and violent, though not permanent, changes in the political state of the vast range of country, which extends from the Thian-schan mountains to the Kuen-luu, the chain of Northern Thibet. During the reigns of the Roman emperors Vespasian and Domitian, a Chinese military expedition overran and oppressed the Hiungnu country, rendered tributary the little kingdoms of Khotan and Kashgar, and carried its victorious arms as far as the eastern coast of the Caspian. This was the great expedition led by the military commander Pantschab, under the Emperor Mingti of the dynasty of the Han. Chinese writers even ascribe to this adventurous and fortunate leader, cotemporaneous with Vespasian and Domitian, a grander plan; they assert that he designed to attack the empire of the Romans (Tathsin); but that the advice of the Persians induced him to change his purpose (286). Thus there arose connections between the coasts of the Pacific, the Shensi, and the region around the Oxus, in which there had been, from very early times, an animated traffic with the neighbourhood of the Black Sea.

The direction in which the great tide of population flowed in Asia was from east to west, as in the New Continent from north to south. A century and a half before our era, near the time of the destruction of Corinth and of Carthage, the attacks of the Hiungnu (a Turkish tribe confounded by De Guignes and Johannes Müller with the Finnish Huns) on the fair-haired and blue-eyed, probably Indo-Germanic, race of the (287) Yueti (Getæ?) and Usun, near the Chinese wall, gave the first impulse to that "migration of nations" which did not reach the borders of Europe until five centuries later. Thus the wave of population flowed (or was propagated) from the upper valley of the Hoangho to the Don and the Danube; and in the northern part of the Old Continent, movements advancing in different directions brought one part of mankind first into hostile collision, and subsequently into peaceful and commercial contact with another. Thus we may regard great currents of population, moving forward like the currents of the ocean between unmoved masses at rest, as facts of cosmical importance.

Under the reign of the Emperor Claudius, the embassy of Rachias came from Ceylon, through Egypt, to Rome. Under Marcus Aurelius Antoninus (called by the historians of the dynasty of the Han, Antun), Roman legates appeared at the Chinese court, having come by water by Tunkin. I here point out the first traces of an extended intercourse between the Roman Empire and China and India for this reason among others, that it is highly probable that through this intercourse the knowledge of the Greek sphere, the Greek zodiac, and the astrological planetary week, extended to the last-named countries in the first centuries of our era. (288) The great Indian mathematicians Warahamihira, Bramagupta, and perhaps even Aryabhatta, are later than the period of which we are treating; (289) but it is also possible that a partial knowledge of discoveries earlier made, in ways distinct and apart in India itself and originally belonging to that anciently civilized nation, may have been conveyed to the countries of the West before Diophantus, through the extensive commercial intercourse which took place under the Lagidæ and the Cæsars. We do not here undertake to distinguish accurately what belongs to each nation and to each epoch; it is enough if we point out the channels which were opened to the communication and interchange of ideas.

The gigantic works of Strabo and of Ptolemy testify in the most lively manner the increase which had taken place in these channels and in general international intercourse. The ingenious geographer of Amasia had not Hipparchus's exactness of measurements or the mathematical views of Ptolemy; but his work surpasses all the geographical writings of antiquity both in grandeur of plan and in the variety and abundance of materials. Strabo, as he takes pleasure in telling us, had seen with his own eyes a considerable part of the Roman empire, "from Armenia to the Tyrrhenian coasts, and from the Euxine to the borders of Ethiopia." After having completed forty-three books of history as a continuation of Polybius, he had the courage in the eighty-third year of his age (290) to commence his great geographical work. He reminds his readers "that in his time the power of the Romans and of the Parthians had opened the world even more than Alexander's expeditions, on which Eratosthenes had rested." The commerce of India was no longer in the hands of the Arabians: Strabo saw in Egypt with surprise the increased number of ships which sailed direct from Myos Hormos to India; (291) and his imagination led him beyond India itself to the eastern coasts of Asia. In the parallel of latitude which passes through the pillars of Hercules and the Island of Rhodes, and in which Strabo believed that a connected chain of mountains traversed the old continent in its greatest breadth, he conjectured the existence of "another continent" between the western coast of Europe and Asia. He says, (292) "it is very possible that there may be, besides the world which we inhabit, in the same temperate zone, about the parallel of Thinæ (or Athens?) which passes through the Atlantic Sea, one or more other worlds inhabited by men different from ourselves." It is surprising that the attention of Spanish writers in the beginning of the sixteenth century, who thought that they found everywhere in the classics traces of a knowledge of the new world, should not have been attracted by this passage.

"Since," as Strabo finely says, "in all works of art which would represent something great, the object is not the finish and completeness of separate parts," so in his "gigantic work" it was his wish to fix his attention primarily on the form of the whole. This predilection for generalisation has at the same time not prevented him from bringing forward a great number of excellent physical observations, and particularly many concerning the structure of the earth. (293) Like Posidonius and Polybius, he discusses the influence of the shorter or longer interval between successive passages of the sun through the zenith under the tropic or the equator upon the maximum of temperature of the air; he treats of the various causes of the changes which the surface of the earth undergoes; of the breaking through of the boundaries of lakes or seas originally closed; of the general level of the sea (already recognised by Archimedes); of its currents; of the eruptions of submarine volcanoes; of petrifactions of shells, and impressions of fishes; and even of the oscillations of the crust of the earth, which last point especially arrests our attention, as it has become the nucleus of modern geology. Strabo says expressly that the alterations of the boundaries between land and sea are to be attributed to the rising and sinking of the land rather than to small inundations; "that not only detached masses of rock, or small or large islands, but even whole continents may be raised up." Like Herodotus, Strabo is also attentive to the descent of nations, and to the diversities of race in mankind; he curiously enough calls man a "land and air animal" who "requires much light" ([294]). We find the ethnological distinctions of races most acutely and accurately marked in the commentaries of Julius Cæsar, as well as in Tacitus's fine eulogium on Agricola.

Unfortunately Strabo's great work, so rich in facts and in the cosmical views which we have here referred to, remained almost unknown in Roman antiquity until the fifth century, and was not even employed by the all-collecting Pliny. Towards the end of the middle ages Strabo's work became influential on the direction of ideas, though in a less degree than the more mathematical and more dry and tabular geography of Claudius Ptolemæus, from which physical views are almost entirely absent. This latter work became the guiding clue of all travellers as late as the sixteenth century;

they imagined that they recognised in it under different names whatever new places they discovered. In the same manner that natural historians long attached to new found plants and animals the marks of the classes of Linnæus, so the earliest maps of the New Continent appeared in the atlas of Ptolemy which Agathodæmon prepared, at the same time that, in the farthest part of Asia, among the highly civilised Chinese, the western provinces of the empire (295) were already marked in forty-four divisions. The universal geography of Ptolemy has, indeed, the merit of presenting to us the whole of the ancient world graphically in outlines, as well as numerically in positions assigned according to longitude, latitude, and length of day; but often as he affirms the superiority of astronomical results over itinerary estimates by land or water, we are unfortunately without any means of distinguishing among these assigned positions, above 2500 in number, the nature of the foundation on which each rests, or the relative probability which may be ascribed to them according to the itineraries then existing.

The entire ignorance of the polarity of the magnetic needle, and of the use of the compass, which 1250 years before the time of Ptolemy, under the Chinese emperor Tschingwang, had been employed in the construction of "magnetic cars" furnishing an index to the road to be followed, rendered the most detailed itineraries of the Greeks and Romans extremely uncertain, from a want of knowledge of the direction or angle with the meridian. (296)

In the better knowledge which has recently been obtained of the Indian and ancient Persian (or Zend) languages, we are struck by the fact that a great part of the geographical nomenclature of Ptolemy may be regarded as an historic monument of the commercial relations between the West and the most distant regions of southern and central Asia. (297) One of the most important geographical results of these relations was the correct opinion of the insulation of the Caspian Sea, which was restored by Ptolemy after the contrary error had lasted five hundred years. The truth on this subject had been recognised both by Herodotus and by Aristotle, the latter having fortunately written his Meteorologica before the Asiatic campaigns of Alexander. The Olbiopolites, from whose lips the father of history had gathered the account which he followed, were familiar with the northern shores of the Caspian between the Kuma, the Volga (Rha), and the Jaik (Ural); and there was nothing there which could give them an idea of an outlet to the Icy Sea. Very different reasons produced the erroneous impression received by the Macedonian army, when, passing through Hecatompylos (Damaghan), they descended into the humid forests of Mazanderan, and, at Zadracarta, a little to the west of the present Asterabad, saw the apparently boundless expanse of the Caspian in the northern direction. Plutarch tells us in his Life of Alexander that this sight first caused the hypothesis that the sea thus seen was a gulf of the Euxine. (298) The Macedonian expedition, although it was upon the whole very favourable to the progress of geographical knowledge, yet gave rise to particular errors which long maintained themselves. The Tanais was confounded with the Jaxartes (the Araxes of Herodotus), and the Caucasus with the Paropanisus (the Hindoo Coosh). Ptolemy, during his residence at Alexandria, was able to obtain certain accounts from countries immediately adjoining he Caspian, (from Albania, Atropatene, and Hyrcania), of the caravan roads of the Aorsi, whose camels carried Indian and Babylonian goods to the Don and to the Black Sea (299). If, contrary to the juster knowledge of Herodotus, Ptolemy believed the length of the Caspian to be greatest in the east and west direction, he may perhaps have been thus misled by some obscure knowledge of the former greater extent of the Scythian Gulf (Karabogas); and the existence of Lake Aral, the first decided notice of which we find in a Byzantine author, Menander, who wrote a continuation of Agathias. (300)

It is to be regretted that Ptolemy, who reclosed the Caspian Sea, (which the hypothesis of four gulfs supposed to be the reflections or counterparts of similar ones in the disk of the moon (301) had long kept open), did not at the same time give up the fable of the "unknown southern land" connecting Cape Prasum with Cattigara and Thinæ, (Sinarum metropolis); therefore connecting eastern Africa with the land of Tsin, or China. This myth, which would make the Indian Ocean an inland sea, was derived from views which may be traced back from Marinus of Tyre to Hipparchus, Seleucus the Babylonian, and even to Aristotle. (302) In these cosmical descriptions of the progressive advance of the knowledge and contemplation of the Universe, it is sufficient to recal by a few examples how in successive fluctuations the already half recognised truth has often been again obscured. The more the increased extent both of navigation and of traffic by land seemed to render it possible to know the whole of the earth's surface, the more actively, especially in the Alexandrian period under the Lagidæ, and under the Roman empire, did the never slumbering Hellenic imagination seek by ingenious combinations to blend all previous conjectures with the newly added stores of actual knowledge, and thus to complete at once the yet scarcely sketched map of the earth.

We have already briefly noticed that Claudius Ptolemæus by his optical researches (which have been preserved to us, although in a very incomplete state, by the Arabians) became the founder of a branch of mathematical physics; which, indeed, according to Theon of Alexandria, (303) had already been touched upon, so far as relates to the refraction of rays, in the Catoptrica of Archimedes. It is a very important step in advance, when physical phenomena, instead of being simply observed and compared with each other,—of which we find memorable examples in Grecian antiquity in the pseudo-Aristotelian problems, which are full of matter, and in Roman antiquity in the writings of Seneca,—are produced at will under altered conditions, and measured. (304) The process thus referred to characterises Ptolemy's researches on the refraction of rays of light when made to pass through media of unequal density. He caused the rays to pass from air into water and glass, and from water into glass, under different angles of incidence. The results of these "physical experiments" were collected by him into tables. This measurement of a physical phenomenon purposely called forth, of a natural process not reduced to a movement of of the waves of light (Aristotle assumed a movement of the medium intervening between the eye and the object seen), is a solitary occurrence in the period of which we are treating. (305) In the investigation of inorganic nature, this period offers in addition only a few chemical experiments by Dioscorides, and, as I have elsewhere observed, the technical art of collecting fluids when passing over in distillation. [306] As chemistry only begins when men have learnt to employ mineral acids as powerful solvents, and as means of liberating substances, the distillation of sea-water, described by Alexander of Aphrodisias, in the reign of Caracalla, is deserving of great attention. It indicates the path by which men gradually arrived at the knowledge of the heterogeneity of substances, their combination in chemical compounds, and their reciprocal attractions or affinities.

We can only cite, as having advanced the knowledge of organic nature, the anatomist Marinus, Rufus of Ephesus who dissected apes and distinguished between nerves of sensation and of motion, and Galen of Pergamos who eclipses all other names. The natural history of animals by Ælian of Præneste, and the poem treating of fishes written by Oppianus of Cilicia, do not contain facts based on the author's own examination, but only scattered notices derived from other sources. It is hardly conceivable how the enormous multitude [307] of rare animals, which, during four centuries, were massacred in the Roman circus,—elephants, rhinoceroses, hippopotamuses, elks, lions, tigers, panthers, crocodiles, and ostriches,—should never have been rendered of any use to comparative anatomy. I have already spoken of the merits of Dioscorides in regard to the knowledge of collected plants: his works exercised a powerful and long-enduring influence on the botany and pharmaceutical chemistry of the Arabians. The botanical garden of the Roman physician Antonius Castor (who lived to upwards of a hundred years of age), imitated, perhaps, from the botanical gardens of Theophrastus and Mithridates, was probably of no greater scientific use than the collection of fossil bones of the Emperor

Augustus, or the collection of objects of natural history which has been ascribed on very feeble grounds to Appuleius of Madaura. (308)

Before we close the description of what the period of the Roman empire contributed towards the advancement of cosmical knowledge, we have still to mention the grand undertaking of a description of the Universe which Caius Plinius Secundus endeavoured to comprise in thirty-seven books. In the whole of antiquity nothing similar had been attempted; and although in the execution of the work it became a kind of encyclopædia of nature and art (the author in his dedication to Titus not scrupling to apply to his work the then more noble Greek expression εγκυκλοπαιδεια), yet it cannot be denied that, notwithstanding the want of an internal connection and coherence of parts, still the whole presents a plan or sketch of a physical description of the Universe.

The Historia Naturalis of Pliny,—termed Historia Mundi in the tabular view which forms what is now called the first book, and in a letter of his nephew's to his friend Macer more finely described as a Naturæ Historia,—embraces the heavens and the earth, the position and course of the heavenly bodies, the meteorological processes of the atmosphere, the forms of the earth's surface, and all terrestrial objects, from the vegetable covering of the land and the molluscæ of the ocean up to the race of man. Mankind are considered according to the variety of their mental dispositions and intellectual powers, and to the cultivation and exaltation of these as manifested in the noblest works of art. I have here named the elements of a general knowledge of nature which lie scattered almost without order in the great work of which we are speaking. "The path in which I propose to walk," says Pliny, with noble confidence in himself, "is untrodden, (non trita auctoribus via); no one among ourselves, no one among the Greeks, has undertaken to treat as one the whole of nature (nemo apud Græcos qui unus omnia tractaverit). If my undertaking is not successful, still it is something fair and noble (pulchrum atque magnificum) to have attempted its accomplishment."

There floated before the mind of Pliny a grand and single image; but diverted from his purpose by specialities, and wanting the living personal contemplation of nature, he was unable to hold fast this image. The execution remained imperfect, not merely from haste and frequent want of knowledge of the objects to be treated, but also from defective arrangement. We may judge thus from those portions of work which are now accessible to us. We recognise in the author a man of rank, full of occupation, who prides himself on labour bestowed on his work in sleepless nights, but who, whilst exercising the functions of government in Spain, and those of superintendent of the fleet in Lower Italy, doubtless too often confided to imperfectly educated dependants the loose web of an endless compilation. This fondness for compilation, *i. e.*, for a laborious collection of separate observations and facts such as the state of knowledge could then afford, is, in itself, by no means deserving of censure; the imperfection in the success of the result arose from the want of capacity fully to master and command the accumulated materials,—to subordinate the descriptions of nature to higher and more general views,—and to keep steadily to the point of view from which the whole should be seen, viz., that of a comparative study of nature. The germs of such higher views, not merely orographic, but truly geognostic, were to be found in Eratosthenes and Strabo; but the works of the former are made use of by Pliny only in one instance, and those of the latter not at all. Nor has he learned from Aristotle's anatomical history of animals, either the division into great classes based upon the principal diversities of internal organisation, or the method of induction, the only safe means of generalisation of results.

Commencing with pantheistic contemplations and considerations, Pliny descends from the celestial spaces to terrestrial objects. Recognising the necessity of presenting the powers and the majesty of nature (naturæ vis atque majestas) as a great and concurrent whole, (I refer here to the motto on the title of my work), he also distinguishes, in the beginning of the third book, between general and special geography; but this distinction is soon again forgotten and neglected when he plunges into the dry nomenclature of countries, mountains, and rivers. The greater part of books viii. to xxvii., xxxiii. and xxxiv., xxxvi. and xxxvii. is filled with catalogues of the three kingdoms of nature. The younger Pliny, in one of his letters, characterises his uncle's work with great justness as "a work learned and full of matter; no less various than nature herself (opus diffusum, eruditum, nec minus varium quam ipsa natura)." Much which has been made a subject of reproach to Pliny as needless and extraneous admixture, I am inclined to regard rather as deserving of praise. I view with particular pleasure the frequent references which he makes, with evident predilection, to the influence of nature on the civilization and mental development of mankind. His points of connection, however, are seldom happily chosen (vii. 24 to 47; xxv. 2; xxvi. 1; xxxv. 2; xxxvi. 2 to 4; xxxvii. 1.) The nature of mineral and vegetable substances, for example, leads to a fragment of the history of the plastic arts; but this fragment has become in the present state of our knowledge of greater interest and importance than almost all which we can gather from his work in descriptive natural history.

The style of Pliny is rather spirited and lively than characterised by true grandeur; he seldom defines picturesquely; and we feel, in reading his work, that the author had derived his impressions from books, and not from the free aspect of nature herself, although he had enjoyed that aspect in various regions of the earth. A grave and melancholy colouring is spread over the whole, and with this sentimental tone there is blended a degree of bitterness whenever man and his circumstances and destiny are touched upon. At such times (almost as in the writings of Cicero, (309) though with less simplicity of diction), the view of the great universal whole of the world of nature is described as reassuring and consolatory.

The conclusion of the Historia Naturalis of Pliny, the greatest Roman memorial bequeathed to the literature of the middle ages, is conceived in the true spirit of a description of the universe. As we now possess it, since 1831, (310) it contains a cursory view of the comparative natural history of countries in different zones; and a laudatory description of Southern Europe between the natural boundaries of the Mediterranean and the Alps, and of the serene heaven of

Hesperia, "where," according to a dogma of the older Pythagoreans, "the soft and temperate climate had early hastened the escape of mankind from barbarism."

The influence of the Roman dominion, as a constant element of union and fusion, deserves to be brought forward, in a history of the contemplation of the universe, with the more detail and force, because we can recognise its consequences even at a period when the union of the empire had been loosened, and in part destroyed, by the assaults and irruptions of the barbarians. Claudian, who, in a late and troubled age, under Theodosius the Great and his sons, came forward with new poetic productiveness in the decline of literature, still sings, in too laudatory strains, of the Roman sovereignty (311) :—

> " Hæc est, in gremium victos quæ sola recepit,
> Humanumque genus communi nomine fovit
> Matris, non dominæ, ritu ; civesque vocavit
> Quos domuit, nexuque pio longinqua revinxit.
> Hujus pacificis debemus moribus omnes
> Quod veluti patriis regionibus utitur hospes" . . . . .

Outward means of constraint, skilfully disposed civil institutions, and long-continued habits of servitude, may indeed produce union, by taking away separate national existence; but the feeling of the unity of mankind, of their common humanity, and of the equal rights of all portions of the human race, has a nobler origin: it is in the inmost impulses of the human mind, and in religious convictions, that its foundations are to be sought. Christianity has preeminently contributed to call forth the idea of the unity of mankind, and has thereby acted beneficently on the "humanizing" of nations, in their manners and institutions. Deeply interwoven from the first with Christian doctrines, the idea of humanity has nevertheless only slowly obtained its just recognition. At the time when, from political motives, the new faith was established at Byzantium as the religion of the state, its adherents were already involved in miserable party strife, whilst intercourse with distant nations had been checked, and the foundations of the empire had been shaken by external assaults. Even the personal freedom of entire classes of men long found no protection in Christian states, and even among ecclesiastical proprietors and corporations.

Such unnatural impediments, and many others which still stand in the way of the intellectual and social advancement and ennoblement of mankind, will gradually vanish. The principle of individual and political freedom is rooted in the indestructible conviction of the equal rights of the whole human race. Thus, as I have already said in another place, ([312]) mankind, as one great brotherhood, advance together towards the attainment of one common object, the free development of their moral faculties. This view of humanity, or at least the tendency towards the formation of this view,—sometimes checked, sometimes advancing with powerful and rapid steps, and by no means a discovery of modern times—by the universality of its direction, belongs most properly to our subject, as elevating and animating cosmical life. In depicting a great epoch in the history of the world, that of the Empire of the Romans and the laws which they originated, and of the beginning of the Christian religion, it was fitting that I should, before all things, recal the manner in which Christianity enlarged the views of mankind, and exercised a mild and enduring, although slowly operating, influence on Intelligence and Civilization.

EPOCHS IN THE HISTORY OF THE CONTEMPLATION OF THE UNIVERSE.—ITS ADVANCEMENT BY THE ARABIANS.

V.

Invasion of the Arabians—Aptitude of this part of the Semitic Race for Intellectual Cultivation—Influence of a Foreign Element on the Development of European Civilization and Culture—Peculiarities of the National Character of the Arabians—Attachment to the Study of Nature and its powers—Science of Materia Medica and Chemistry—Extension of Physical Geography to the Interior of Continents, and Advances in Astronomy and in the Mathematical ciences.

In my sketch of the history of the physical contemplation of the universe, I have already enumerated four leading epochs in the gradual development of the recognition of the universe as a whole. These included, firstly, the period when the inhabitants of the coasts of the Mediterranean endeavoured to penetrate eastward to the Euxine and the Phasis, southward to Ophir and the tropical gold lands, and westward through the Pillars of Hercules into the "all-surrounding ocean;" secondly, the epoch of the Macedonian expeditions under Alexander the Great; thirdly, the period of the Lagidæ; and fourthly, that of the Roman Empire of the World. We have now to consider the powerful influence exercised by the Arabians, whose civilization was a new element foreign to that of Europe,—and, six or seven centuries later, by the maritime discoveries of the Portuguese and Spaniards,—on the general physical and mathematical knowledge of nature, in respect to form and measurement on the earth and in the regions of space, to the heterogeneity of substances, and to the powers or forces resident therein. The discovery and exploration of the New Continent, with its lofty Cordilleras and their numerous volcanoes, its elevated plateaus with successive stages of climate placed one above another, and its various vegetation ranging through 120 degrees of latitude, mark incontestably the period in which there was offered to the human mind, in the smallest space of time, the greatest abundance of new physical perceptions.

Henceforward the extension of cosmical knowledge has no longer been connected with political events acting within definite localities. From that period the human intellect has brought forth great things by virtue of its own proper strength; and instead of being principally incited thereto by the influence of extraneous events, it now works simultaneously in many directions: by new combinations of thought it creates for itself new organs, wherewith to examine, on the one hand, the wide regions of celestial space, and, on the other, the delicate tissues of animal and vegetable structure which form the substratum of life. The whole of the seventeenth century, brilliantly opened by the great discovery of the telescope and by the more immediate fruits of that discovery,—from Galileo's observations of Jupiter's satellites, the crescent form of the disk of Venus, and the solar spots, to Newton's theory of gravitation,—is distinguished as the most important epoch of a newly created "physical astronomy." We here find, therefore, once more a marked epoch, characterised by unity in the endeavours devoted to the observation of the heavens, and to mathematical research; it forms a well-defined section in the great process of intellectual development, which since that period has advanced uninterruptedly forward.

Nearer to our own time it becomes so much the more difficult to distinguish particular epochs, as the intellectual activity of mankind has moved forward simultaneously in many directions, and as with a new order of social and political relations a closer bond of union now subsists between the different sciences. In the separate studies the development of which belongs to the "history of the physical sciences," in chemistry and descriptive botany, it is still quite possible, even up to the most recent time, to distinguish insulated periods in which the greatest advances were made, or in which new views suddenly prevailed; but in the "history of the contemplation of the universe,"—which, according to its essential character, ought to borrow from the history of separate studies only that which relates most immediately to the extension of the idea of the Cosmos,—connection with particular epochs becomes unsafe and impracticable, since that which we have just termed an intellectual process of development supposes an uninterrupted simultaneous advance in all departments of cosmical knowledge. Having now arrived at the important point of separation, at which, after the fall of the Roman Empire of the World, there appears a new and foreign element of cultivation received by our continent for the first time direct from a tropical country, it may be useful to cast a general glance at the path which yet remains to be travelled over.

The Arabians, a primitive Semitic race, partially dispelled the barbarism which for two centuries had overspread the face of Europe, after it had been shaken to its foundations by the tempestuous assaults of the nations by whom it was overrun. The Arabians not only contributed to preserve scientific cultivation, by leading men back to the perennial sources of Greek philosophy, but they also extended that cultivation, and opened new paths to the investigation of nature. The desolation of our continent by the overwhelming torrent of invading nations commenced in the reign of Valentinian I., in the last quarter of the 4th century, when the Huns (of Finnish not Mongolian origin) crossed the Don, and oppressed the Alani, and later with the help of these, the Ostrogoths. Far off in eastern Asia, the torrent of migrating nations had been set in motion several centuries before our era. The first impulse was given, as we have already said, by the attack of the Hiungnu (a Turkish tribe), on the fair haired and blue-eyed, perhaps Indo-germanic, population of the Usün, dwelling adjacent to the Yueti (Getæ?), in the upper valley of the Hoangho in North-western China. This desolating torrent, propagated from the great wall erected against the Hiungnu (214 B.C.) to the most western parts of Europe, moved through central Asia north of the chain of the Himalaya. These Asiatic hordes were not animated by any religious zeal before they came in contact with Europe; it has even been shown that they were not yet Buddhists (313) when they arrived as conquerors in Poland and Silesia. Causes of an entirely different kind gave to the warlike outbreak of a southern people, the Arabians, a peculiar character.

In the generally compact and unbroken continent of Asia, (314) the almost detached peninsula of Arabia, between the Red Sea and the Persian Gulf, the Euphrates and the Syrian part of the Mediterranean, forms a remarkably distinct feature. It is the westernmost of the three peninsulas of southern Asia, and its proximity to Egypt and to a European sea, render its geographical position a very favourable one, both politically and commercially. In the central parts of the Arabian peninsula lived the population of the Hedjaz, a noble and powerful race, uninformed but not rude, imaginative, and yet devoted to the careful observation of all the phenomena presenting themselves to their eyes in the open face of nature, on the ever clear vault of heaven, and on the surface of the earth. After this race had lived for thousands of years almost without contact with the rest of the world, and leading for the most part a nomadic life, they suddenly broke forth, became polished and informed by mental contact with the inhabitants of the ancient seats of cultivation, and subdued, proselytised, and ruled over nations from the Pillars of Hercules to the Indus, as far as the point where the Bolor chain intersects that of the Hindoo Coosh. Even from the middle of the ninth century they maintained commercial relations at once with the northern countries of Europe and with Madagascar, with East Africa, India and China; they diffused their language, their coins, and the Indian system of numbers, and founded a powerful combination of countries held together by the ties of a common religious faith. It often happened that great provinces were only temporarily overrun. The swarming troop, threatened by the natives, encamped, according to a comparison of their native poets, "like groups of clouds which are soon scattered anew by the wind." No national movement ever offered more animated phenomena; and the mind-repressing spirit which appears to be inherent in Islam, has manifested itself, on the whole, far less under the Arabian empire than among the Turkish races. Religious persecution was here as elsewhere (among Christian nations also), rather the effect of a boundless dogmatising despotism, (315) than of the original faith and doctrine or of the religious contemplation of the nation. The severity of the Koran is principally directed against idolatry, and especially against the worship of idols by Aramean races.

As the life of nations is determined not only by their internal mental dispositions, but also by many external conditions of soil, climate, proximity of the sea, &c., we should first recal the diversities of form presented by the Arabian peninsula. Although the first impulse which led to the great changes which the Arabians wrought in the three continents proceeded from the Ismaelitish Hedjaz, and owed its principal strength to a solitary pastoral tribe, yet the coasts of the other parts of the peninsula had for thousands of years enjoyed some portion of intercourse with the rest of the world. In order to obtain an insight into the connection and necessary conditions of great and singular events, we must ascend to the causes which gradually prepared the way for them.

Towards the south west, near the Erythrean Sea, is situated the fine fruitful and agricultural country of the Joctanides, (316) Yemen, the ancient seat of civilization (Saba). It produces incense (lebonah of the Hebrews, perhaps Boswellia thurifera, Colebr.), (317) myrrh (a kind of Amyris, first exactly described by Ehrenberg), and what is called the balsam of Mecca (Balsamodendron gileadense, Kunth): all of which formed articles of a considerable trade with neighbouring nations, and were carried to the Egyptians, to the Persians and Indians, and to the Greeks and Romans. It was from these productions that the geographical denomination of Arabia Felix, which we find first employed by Diodorus and Strabo, was given. On the south-east of the peninsula, on the Persian Gulf, the town of Gerrha, situated opposite to the Phœnician settlements of Aradus and Tylus, formed an important mart for the traffic in Indian goods. Although almost the whole of the interior of Arabia may be termed a treeless sandy desert, yet there are in Oman (between Jailan and Batna), a whole chain of oases, watered by subterranean canals; and we owe to the activity of the meritorious traveller Wellsted, (318) the knowledge of three mountain chains, of which the loftiest summit, Djebel Akhdar, rises, clothed with forests, to an elevation of more than six thousand feet above the level of the sea. There are also in the mountain country of Yemen, east of Lopeia, and in the littoral chain of Hedjaz in Asyr, as well as east of Mecca near Tayef, elevated plains, of which the constantly low temperature was known to the geographer Edrisi. (319)

The same variety of mountain landscape characterises the peninsula of Sinai, the "copper land" of the Egyptians of the "ancient kingdom" (before the time of the Hyksos), and the rocky valleys of Petra. I have already spoken, in a preceding section, (320) of the Phœnician trading settlements on the most northern part of the Red Sea, and the voyages to Ophir of the ships of Hiram and Solomon, which sailed from Ezion Geber. Arabia, and the adjacent island of Socotora (the Island of Dioscorides), inhabited by Indian settlers, were the intermediate links of the traffic of the world with India and the east coast of Africa. The productions of these countries were commonly confounded with those of Hadramaut and Yemen. We read in the prophet

Isaiah, "they (the dromedaries of Midian) shall come from Saba, they shall bring gold and incense." (321) Petra was the emporium for the valuable goods designed for Tyre and Sidon, and a principal seat of the once powerful commercial nation of the Nabateans, supposed by the learned Quatremère to have had their original dwelling-place in the Gerrha mountains, near the lower Euphrates. This northern part of Arabia, by its proximity to Egypt, by the spreading of Arabian tribes into the mountains bounding Syria and Palestine and into the countries near the Euphrates, as well as by the celebrated caravan road from Damascus through Emesa and Tadmor (Palmyra) to Babylon, had come into influential contact with other civilised states. Mahomet himself, sprung from a noble but impoverished family of the tribe of Koreish, in the course of his trading occupations, before he came forward as an inspired prophet and reformer, had visited the fair of Bosra on the Syrian border, the fair held in Hadramaut the land of incense, as well as the twenty days' fair of Okadh near Mecca, where poets, chiefly Bedouins, assembled for lyrical contests. I allude to these particulars of the Arabian commerce, and the circumstances thence arising, in order to give a more vivid picture of that which prepared great revolutions in the world.

The spreading of the Arabian population towards the north, reminds us of two events, the circumstances of which are indeed veiled in obscurity, but which afford evidence that ages before Mahomet the inhabitants of the peninsula had mixed in the affairs of the world by outbreaks to the west and east, towards Egypt and the Euphrates. The Semitic or Aramaic descent of the Hyksos, who, under the twelfth dynasty, 2200 years before our era, put an end to the "ancient kingdom" of Egypt, is now received by almost all historic investigators. Manetho even had said, "some maintain that these shepherds were Arabians." In other sources of historical knowledge they are called Phœnicians—a name which in antiquity was extended to the inhabitants of the valley of the Jordan, and to all the Arabian tribes. The acute Ewald refers particularly to the Amalekites (Amalekalians), who originally dwelt in Yemen, and then spread themselves by Mecca and Medina to Canaan and Syria, and are said, in early Arabian historical works, to have had power over Egypt in the time of Joseph. (322) It still must appear remarkable how the nomadic tribes of the Hyksos should have been able to overthrow the powerful and well-established "ancient kingdom" of Egypt. Men accustomed to freedom fought with success against men habituated to a long course of servitude, even though at that period the victorious Arabian invaders were not, as they subsequently were, animated by religious enthusiasm. From fear of the Assyrians (races of Arpachsad), the Hyksos established the fortress of Avaris as a place of arms on the eastern branch of the Nile. Perhaps this circumstance may indicate a succession of advancing warlike masses, or a movement of nations directed towards the west. A second event, which occurred fully 1000 years afterwards, is that which Diodorus (323) relates from Ctesias. Ariæus, a powerful Himyarite prince, entered into alliance with Ninus on the Tigris, and with him, defeated the Babylonians, and returned to his home in southern Arabia laden with rich spoils. (324)

Although, on the whole, the prevailing mode of life in Hedjaz, and that followed by a large and powerful portion of the people, was a free and pastoral one, yet even then the towns of Medina and Mecca (the latter with its highly ancient and enigmatical sacred Kaaba) were distinguished as places of importance visited by foreign nations. In districts adjacent to the sea, or to the caravan roads which act as river vallies, the complete savage wildness engendered by entire insulation never prevailed. Gibbon, whose conception of the different circumstances of mankind is always so clear, notices the important distinction to be drawn between the nomadic life of the inhabitants of the Arabian peninsula, and that of the Scythians described by Herodotus and Hippocrates; since among the latter, no part of the pastoral population ever settled in towns, whereas in the great Arabian peninsula, the inhabitants of the country have always kept up intercourse with the inhabitants of the towns, who they regard as descended from the same original race as themselves. (325) In the Kirghez Steppe, a portion of the plains inhabited by the ancient Scythians (Scoloti and Sacæ) and exceeding Germany in superficial extent, (326) no town has existed for thousands of years; yet at the time of my Siberian journey, the number of tents (yourtes or kibitkos) in the three wandering hordes still exceeded 400,000, indicating a nomadic population of two millions. I need not enter more fully on the influence which such differences, in regard to the greater or less insulation of nomadic life, must have exercised on the national aptitude for mental cultivation, even supposing an equality of original disposition and capacity.

In the noble and richly-gifted Arab race, the internal disposition and aptitude for mental cultivation concur with the external circumstances to which I have adverted,—I mean the natural features of the country, and the ancient commercial intercourse of the coasts with highly-civilised neighbouring states,—in explaining how the irruptions into Syria and Persia, and at a later period the possession of Egypt, could have so rapidly awakened in the conquerors a love for the sciences, and a disposition to original investigation. We may perceive that, in the wonderful arrangement of the order of the world, the Christian sect of the Nestorians, who had exerted a very important influence on the diffusion of knowledge, became also of use to the Arabians before the latter came to the learned and controversial city of Alexandria; and even that Nestorian Christianity was enabled to penetrate far into eastern Asia under the protection of armed Islam. The Arabians were first made acquainted with Greek literature through the Syrians, (327) a cognate Semitic race, who had received this knowledge hardly a century and a half before from the Nestorians. Physicians trained in Grecian establishments of learning, or in the celebrated medical school founded at Edessa in Mesopotamia by Nestorian Christians, were living at Mecca in the time of Mahomet, and connected by family ties with himself and Abu-Bekr.

The school of Edessa, a prototype of the Benedictine schools of Monte-Cassino and Salerno, awakened a disposition for the pursuit of natural history, by the investigation of "healing substances in the mineral and vegetable kingdoms." When this school was dissolved from motives of fanaticism under Zeno the Isaurian, the Nestorians were scattered into Persia, where they soon obtained a political importance, and founded a new and much-frequented medicinal institution at Chondisapur, in Khusistan. They succeeded in carrying both their scientific and literary knowledge and their religion as far as China, under the dynasty of the Thang, towards the middle of the seventh century, 572 years after Buddhism had arrived there from India.

The seeds of western cultivation scattered in Persia by learned monks, and by the philosophers of the school of the later Platonists at Athens persecuted by Justinian, had exercised a beneficial influence on the Arabians during their Asiatic campaigns. However imperfect the scientific knowledge of the Nestorian priests may have been, yet, by its particular medico-pharmaceutical direction, it was the more effectual in stimulating a race of men who had long lived in the enjoyment of the open face of nature, and preserved a fresher feeling for every kind of natural contemplation, than the Greek and Italian inhabitants of cities. That which gives to the epoch of the Arabians the cosmical importance which we are endeavouring to illustrate, is very much connected with this feature of the national character. The Arabians are, we repeat, to be regarded as the proper founders of the *physical sciences,* in the sense which we are now accustomed to attach to the term.

In the world of ideas, the internal connection and enchainment of all thought renders it indeed always difficult to attach an absolute beginning to any particular period of time. Separate points of knowledge, as well as processes by which knowledge may be attained, are, it is true, to be seen scattered in rare instances at an earlier period. How wide is the difference between Dioscorides who separated mercury from cinnabar and the Arabian chemist Djeber; and between Ptolemy as an investigator of optics and Alhazen! But the foundation of physical studies, and of the natural sciences themselves, first begins when newly opened paths are pursued by many at once, although with unequal success. After the simple contemplation of nature, after the observation of such phenomena on the surface of the earth or in the heavens as present themselves spontaneously to the eye, comes investigation, the seeking after that which exists, the measurements of magnitudes and of the duration of motion. The earliest epoch of such an investigation of nature, chiefly limited, however, to the organic world, was that of Aristotle. In the progressive knowledge of physical phenomena, in the searching out of the powers of nature, there still remains a third and higher stage,—that of the knowledge of the action of these powers or forces in producing new forms of matter, and of the substances themselves which are set at liberty in order to enter into new combinations. The means which lead to this liberation belong to the calling forth at will of phenomena, or to "experiment."

It is on this last stage, which was almost wholly untrodden by the ancients, that the Arabians principally distinguished themselves. Their country enjoys throughout the climate necessary for the growth of palms, and in its larger portion possesses a tropical climate, as the tropic of Cancer crosses the peninsula nearly from Maskat to Mecca;—it is therefore a part of the world in which the higher vital energy of the vegetable kingdom offers an abundance of aromas, of balsamic juices, and of substances injurious as well as beneficial to man. The attention of the people must have been early directed to the productions of their native soil, and to those obtained by commerce from the coasts of Malabar, Ceylon, and eastern Africa. In these portions of the torrid zone organic forms are "individualised" in the smallest geographical spaces, each of which offers peculiar productions, —and thus incitements to the intercourse of men with nature were increased and multiplied. Great desire was felt to become acquainted with articles so precious and so important to medicine, industry, and the luxury of the temple and the palace; to distinguish them carefully from each other; and to find out their native place, which was often artfully concealed from motives of covetousness. Numerous caravan roads, departing from the commercial mart, Gerrha, on the Persian Gulf, and from the incense district of Yemen, traversed the whole interior of Arabia to Phœnicia and Syria; and thus the names of these much-desired productions, and the interest felt in them, became generally diffused.

The science of materia medica, the foundation of which was laid in the Alexandrian school by Dioscorides, is, in its scientific form, a creation of the Arabians, who, however, had previously access to a rich source of instruction, the most ancient of all, that of the Indian physicians. (328) The apothecary's art was indeed formed by the Arabians, and the first official authoritative rules for the preparation of medicines were taken from them, and were diffused through southern Europe by the school of Salerno. Pharmacy and the materia medica, the first requirements of the healing art, conducted to the studies of botany and chemistry. From the confined sphere of utility and of single application, the study of plants gradually expanded into a wider and freer field: it examined the structure of organic tissues; the connection of this structure with the laws of their development; and the laws according to which vegetable forms are distributed geographically over the earth's surface, according to differences of climate and of elevation.

After the Asiatic conquests, for the maintenance of which

Bagdad subsequently became a central point of power and civilisation, the Arabs, in the short space of seventy years, extended their conquests over Egypt, Cyrene, and Carthage, and through the whole of northern Africa to the farthest Iberian peninsula. The low state of cultivation of the armed masses and of their leaders, may indeed render the occurrence of any outbreak of a rude spirit not altogether improbable. The tale of the burning of the Alexandrian library by Amru, 40,000 baths being heated for six months by its contents, rests, however, solely on the testimony of two writers who lived 580 years after the supposed event. ([329]) We need not here describe how, in more peaceful times, but without the mental cultivation of the mass of the nation having attained any free development, in the brilliant epoch of Al-Mansur, Harun Al-Raschid, Mamun, and Motasem, the courts of princes and the public scientific institutions were able to assemble a considerable number of highly distinguished men. We cannot attempt in these pages to characterise the extensive, varied, and unequal Arabic literature; or to distinguish that which springs from the hidden depths of the particular organisation of a race and the natural unfolding of its faculties, from that which is dependent on external incitements and accidental conditions. The solution of this important problem belongs to a different sphere of ideas. Our historical considerations are limited to a fragmentary notice of what the Arabian nation has contributed, by mathematical and astronomical knowledge, and by the physical sciences, to the more general contemplation of the Universe.

The true results of investigation are indeed here, as elsewhere in the middle ages, alloyed by alchemy, supposed magical arts, and mystic fancies; but the Arabians, incessant in their own independent endeavours, as well as laborious in appropriating to themselves by translations the fruits of earlier cultivated generations, have produced much which is truly their own, and have enlarged the view of nature. Attention has been justly called (330) to the different circumstances in respect to cultivation of the invading and immigrating Germanic and Arabic races. The former became civilized after their immigration; the latter brought with them from their native country not only their religion, but also a highly polished language, and the tender blossoms of a poetry which has not been altogether without influence on the Provençal poets and the Minnesingers.

The Arabs possessed qualities which fitted them in a remarkable manner for obtaining influence and dominion over, and for assimilating and combining different nations, from the Euphrates to the Guadalquivir, and southward to the middle of Africa: they possessed a mobility unexampled in the history of the world; a disposition, very different from the repellent Israelitish spirit of separation, to effect a fusion with the conquered nations; and yet, notwithstanding perpetual change of place, to preserve unimpaired their own national character, and the traditional remembrances of their original home. No nation can shew examples of more extensive land journies undertaken by individuals, not always for commercial objects, but also for collecting knowledge: even the Buddhistic priests from Thibet and China, even Marco Polo, and the Christian missionaries who were sent to the Mogul princes, moved over a smaller range of geographical space. Through the many relations subsisting between the Arabs and India and China, for their conquests had extended under the Caliphate of the Ommaiades by the end of the seventh century ([331]) to Kashgar, Caubul, and the Punjab), important portions of Asiatic knowledge reached Europe. The acute researches of Reinaud have shewn how much may be derived from Arabic sources, for the knowledge of India. Although the invasion of China by the Moguls for a time disturbed the communications across the Oxus, ([332]) the Moguls themselves soon became a uniting link to the Arabs, who, by their own observations, and by laborious researches, have illustrated the knowledge of the earth's surface from the coasts of the Pacific to those of Western Africa, and from the Pyrenees to Edrisi's marsh-land of Wangara in the interior of Africa. The geography of Ptolemy was translated into Arabic, according to Frähn, by the command of the Caliph Mamun between 813 and 833; and it is even not improbable that some fragments of Marinus of Tyre which have not come down to us may have been used in the translation. ([333])

Of the long series of distinguished geographers which Arabic literature affords, it is sufficient to name the earliest and the latest:—El-Istachri, ([334]) and Alhassan (Johannes Leo Africanus). At no period before the discoveries of the Portuguese and Spaniards, did the knowledge of the earth's surface receive a larger accession. Only fifty years after the death of Mahomet the Arabs had reached the extreme western coast of Africa at the harbour of Asfi. Whether, subsequently, when the adventurers known under the name of Almagrurin navigated the "mare tenebrosum," the islands of the Guanches were visited by Arab ships, as I long thought probable, has recently been rendered again doubtful. ([335]) The quantity of Arabic coins found buried in the countries about the Baltic, and in the extreme North in Scandinavia, are not to be attributed to commerce by sea properly so called, but to the far extended inland traffic of the Arabs. ([336])

Geography did not continue to be restricted to the enumeration of countries and their boundaries, and to positions in latitude and longitude, (which were multiplied by Abul-Hassan); ([337]) it led a people familiar with nature to consider the organic productions of different places, and more especially those of the vegetable world. The horror which the followers of Islam have for anatomical examinations prevented all progress in the natural history of animals. They were content with appropriating to themselves by translation what they could find in Aristotle ([338]) and Galen; yet Avicenna's history of animals, (which is in the Royal Library at Paris), ([339]) differs from that of Aristotle. As a botanist we may name Ibn-Baithar of Malaga, ([340]) who, from his journies into Greece, Persia, India, and Egypt, may also be regarded as an example of the endeavour to compare by direct observation the productions of different regions,—of the East and of the West. The study of medicines was, however, always the point from which these endeavours proceeded; it was through it that the Arabs long swayed the schools of Christendom, and for its improvement and completion Ibn-Sina (Avicenna), a native of Affschena near Bokhara, Ibn-Roschd (Averroes) of Cordova, the younger Serapion of Syria, and Mesue of Maridin on the Euphrates, availed themselves of all the materials furnished by the Arabian caravan and sea traffic. I have purposely cited these widely-separated birth-places of celebrated and learned Arabs, because they bring vividly before us the manner in which, by the peculiar disposition of this race of men, natural knowledge was spread over a large portion of the earth's surface, and the circle of ideas enlarged by simultaneous efforts proceeding from many quarters.

The knowledge possessed by a more anciently cultivated people, the Indians, was also drawn into the same circle: several important works, probably those known under the semifabulous names of Tscharaka and Susruta, (341) were translated from Sanscrit into Arabic. Avicenna, a man of comprehensive mind, and who has often been compared to Albertus Magnus, affords in his Materia Medica a very striking instance of this influence of Indian literature, in showing himself acquainted, as the learned Royle remarks, with the Deodara (Cedrus deodvara) (342) of the snowy Himalayan Alps, which, in the 11th century, had assuredly never been visited by any Arabian traveller: he calls it by its true Sanscrit name, and speaks of it as a lofty species of juniper, from which oil of turpentine was obtained. The sons of Averroes lived at the court of the Emperor Frederic II., the great prince of the house of Hohenstauffen, who was indebted for part of his knowledge of natural history to communication with learned Arabs and Spanish Jews. (343) The Caliph Abderrahman established a botanical garden at Cordova, (344) and sent travellers into Syria and other parts of Asia to collect rare plants. He planted, near the palace of Rissafah, the first date tree, which he celebrates in strains full of tender regrets and longings for his native home, Damascus.

But the most important influence exerted by the Arabians on the general knowledge of nature, was in the progress of chemistry; with their labours commenced a new epoch in that science. Alchemistic and new Platonic fancies were, it is true, as nearly allied with their chemistry, as was astrology with their astronomy; but the demands of pharmacy, and the equally pressing requirements of the technical arts, led to discoveries which were favoured sometimes by design, and sometimes, through a happy accident, by metallurgic attempts connected with alchemy. The labours of Geber, or rather Djaber (Abu-Mussah-Dschafar al-Kufi), and the much later ones of Razes (Abu-Bekr Arrasi), have had the most important results. This epoch is marked by the preparation of sulphuric and nitric acids, (345) aqua regia, preparations of mercury and other metallic oxides, and by the knowledge of alcoholic (346) processes of fermentation. The first foundation and earliest advances of the science of chemistry are of so much the greater importance in the history of the contemplation of the universe, because thereby the heterogeneity of substances, and the nature of forces or powers not manifested visibly by motion, were first recognised; and the students of nature, no longer looking exclusively to the Pythagorean Platonic perfection of form, perceived that composition was also deserving of regard. Differences of form and differences of composition are the elements of all our knowledge of matter; they are the abstractions by which, through measurement and analysis, we believe that we can form a conception of the entire universe.

It would be difficult to determine at present what portion of knowledge the Arabian chemists may have derived, either from their acquaintance with Indian literature (writings on the Rasayana), (347) from the primitive technical arts of the ancient Egyptians, from the comparatively modern alchemistic rules of the Pseudo-Democritus and the Sophist Synesius, or even from Chinese sources through the medium of the Mogols. According to the most recent and very careful investigations of a celebrated orientalist, Reinaud, the invention of gunpowder, (348) and its application to projectiles, are not to be ascribed to the Arabians: Hassan Al-Rammah, who wrote between 1285 and 1295, was not acquainted with this application; while, as early as the twelfth century, 200 years therefore before Berthold Schwarz, a kind of gunpowder was used at Rammelsberg, in the Harz, for blasting rocks. The invention of an air thermometer has been ascribed to Avicenna, on the strength of a notice by Sanctorius; but this notice is very obscure, and six centuries elapsed before Galileo, Cornelius Dreddel, and the Academia del Cimento, by the establishment of an exact measure of temperature, created the important means of penetrating into a world of almost unknown phænomena, whose regularity and periodicity excite our astonishment; and of recognising the cosmical connection of effects taking place in the atmosphere, in the superimposed aqueous strata of the ocean, and in the interior of the earth. Among the advances which physical science owes to the Arabians, it will be sufficient to name Alhazen's work on the refraction of rays, which may indeed have been partially derived from Ptolemy's optical researches; and the knowledge and first application of the pendulum as a measure of time (349) by the great astronomer Ebn-Junis.

The purity and rarely disturbed transparency of the Arabian sky had in a peculiar manner drawn the attention of the Arab race, in their earliest uncultivated state in their native land, to the motions of the heavenly bodies; for we find that, besides the worship of the planet Jupiter among the Lachmites, the tribe of the Asedites worshipped the planet Mercury, which, from its proximity to the solar orb, is rarely visible. Notwithstanding this, however, the distinguished scientific activity of the civilised Arabians in all departments of practical astronomy, is rather to be ascribed to Chaldean and Indian influences. Atmospheric conditions can only encourage and favour such pursuits where a disposition towards them has been produced by the original mental endowments of richly gifted races, or by intercourse with more highly civilised neighbouring nations. How many districts of tropical America, as Cumana, Coro, and Payta, where rain never falls, enjoy an atmosphere even more transparent than that of Egypt, Arabia, or Bokhara! The climate of the tropics, and the eternal serenity of the vault of heaven, resplendent with stars and nebulæ, are indeed never without some influence on the dispositions of men; but they are fruitful in intellectual results, and incite the human mind to labour in the development of mathematical ideas, only where an impulse is given, independently of climate, by other causes belonging either to the character of the race, or to external circumstances; as, for example, where the exact division of time becomes an object of social necessity for the satisfaction of religious or of agricultural requirements. Among calculating commercial nations like the Phœnicians, and nations like the Egyptians and Chaldeans fond of architecture and constructions of all kinds, and much accustomed to ground surveys and measurements, empirical rules of arithmetic and of geometry were early discovered; but these can only prepare the way for the mathematical and astronomical sciences. Nor is it until cultivation has reached a still higher point, that the regularity and subjection to laws, which characterise the movements of the heavenly bodies, are seen to be, as it were, reflected in terrestrial phænomena, and that men seek to discover in these also, to use the expression of our great poet, the "fixed unchanging pole." In all climates, the conviction of the regularity of the planetary movements, and of their subjection to law and order, has contributed more than any thing else to lead men to seek the same subjection to law and order, in the undulations of the aerial ocean, in the oscillations of the sea, in the periodical march of the magnetic needle, and in the distribution of vegetable and animal life on the surface of the globe.

The Arabians were in possession of Indian planetary tables (350) as early as the end of the eighth century. I have already mentioned that the Susruta, the ancient epitome comprising all the medicinal knowledge of the Indians, was translated by learned men belonging to the court of the Caliph Haroun Al-Raschid,—a proof of the early introduction of Sanscrit literature. The Arabian mathematician Albyruní went himself to India to study astronomy there. His writings, which have only very lately become accessible to us, shew how well he was acquainted with the country, the traditions, and the extensive knowledge of the Indians. (351)

But however much the Arabian astronomers may have owed to earlier civilized nations, and especially to the Indian and Alexandrian schools, they still must be regarded as having considerably enlarged the domain of astronomy, by their peculiar practical turn of mind, by the great number and the direction of their observations, by their improvements in instruments for angular measurements, and by their zealous endeavours to correct the earlier tables by careful comparison with the heavens. Sedillot has recognised in the seventh book of the Almagest of Abul-Wefa the important inequality in the moon's motion, which vanishes at the Syzygies and Quadratures, and has its greatest value at the Octants, and which under the name of "variation" has long been regarded as a discovery of Tycho Brahe. (352) Ebn-Junis's observations at Cairo have become particularly important for the perturbations and secular changes of the orbits of the two great planets, Jupiter and Saturn. (353) A measurement of a degree of the meridian, executed by the orders of the Caliph Al-Mamun in the great plain of Sindschar between Tadmor and Rakka, by observers whose names have been preserved to us by Ebn-Junis, is less important for its result than for the evidence which it affords of the scientific cultivation of the Arabian race.

We must also attribute to this cultivation, in the West, the astronomical congress held in Toledo in Christian Spain under Alphonso of Castile, in which the Rabbi Isaac Ebn Sid Hazan occupied a prominent place; and in the far East, the Observatory provided with many instruments established by Ilschan Holagu, the grandson of Ghengis-khan, on a mountain near Meragha, in which Nassir-Eddin of Tus in Khorasan made his observations. These details are deserving of notice in the history of the contemplation of the Universe, because they remind us in a lively manner of what the Arabians have effected in the extension of knowledge over wide portions of the earth's surface, and in the accumulation of numerical results: results which contributed materially in the great epoch of Kepler and Tycho Brahe to the foundation of theoretical astronomy, and to a correct view of the motions of the heavenly bodies in space. The light kindled in the part of Asia inhabited by Tatar nations extended in the fifteenth century to the westward as far as Samarcand, where Ulugh Beig a descendant of Timour established an astronomical observatory, and a gymnasium of the class of the Alexandrian Museum, and caused a star catalogue to be prepared founded entirely on new and independent observations. (354)

Besides the tribute of praise which we have here paid to the advances made by the Arabians in the knowledge of nature, both in the terrestrial and celestial spheres, we have still to allude to the additions which, in the solitary paths of the development of ideas, they made to the treasury of pure mathematical knowledge. According to the most recent works written in England, France, and Germany (355) on the history of mathematics, the algebra of the Arabians is to be regarded as "having originated from the confluence of two streams which had long flowed independently of each other, one Indian and one Greek." The compendium of algebra written by the command of the Caliph Al-Mamun by the Arabian mathematician Mahommed Ben-Musa (the Chowarezmian) is based, as my deceased learned friend Friedrich Rosen has shewn, (356) not on the works of Diophantus, but on Indian knowledge; and even as early as under Almansor at the end of the eighth century Indian astronomers were called to the brilliant court of the Abassides. According to Castri and to Colebrooke, Diophantus was not translated into Arabic until the end of the tenth century by Abul-Wefa Buzjani. The Arabians were indebted to the Alexandrian school for that which we miss in the old Indian algebraists, namely, the establishment of a conclusion by the successive advance from proposition to proposition. This fair inheritance, yet farther increased by their own exertions, passed in the twelfth century from the Arabs to the European literature of the middle ages through Johannes Hispalensis and Gerard of Cremona. (357) "In the algebraical works of the Indians we find the general solution of indeterminate equations of the first degree, and a far more highly finished treatment of those of the second degree, than in the writings of the Alexandrian school which have come down to us; there is therefore no doubt, that if the works of the Indian writers had been made known to Europeans two centuries earlier, instead of only in our own time, they must have aided the development of modern analysis."

The Arabs in Persia and on the Euphrates, as well as in Arabia, received in the 9th century, the knowledge of the Indian numerical characters, through channels similar to those which had led to their acquaintance with Indian algebra. Persians were employed at that period as revenue collectors on the Indus; and the use of Indian numbers became general amongst the Arab revenue officers, and extended to Northern Africa, opposite to the coast of Sicily. Nevertheless, the profound and important historical investigations to which a distinguished mathematician, M. Chasles, was led, by his correct interpretation of the so-called Pythagorean table in the geometry of Boethius, (358) render it more than probable that the Christians in the West were acquainted even earlier than the Arabians with the Indian system of numeration; the use of the nine figures, having their values determined by position, being known by them under the name of the system of the Abacus.

The present work is not the place for entering more fully on this subject, which was treated by me many years ago in two memoirs presented in 1819 and in 1829 to the Académie des Inscriptions at Paris, and the Akademie der Wissenschaften at Berlin; (359) but in an historical problem, in which much still remains to be discovered, the question arises, whether the highly ingenious artificial idea of value by position, which appears both in the Tuscan Abacus and in the Suan-pan of the interior of Asia, was separately discovered in the East and in the West; or whether, through the direction of the commerce of the world under the Lagidæ, it made its way from the western peninsula of India to Alexandria, and subsequently, in the renewal of the dreams of the Pythagoreans, was represented as a discovery of their founder. We need not dwell on the mere possibility of ancient relations with which we are entirely unacquainted having subsisted prior to the 60th Olympiad. Why may we not suppose that, under a sense of similar wants, the same combinations of ideas may have presented themselves separately to highly-gifted nations of different races?

The algebra of the Arabians, including what they had received from the Greeks and the Indians and what they had themselves originated, notwithstanding its great deficiency in symbolic notation, exercised a beneficial influence during the brilliant period of the Italian mathematicians of the middle ages; the Arabians have also the merit of having by their writings, and by their extensive commercial intercourse, accelerated the use of the Indian system of numbers from Bagdad in the East to Cordova in the West. Both circumstances contributed powerfully, although in different ways, to advance the mathematical part of natural knowledge, and to facilitate the access to fields which without these aids must have remained unopened, in astronomy, in optics, in physical geography, in thermometrics, and in the theory of magnetism.

In studying the history of nations, the question has often been raised, what would have been the effect on the course of events if Carthage had conquered Rome, and had subjected to its dominion the European West: Wilhelm von Humboldt (360) has remarked, that "we might ask with equal justice, what would have been the state of our present intellectual cultivation, if the Arabs had continued the exclusive possessors of science as they were for a long period, and had spread themselves permanently over the West? In both cases it appears to me we can scarcely doubt that the result would have been less favourable. It is to the same causes which led to the Roman universal empire, namely, to the Roman mind and character and not to external accidents, that we owe the influence of the Romans on our civil institutions, our laws, our languages, and our civilization. Through this beneficial influence, and in consequence of our belonging to a kindred race, we have been enabled to receive the impression of the Grecian mind and Grecian language; whereas the Arabians only attached themselves to the scientific results of Greek investigation in natural history, physics, astronomy, and pure mathematics." The Arabians, by sedulous care in preserving the purity of their native idiom, and by the ingenuity of their figurative modes of speech, knew how to lend to the expression of their feelings, and to the enunciation of noble and sage maxims, the grace of poetic colouring; but judging from what they were under the Abassides, even if they had built on the same foundation of classical antiquity with which we find them familiar, they yet could never have produced those works of sublime poetry and creative art which are the boast of our European cultivation.

VI

EPOCHS IN THE HISTORY OF THE CONTEMPLATION OF THE UNIVERSE.—OCEANIC DISCOVERIES.

Epoch of the Oceanic Discoveries—Opening of the Western Hemisphere—Events, and Extension of different Branches of Scientific Knowledge, which prepared the way for the Oceanic Discoveries. —Columbus, Sebastian Cabot, and Vasco de Gama—America and the Pacific Ocean—Cabrillo, Sebastian Vizcaino, Mendaña, and Quiros.—The rich abundance of materials for the foundation of Physical Geography offered to the nations of Europe.

The fifteenth century belongs to those rare epochs in the history of the world, in which all the efforts of the human mind are invested with a determinate and common character, and manifest an unswerving direction towards a single object. The unity of these endeavours, the success with which they were crowned, and the vigour and activity displayed by entire nations, give grandeur and enduring splendour to the age of Columbus, of Sebastian Cabot, and of Vasco de Gama. Intervening between two different stages of cultivation, the fifteenth century forms a transition epoch belonging at once to the middle ages and to the commencement of modern times. It is the epoch of the greatest discoveries in geographical space, comprising almost all degrees of latitude, and almost every gradation of elevation of the earth's surface. To the inhabitants of Europe it doubled the works of Creation, while at the same time it offered to the intellect new and powerful incitements to the improvement of the natural sciences in their physical and mathematical departments. (361)

The world of objects, now as in Alexander's campaigns but with yet more preponderating power, presented to the combining mind the separate forms of sensible objects, and the concurrent action of animating powers or forces. The scattered images offered to the contemplation of the senses, notwithstanding their number and diversity, were gradually fused into a concrete whole; terrestrial nature was conceived in its generality, no longer according to mere presentiments or conjectures floating in varying forms before the eye of fancy, but as a result of actual observation. The vault of heaven also offered to the yet unarmed eye new regions, adorned with constellations before unseen. As I have already remarked, at no period has there been offered to mankind a greater abundance of new facts, or fuller materials for the foundation of comparative physical geography. I may add, that never were geographical or physical discoveries more influential on human affairs. A larger field of view was opened, commerce was stimulated by a great increase in the medium of exchange, as well as by a large accession to the number of natural productions valued for use or enjoyment; above all, there were laid the foundations of colonies, of a magnitude never before known: and through the agency of all these causes, extraordinary changes were wrought in manners and customs, in the condition of servitude long experienced by a portion of mankind, and in their late awakening to political freedom.

When a particular epoch thus stands out in the history

of mankind, as marked by important intellectual progress, we shall find on examination that preparations for this progress had been made during a long series of antecedent centuries. It does not appear to belong to the destinies of the human race that all portions of it should suffer eclipse or obscuration at the same time. A preserving principle maintains the ever living process of the progress of reason. The epoch of Columbus attained the fulfilment of its objects so rapidly, because their attainment was the development of fruitful germs, which had been previously deposited by a series of highly gifted men, who formed as it were a long beam of light which we may trace throughout the whole of what have been called the dark ages. A single century, the thirteenth, shows us Roger Bacon, Nicolaus Scotus, Albertus Magnus, and Vincentius of Beauvais. The subsequent more general awakening of mental activity soon bore fruit in the extension of geographical knowledge. When, in 1525, Diego Ribero returned from the geographico-astronomical congress which was held at the Puente de Caya near Yelves, for the termination of differences respecting the boundaries of the two great empires of the Portuguese and Spanish monarchies, the outlines of the New Continent had already been traced from Terra del Fuego to the coasts of Labrador. On the western side, opposite to Asia, the advances were naturally less rapid; yet in 1543 Rodriguez Cabrillo had already penetrated north of Monterey; and after this great and adventurous navigator had met his death off New California, in the Channel of Santa Barbara, the pilot Bartholomew Ferreto still led the expedition as far as the 43d degree of latitude, where Vancouver's Cape Oxford is situated. The emulative activity of the Spaniards, English, and Portuguese, was then so great, that half a century sufficed to determine the outline or the general direction of the coasts of the Western Continent.

Although the acquaintance of the nations of Europe with the western hemisphere is the leading subject to which this section is devoted, and around which are grouped the numerous results which flow from it of juster and grander views of the Universe, yet we must draw a strongly marked line of separation between the first discovery of America in its more northern portions, which is certainly to be ascribed to the Northmen, and the re-discovery of the same Continent in its tropical portions. Whilst the Caliphate of Bagdad still flourished under the Abassides, and while the Samanides whose reign was so favourable to poetry bore sway in Persia, America was discovered in the year 1000, by a northern route, as far south as 41½° north latitude, by Leif, the son of Eric the Red. (362) The first but accidental step towards this discovery was made from Norway. In the second half of the ninth century, Naddod, having sailed for the Färoe Islands, which had previously been visited from Ireland, was driven by storms to Iceland, and the first Norman settlement was established there by Ingolf, in 875. Greenland, the eastern peninsula of a land which is everywhere separated by the sea from America proper, was early seen, (363) but was first peopled from Iceland a hundred years later, in 983. The colonization of Iceland, which had been first called by Naddod, Snowland (Snjoland), now conducted, in a south-westerly direction, passing by Greenland, to the New Continent.

The Färoe Islands and Iceland must be regarded as intermediate stations, and as points of departure for enterprises to Scandinavian America. In a similar manner the settlement of the Tyrians at Carthage had aided them to reach the Straits of Gadeira and the port of Tartessus, and Tartessus itself conducted this enterprising race from station to station to Cerne, the Gauleon (ship island) of the Carthaginians. [364]

Notwithstanding the proximity of the opposite coast of Labrador (Helluland it mikla or the great), 125 years elapsed from the first settlement of Northmen in Iceland, to Leif's great discovery of America; so small were the means which, in this remote and desolate part of the globe, a noble, energetic, but not wealthy race, were able to devote to naval enterprises. The line of coast called Vinland, from wild vines which were found there by the German Tyrker, charmed its discoverers by the fertility of its soil and the mildness of its climate, compared with Iceland and Greenland. The tract which received from Leif the name of Vinland it goda (Vinland the good), comprised the coast line between Boston and New York; therefore parts of the present states of Massachusetts, Rhode Island, and Connecticut, between the parallels of Civita Vecchia and Terracina, but which corresponded there to mean annual temperatures of 51·8° and 57·2° of Fahr. [365] This was the principal settlement of the Northmen. The colonists had frequently to contend with a very warlike tribe of Esquimaux, then extending much farther to the south, under the name of Skrälinger. The first bishop of Greenland, Eric Upsi, an Icelander, undertook, in 1121, a Christian mission to Vinland; and the name of the colonised country has even been met with in old national songs of the natives of the Färoe Islands. [366]

The activity, courage, and enterprising spirit of the adventurers from Iceland and Greenland, is manifested by the fact, that after they had settled themselves so far south as 41½° N. latitude, they prosecuted their researches to the latitude of 72° 55′ on the east coast of Baffin's Bay; where, on one of the Women's Islands, (367) north-west of the present most northern Danish settlement of Upernavik, they set up three stone pillars marking the limit of their discoveries. The Runic inscription on the stone discovered there in the autumn of 1824, contains, according to Rask and Finn Magnusen, the date 1135. From this eastern coast of Baffin's Bay the colonists very regularly visited Lancaster Sound, and a part of Barrow's Strait, for purposes of fishing, more than six centuries before the adventurous voyage of Parry. The locality of the fishery is very distinctly described, and priests from Greenland from the bishopric of Gardar conducted the first voyage of discovery (1266). This north-westernmost summer station is called the Kroksfjardar-Heide. Mention is made of the drift-wood (doubtless from Siberia) which was collected there, and of the abundance of whales, seals, walruses, and sea-bears. (368)

Our accounts of the communications of the extreme north of Europe, and of Iceland and Greenland, with the American Continent properly so called, only extend to the middle of the 14th century. In 1347, a ship was sent from Greenland to Markland (Nova Scotia), to bring building timber and other necessary articles. In returning from Markland the ship was driven by tempests and forced to take refuge in Straumfiord, in the West of Iceland. This is the latest notice having reference to America, preserved to us in ancient Scandinavian writings. (369)

I have hitherto kept strictly on historic ground. By the

critical and highly praiseworthy labours of Christian Rafn, and of the Royal Society established at Copenhagen for the study of northern antiquities, the Sagas and original sources of information respecting the voyages of the Northmen to Helluland (Newfoundland), to Markland, the mouth of the St. Lawrence, and Nova Scotia, and to Vinland (Massachusetts), have been severally printed, and satisfactorily commented on. (370) The duration of the voyage, the course, and the times of sunrise and sunset, are all expressly given.

There is less certainty respecting the traces which have been supposed to be found of a discovery of America from Ireland previous to the year 1000. The Skrälinger related to the Northmen settled in Vinland, that farther to the south, beyond Chesapeake Bay, there were "white men, wearing long white garments, who carried before them poles with pieces of cloth fastened to them, and who called with a loud voice." This account was interpreted by the Christian Northmen to indicate processions, with banners and singing. In the oldest Sagas, in the historical narratives of Thorfinn Karlsefne, and the Icelandic Landnama-books, these southern coasts between Virginia and Florida are designated by the name of White Men's Land. They are also called Great Ireland (Irland it mikla), and it is asserted that they were peopled from Ireland. According to testimonies which go back as far as 1064, before Leif discovered Vinland, probably about the year 982, Ari Marsson, of the powerful Icelandic family of Ulf the squint-eyed, on a voyage from Iceland to the southward, was driven by storms to the coast of White Men's Land, and was there baptized a Christian; and not being permitted to go away, was recognised there by men from there Orkney Islands and Iceland. (371)

Some northern antiquaries are of opinion that as in the oldest Icelandic documents the first inhabitants of the island are called "West men who arrived by sea," (and settled themselves at Papyli on the south-east coast and on the adjacent small island of Papar), Iceland must have been first peopled not directly from Europe, but from Virginia and Carolina, that is to say from Irland it mikla or White Men's land, which had received its inhabitants from Ireland at a still earlier period. The important treatise entitled "de Mensurâ Orbis Terræ" by the Irish monk Dicuil, which was written in 825, being 38 years before Iceland was discovered by Northman Naddod, does not, however, confirm this opinion.

Christian anchorites in the north of Europe, and Buddhist monks in the interior of Asia, have explored and opened to civilisation regions which were supposed to be inaccessible. The desire of extending religious dogmas has led sometimes to warlike enterprises, and sometimes has prepared the way to peaceful ideas and to commercial relations. In the first half of the middle ages geography was advanced by enterprises dictated by the religious zeal, strongly contrasted with the indifference of the polytheist Greeks and Romans, of Christians, Buddhists, and Mahometans. Letronne, in his commentary on Dicuil, has with much ingenuity and acuteness made it appear probable that after the Irish missionaries were expelled from the Färoe Islands by the Northmen, they began about the year 795 to visit Iceland. When the Northmen first landed in Iceland they found there Irish books, Mass bells, and other objects which had been left behind by earlier visitors called Papar: these papæ (fathers) were the clerici of Dicuil. (372) If then, as we may suppose from the testimony here referred to, these objects belonged to Irish monks (papar) who had come from the Färoe Islands, why should they have been termed in the native Sagas, "West men" (Vestmen), "who had come over the sea from the westward" (kommir til vestan um haf)? All that relates to the supposed voyage of the Gaelic chieftain Madoc the son of Owen Gwyneth, is as yet veiled in profound obscurity: the supposed race of Celto-Americans, which credulous travellers thought they had discovered in several parts of the United States, is gradually disappearing since the introduction of strict ethnological comparison, founded not on accidental resemblances of words, but on organic structure and grammatical forms. (373)

That this first discovery of America in or before the eleventh century was not productive of a great and permanent enlargement of the physical contemplation of the Universe, as was the re-discovery of the same continent by Columbus at the close of the fifteenth century, is an almost necessary consequence of the uncultivated condition of the race by whom the first discovery was made, and of the nature of the regions to which it remained limited. The Scandinavians were not prepared by any scientific knowledge to explore the lands in which they settled farther than appeared necessary for the supply of their most immediate wants. Greenland and Iceland, which must be regarded as the true mother countries of those new colonies, are regions in which man has to cope with all the difficulties and hardships of an inhospitable climate. The wonderfully organised Icelandic Free State did, indeed, preserve its independence for three centuries and a half, until the destruction of civil freedom, and the subjection of the country to the Norwegian king, Haco VI. The flower of the Icelandic literature, the historical writings, the collection of Sagas and of the songs of the Edda, belong to the twelfth and thirteenth centuries.

It is a remarkable phenomenon in the history of the intellectual cultivation of nations, that when the national treasures of the oldest documents belonging to the North of Europe were placed in jeopardy by the unquiet state of their own country, they should have been conveyed to Iceland and there carefully preserved, and thus rescued for posterity. This rescue, the remote consequence of Ingolf's first settlement in Iceland in 875, became, amidst the undefined and misty forms of the Scandinavian world of myths and of figurative cosmogonies, an event of much importance in respect to the fruits of the poetical and imaginative faculties of men: it was only natural knowledge which gained no enlargement. Travellers from Iceland visited the learned institutions of Germany and Italy; but the discoveries made from Greenland towards the south, and the inconsiderable intercourse maintained with Vinland, the vegetation of which did not present any striking peculiarity of character, had so little power to divert settlers and mariners from their wholly European interest, that no tidings of these newly settled countries spread among the cultivated nations of Southern Europe. Even in Iceland itself no notice respecting them appears to have reached the ears of the great Genoese navigator. Iceland and Greenland had then been already separated from each other for more than two centuries, as in 1261 Greenland had lost its republican constitution, and as a possession of the crown of Norway had been formally interdicted from all intercourse with foreigners, and even with Iceland. In a now very rare work of Columbus "on the Five Habitable Zones of the Earth," he mentions having visited Iceland in the month of February 1477, and adds that "the sea was not then covered with ice, (374) and that the country was visited by many traders from Bristol." If he had heard there of the former colonisation on an opposite coast of an extensive connected territory—of Helluland it mikla, of Markland, and of "the good Vinland"—and had connected this knowledge of a neighbouring continent with the projects with which he had already been occupied since 1470 and 1473,—his visit to Thule (Iceland) would no doubt have been more spoken of in the celebrated lawsuit respecting the merit of the first discovery, which was not concluded until 1517; for the suspicious Fiscal even mentions a chart (mappa mundo) which Martin Alonso Pinzon had seen at Rome, on which the New Continent was said to have been laid down. If Columbus had designed to seek for a land of which he had obtained information in Iceland, he would certainly not have steered a south-westerly course from the Canaries in his first voyage of discovery. Between Bergen and Greenland, however, commercial relations still subsisted in 1484, seven years after Columbus's voyage to Iceland.

Very different from the first discovery of the new continent in the eleventh century, in its results on the history of the world, and in its influence on the enlargement of the physical contemplation of the Universe, was the re-discovery of America,—the discovery of its tropical lands,—by Columbus. Although in conducting his great enterprise he had by no means in view the discovery of a new part of the world; although it is even certain that both Columbus and Amerigo Vespucci died in the firm persuasion ([375]) that the lands which they had seen were merely portions of Eastern Asia, yet his voyage has all the character of the execution of a plan founded on scientific combinations. The expedition steered confidently onward to the west through the gate which the Tyrians and Colæus of Samos had opened, through the "immeasurable sea of darkness" (mare tenebrosum) of the Arabian geographers; they pressed forwards towards an object of which they thought they knew the distance: the mariners were not accidentally driven by tempests, as were Naddod and Gardar to Iceland, and Gunnbiorn the son of Ulf Kraka to Greenland, nor were the discoverers conducted onward by intervening stations. The great Nuremberg cosmographer, Martin Behaim, who accompanied the Portuguese Diego Cam on his important expeditions to the west coast of Africa, lived four years (1486-1490) at the Azores; but it was not from these islands, situated at $\frac{2}{5}$ths of the distance of the Iberian coast from that of Pensylvania, that America was discovered. The determined purpose of the act is finely celebrated in the stanzas of Tasso. He sings of that which Hercules dared not attempt:—

> Non osò di tentar l'alto Oceano
> Segnò le mete, e'n troppo brevi chiostri
> L'ardir ristrinse dell' ingegno umano . . . . . .
> Tempo verrà che fian d'Ercole i segni
> Favola vile ai naviganti industri . . . . . .
> Un uom della Liguria avrà ardimento
> All' incognito corso esporsi in prima . . . . . .
>
> Tasso, xv. st. 25, 30 and 31.

And yet all that the great Portuguese historical writer John Barros, ([376]) whose first decade appeared in 1552, has to say of this "uom della Liguria," is, that he was a vain and fantastic talker: (homem fallador e glorioso em mostrar suas habilidades e mais fantastico, e de imaginaçoes com sua Ilha Cypango.) It is thus that, throughout all ages and all degrees of civilization yet attained, national animosity has endeavoured to obscure the brightness of glorious names.

In the history of the contemplation of the Universe, the discovery of tropical America by Christopher Columbus, Alonso de Hojeda, and Alvarez Cabral, must not be regarded as an isolated event. Its influence on the extension of physical knowledge, and on the enrichment of the world of ideas, cannot be justly apprehended, without casting a brief glance on the preceding centuries, which separate the age of the great nautical enterprises from the period when the scientific cultivation of the Arabians flourished. That which gave to the era of Columbus its distinctive character, as a series of uninterrupted and successful exertions for the attainment of new geographical discoveries or of an enlarged knowledge of the earth's surface, was prepared beforehand, slowly, and in various ways. It was so prepared by a small number of courageous men, who roused themselves at once to general freedom of independent thought, and to the investigation of particular natural phænomena;—by the influence exerted on the most profound springs of intellectual life by the renewed acquaintance formed in Italy with the works of Greek and Roman literature;—by the discovery of an art which lends to thought at once wings for rapid transmission and indefinitely multiplied means of preservation;—and by the more extensive knowledge of Eastern Asia, which travelling merchants, and the monks who had been sent as ambassadors to the Mogul princes, circulated amongst those nations of south-western Europe who were most disposed to distant commerce and intercourse, and most eagerly desirous of discovering a shorter route to the Spice Islands. The fulfilment of the wishes which all these causes contributed to excite was in the most important degree facilitated towards the close of the 15th century, by advances in the art of navigation, the gradual improvement of nautical instruments, magnetical as well as astronomical; and finally, by the introduction of new methods of determining the ship's place, and by the more general use of the ephemerides of the sun and moon prepared by Regiomontanus.

Without entering into details in the history of the sciences which do not belong to the present work, we must cite among those who had prepared the way for the epoch of Columbus and Gama, three great names, Albertus Magnus, Roger Bacon, and Vincent of Beauvais. I have given these three in the order of time,—but the name of most importance, and which belongs to the most comprehensive genius, is unquestionably that of Roger Bacon, a Franciscan monk of Ilchester, who studied in Oxford and in Paris. All three were in advance of their age, and acted powerfully upon it. In the long and for the most part unfruitful contests of dialectic speculations, and of the logical dogmatism of a philosophy which has been designated by the vague and equivocal term of scholastic, we cannot overlook the advantage derived from what might be called the after-action of the influence of the Arabians. The peculiarity of their national character, described in the preceding section, and their attachment to the contemplation and study of nature, had procured for the newly translated writings of Aristotle an extensive reception, which was intimately connected with the predilection for the experimental sciences, and highly conducive to the gradual establishment of a basis on which they might hereafter be solidly built. Until the end of the twelfth and the commencement of the thirteenth centuries, misunderstood doctrines of the Platonic philosophy prevailed in the schools. The Fathers of the Church (377) had thought they discovered in them types of their own religious contemplations. Many of the symbolising physical fancies of the Timæus were accepted with enthusiasm; and thus confused and erroneous ideas respecting the Cosmos, of which the Alexandrian mathematical school had long since shown the groundlessness, were revived by Christian authority. Thus the predominance of the Platonic philosophy, or, to speak more correctly, of the new modifications of Platonism, was propagated under varying forms from Augustine to Alcuin, John Scotus, and Bernard of Chartres. (378)

When, on the other hand, the Aristotelian philosophy gained the ascendancy, it influenced the minds of its students at once towards the researches of speculative philosophy, and the philosophical elaboration of natural knowledge by way of experiment. Of these two directions the first might appear to be but little connected with the object of the present work; yet it must not be left without allusion, because, in the middle of the period of dialectic scholastics, it tended to incite a few noble and highly gifted minds to the exercise of free and independent thought, in the most different departments of knowledge. An enlarged physical contemplation of the universe not only requires a rich abundance of observations to afford a satisfactory basis for the generalisation of ideas; but also a preparatory invigorating training of men's minds, and this, as well as for other and more obvious reasons, in order that in the often awakened contest between knowledge and faith, they might not be deterred by threatening forms, which even in modern times have been unwisely regarded as forbidding access to certain departments of experimental science.

When touching on intellectual devolopment, we may not separate the animating influences of the conciousness of man's just privilege of intellectual freedom, and the long unsatisfied desire of wider fields of knowledge, embracing the more distant regions of the surface of the earth. A series of such independent thinkers might be named, beginning in the middle ages with Duns Scotus, William of Occam, Nicholas of Cusa, and continued through Ramus, Campanella and Giordano Bruno to Descartes. ([379])

The apparently impassable "gulf between thinking and being, thought and actual existence;—the relation between the mind which recognises and the object recognised" divided the Dialecticians into the two celebrated schools of the Realists and the Nominalists. The almost forgotten contests of these schools of the middle ages are here referred to, because they exerted a material influence on the final establishment of the basis of the experimental sciences. After many fluctuations in the success of the two parties, the victory finally remained in the 14th and 15th centuries with the Nominalists, who allow to external nature only a subjective existence in the human mind. From their greater aversion to empty abstractions they first urged the necessity of experience, and the propriety of augmenting the bases of knowledge, or recognition through the medium of the senses. Thus, this direction of men's thoughts was at least indirectly influential on the cultivation of experimental natural knowledge; but even where the views of the Realists were still exclusively prevalent, acquaintance with Arabian literature had fostered a love for natural knowledge, and aided it to assert its place successfully, amidst the exclusively absorbing tendency of theological studies. Thus, we see that in the different periods of the middle ages, to which too great a unity of character is perhaps usually ascribed, the way for the great work of discoveries over the surface of the earth, and for their successful employment in the enlargement of the circle of cosmical ideas, was gradually prepared through wholly different trains of thought, the one purely ideal and the other empirical.

Natural knowledge was intimately connected among the learned Arabians with the study of medicines and with philosophy; and in the Christian middle ages with philosophy and with dogmatic theological studies. The latter from their tendency to claim exclusive dominion repressed empirical investigation in physics, organic morphology, and astronomy, which indeed was for the most part allied to astrology. The study of the works of the all-embracing mind of Aristotle, which had been brought in by Arabs and Jewish Rabbis, (380) had tended to produce a philosophical fusion of different branches of study; and thus Ibn-Sina (Avicenna) and Ibn-Roschd (Averroes), Albertus Magnus and Roger Bacon, passed as the representatives of all human knowledge possessed by their age. We may hence estimate the fame which in the middle ages surrounded the names of these eminent men.

Albertus Magnus, of the family of the Counts of Bollstadt, must be cited as himself an observer in the domain of analytical chemistry. His hopes were indeed directed to the transmutation of metals; but in seeking the fulfilment of these hopes, he not only materially improved the practical manipulation and treatment of ores, but also gained additional insight into the general mode of operation of the chemical forces of nature. His works contain some exceedingly acute detached remarks on the organic structure and physiology of plants. He was acquainted with the sleep of plants, with the periodical opening and closing of blossoms, with the diminution of sap during evaporation from the cuticle of the leaves, and with the influence of the distribution of the bundles of vessels on the indentations of the leaves. He wrote a commentary upon the whole of Aristotle's works on physics and natural history, following, however, in the history of animals, only the Latin translation of Michael Scot from the Arabic. (381) A work of Albertus Magnus bearing the title of Liber Cosmographicus de Natura Locorum is a species of physical geography. I have found in it considerations on the dependence of temperature concurrently on latitude and elevation, and on the effect of different angles of incidence of the sun's rays in heating the ground, which have excited my surprise. He owes perhaps his having been celebrated by Dante, less to himself than to his beloved scholar Thomas Aquinas, who he took with him from Cologne to Paris in 1245, and brought back to Germany in 1248.

> Questi, che m' è a destra più vicino,
> Frate e maestro fummi; ed esso Alberto
> E' di Cologna, ed io Thomas d' Aquino.
>
> Il Paradiso, x. 97—99.

In all that relates immediately to the extension of the

natural sciences, to their mathematical foundation, and to the intentional production of phenomena in the way of experiment, Albert von Bollstadt or Albertus Magnus, the cotemporary of Roger Bacon, holds the foremost place in the middle ages. These two men occupy between them almost the entire thirteenth century; but to Roger Bacon belongs the praise, that the influence exerted by him on the form and treatment of the study of nature was more beneficial and more permanent in its operation, than the several discoveries which have been with more or less correctness ascribed to him. Awakened himself to independent thought, he condemned strongly the blind faith in the authorities of the schools; yet far from being indifferent to the investigation of Grecian antiquity, he at the same time appreciated and valued a thorough study of that language, (382) the application of mathematics, and the "Scientia experimentalis," to which last he devoted a particular section of the Opus Majus. (383) Protected and favoured by one pope (Clement IV.), and accused of magic and imprisoned by two others (Nicholas III. and IV.), he experienced the alternations of fortune to which in all ages great minds have frequently been subject. He was acquainted with Ptolemy's Optics, (384) and with the Almagest. As, like the Arabians, he always calls Hipparchus Abraxis, we may infer that he too only made use of a Latin translation derived from the Arabic. Next to his chemical experiments on combustible explosive mixtures, his theoretico-optical works on perspective, and on the position of the focus in concave mirrors, are the most important. His Opus Majus, which is full of thought, contains proposals and plans of possible execution, but no clear traces of success in optical discoveries. Nor are we to ascribe to him profound mathematical knowledge. His characteristic is rather a certain liveliness of imagination, which the impression of so many great and unexplained natural phenomena, the long and painful search for the solution of mysterious problems, had raised to a degree of morbid intensity among those of the mediæval monks whose minds were directed to the study of philosophy. The difficulties which, before the invention of printing, the expense of copyists opposed to the assemblage of many separate manuscripts, produced in the middle ages, when after the thirteenth century the circle of ideas began to enlarge, a great predilection for Encyclopædic works. These works are deserving of particular attention in this place, because they led to the generalisation of views. There appeared in succession, one work being in great measure founded on its predecessors, the twenty books de rerum natura of Thomas Cantipratensis, Professor at Louvain in 1230; the mirror of nature (Speculum naturale) which Vincent of Beauvais (Bellovacensis) wrote for St Lewis and his consort Margaret of Provence in 1250; the "book of nature" of Conrad of Meygenberg, a priest at Regensburg in 1349; and the "picture of the world" (Imago Mundi) of Cardinal Petrus de Alliaco, Bishop of Cambray, in 1410. These Encyclopædias were the precursors of the great Margarita philosophica of Father Reisch, the first edition of which appeared in 1486, and which for half a century promoted in a remarkable manner the extension of knowledge. We must here dwell a little more particularly on the Imago Mundi of Cardinal Alliacus (Pierre d'Ailly). I have shewn elsewhere that this work was more influential on the discovery of America, than was the correspondence with the learned Florentine Toscanelli ([385]). All that Columbus knew of Greek and Roman writers, all the passages of Aristotle, Strabo, and Seneca, on the nearness of Eastern Asia to the pillars of Hercules, which, as his son Don Fernando tells us, were what principally incited his father to the discovery of Indian lands, (autoridad de los escritores para mover al Almirante a' descubrir las Indias) were derived by the Admiral from the writings of Alliacus. Columbus carried these writings with him on his voyages; for, in a letter written to the Spanish monarchs in October 1498 from Hayti, he translates word for word a passage from the Cardinal's treatise, "de quantitate terra habitabilis," by which he had been profoundly impressed. He probably did not know that Alliacus had on his part transcribed word for word from another earlier book, Roger Bacon's Opus Majus. ([386]) Singular period, when a mixture of testimonies from Aristotle and Averroes (Avenryz), Esdras and Seneca, on the small extent of the ocean compared with the magnitude of continental land, afforded to monarchs guarantees for the safety and expediency of costly enterprises!

I have noticed the appearance, at the close of the thirteenth century, of a decided predilection for the study of the powers or forces of nature, and of a progressively increasing philosophical tendency in the form assumed by that study, in its establishment on a scientific experimental basis. It still remains to give a brief description of the influence which, from the end of the fourteenth century, the awakening attention to classical literature exercised on the deepest springs of the intellectual life of nations, and thus upon the general contemplation of the Universe. The individual intellectual character of a few highly gifted men had contributed to the augmentation of the riches of the world of ideas. The susceptibility to a more free intellectual development existed at the period when Grecian literature, favoured by many apparently accidental relations, oppressed and driven from its ancient seats, sought a more secure resting-place in western lands. The Arabians in their classical studies had remained strangers to all that belongs to the inspiration of language. Those studies were limited to a very small number of ancient writers; and in accordance with the strong national predilection for the pursuit of natural knowledge, were principally directed to Aristotle's books of Physics, Ptolemy's Almagest, the botany and chemistry of Dioscorides, and the cosmological phantasies of Plato. The dialectics of Aristotle were associated by the Arabians with physical, as they were by the earlier portion of the Christian middle ages with theological, studies. In both cases, men borrowed from the ancients what they judged available for particular applications; but they were far indeed from apprehending the genius of Greece as a whole, from penetrating the organic structure of its language, from delighting in its poetic creations, and from searching out its admirable treasures in the fields of oratory and historical writing.

Almost two centuries before Petrarch and Boccaccio, John of Salisbury and the platonising Abelard had exercised a beneficial influence in reference to acquaintance with some of the works of classical antiquity. Both felt the beauty and the charm of writings in which nature and mind, freedom and subjection to measure, order, and harmony, are ever found conjoined; but the influence of the æsthetic feeling thus awakened in them vanished without leaving farther traces; and the praise of having prepared in Italy a permanent resting place for the exiled Grecian Muses, of having laboured most powerfully for the restoration of classical literature, belongs to two poets intimately linked with each other in the bonds of friendship—Petrarch and Boccaccio. They had both received lessons from a Calabrian monk named Barlaam, who had long lived in Greece enjoying the favour of the Emperor Andronicus. (387) They first commenced the careful collection of Roman and Grecian manuscripts; and even an historical eye for the comparison of languages had been awakened in Petrarch, (388) whose philological acuteness seemed to tend towards a more general contemplation of the Universe. Emanuel Chrysoloras, who was sent as ambassador from Greece to Italy and to England in 1391, Cardinal Bessarion of Trebizond, Gemistus Pletho, and the Athenian, Demetrius Chalcondylas, to whom is owing the first printed edition of Homer, (389) were all important agents in promoting acquaintance with Grecian literature. All these came from Greece before the eventful taking of Constantinople on the 29th of May, 1453; it was only Constantine Lascaris, whose ancestors had once sat on the throne of the eastern empire, who came later to Italy; he brought with him a precious collection of Greek manuscripts, which is now buried in the seldom-used library of the Escurial. (390) The first Greek book was printed only fourteen years before the discovery of America, although the art of printing was discovered (probably simultaneously, and quite independently (391) by Guttenberg in Strasburg and Mayence, and by Lorenz Jansson Koster in Haarlem), between 1436 and 1439, or in the fortunate epoch of the first immigration of learned Greeks into Italy.

Two centuries before the fountains of Grecian literature were open to the nations of the west, and a quarter of a century before the birth of Dante, who formed one of the great epochs in the history of the intellectual cultivation of southern Europe, events were taking place in the interior of Asia, as well as in the East of Africa, which, by extending commercial intercourse, accelerated the arrival of the period of the circumnavigation of Africa and of the expedition of Columbus. The armies of the Moguls in the course of twenty-six years spread the terror of their name from Pekin and the Chinese wall as far as Cracow and Leignitz, and produced a feeling of alarm throughout Christendom. A number of able monks were sent both in a religious and diplomatic capacity; —John de Plano Carpini and Nicolas Ascelin to Batu Khan, and Ruisbroeck (Rubruquis) to Mangu Khan to Karakorum. The last named of these missionaries has left us some acute and important remarks on the geographical extension of different families of nations and of languages in the middle of the thirteenth century. He was the first to recognize that the Huns, the Bashkirs (inhabitants of Paskatir, Basch-gird of Ibn-Fozlan), and the Hungarians, were of Finnish or Uralian race; and he found Gothic tribes, still preserving their language, in the strong holds of the Crimea. ([392]) The accounts given by Rubruquis of the immeasurable riches of Eastern Asia excited the cupidity of two powerful maritime nations of Italy, the Venetians and the Genoese. Rubruquis knew "the silver walls and golden towers of Quinsay," though he does not name that great commercial city, (the present Hangtcheufu), which twenty-five years later acquired such celebrity through the accounts of the greatest of land travellers, Marco Polo ([393]). Truth and naive error are curiously intermingled in the accounts given by Rubruquis of his travels, and preserved to us by Roger Bacon. "Near Cathay, which is bounded by the Eastern Ocean," he describes a happy land "where men and women arriving from other countries cease to grow old" ([394]). Still more credulous than the monk of Brabant, and for that reason much more extensively read, was the English knight, Sir John Mandeville. He describes India and China, Ceylon and Sumatra. The variety and personal interest of his narrative have, (like the itineraries of Balducci Pegoletti, and the narrative of Ruy Gonzalez de Clavijo), contributed not a little to increase the disposition towards intercourse with distant countries.

It has been often and with singular decision asserted, that the excellent work of the truth-loving Marco Polo, and particularly the knowledge which he gave of the Chinese ports and of the Indian archipelago, had great influence on Columbus, and that he even had a copy of Marco Polo's travels with him on his first voyage of discovery. ([395]) I have shown that both Columbus himself, and his son Fernando, speak of Æneas Sylvius's (Pope Pius II.) geography of Asia, but never name Marco Polo or Mandeville. What they knew of Quinsay, Zaitun, Mango and Zipangu, may have been gained, without any immediate acquaintance with chapters 68 and 77 of the second book of Marco Polo, from the celebrated letter of Toscanelli, in 1474, on the facility of reaching Eastern Asia from Spain, and from the accounts of Nicolo de Conti, who travelled for 25 years through

India and Southern China. The oldest printed edition of Marco Polo's travels is a German translation made in 1477, and this certainly would not have been intelligible to Columbus and Toscanelli. The possibility of Columbus having seen a manuscript written by the Venetian traveller between the years 1471 and 1492, in which he was occupied with the project of sailing "to the East by the West" (buscar el levante por el poniente, pasar a donde nacen las especerias, navegando al occidente), cannot certainly be denied; (396) but if so, why, in the letter which he wrote to the monarchs from Jamaica, June 7, 1503,—in which he describes the coast of Veragua as a part of the Asiatic Ciguare, and hopes to see horses with golden trappings,—does not he refer to the Zipangu of Marco Polo rather than to that of Papa Pio?

At the period when the extension of the great Mogul empire from the Pacific to the Volga rendered the interior of Asia accessible, the maritime nations of Europe acquired a knowledge of Cathay and Zipangu (China and Japan), through the diplomatic missions of the monks, and through mercantile enterprises conducted by means of land journies. By an equally remarkable concatenation of circumstances and events, the mission of Pedro de Covilham and Alonso de Payva, sent in 1487 by King John II. to seek for "the African Prester John," prepared the way, not indeed for Bartholomew Diaz, but for Vasco de Gama. Confiding in reports brought by Indian and Arabian pilots to Calicut, Goa, and Aden, as well as to Sofala on the east coast of Africa, Covilham sent word to King John, by two Jews from Cairo, that if the Portuguese prosecuted their voyages of discovery on the western coast of Africa towards the south, they would arrive at the extremity of that continent; from whence the navigation to the Moon Island (the Magastar of Polo), to Zanzibar, and to Sofala rich in gold, would be found extremely easy. Long before these tidings reached Lisbon, however, it had been known there that Bartholomew Diaz had not only discovered the Cape of Good Hope (Cabo Tormentoso), but had already sailed round it, though only for a very short distance. (397) Accounts of the Indian and Arabian trading stations on the eastern coast of Africa, and of the configuration of the southern extremity of the continent, may, indeed, have reached Venice very early in the middle ages, through Egypt, Abyssinia, and Arabia. The triangular form of Africa is distinctly laid down in the planisphere of Sanuto (398) as early as 1306; in the Genoese Portulano della Mediceo-Laurenziana of 1351 discovered by Count Baldelli, and in the map of the world by Fra Mauro. It is fitting that the history of the contemplation of the Universe should indicate by a passing allusion the epochs when the general form of the great continental masses was first recognised.

Whilst the gradually advancing knowledge of geographical relations led men to think of new and shorter maritime routes, the means of improving practical navigation by the application of mathematics and astronomy, by the invention of new measuring instruments, and by the more skilful use of the magnetic forces, were also rapidly increasing. It is highly probable that Europe owes the adaptation of the directing powers of the magnet to the purposes of navigation —or the use of the mariner's compass—to the Arabians, and that they again were indebted for it to the Chinese. In a Chinese work, (the historic Szuki of Szumathsian, a writer belonging to the first half of the second century before our era), mention is made of "magnetic cars" given, more than 900 years before, by the emperor Tschingwang of the old dynasty of the Tscheu to the ambassadors from Tunkin and Cochin China, that they might not miss their way on their homeward journey by land. In Hiutschin's dictionary Schuewen, written in the third century under the dynasty of the Han, a description is given of the manner in which the property of pointing with one extremity to the south is communicated to an iron bar: navigation being then most usually directed to the south, the end of the magnet which pointed southwards was the one always referred to. A century later, under the dynasty of the Tsin, Chinese ships used the south magnetic direction to guide their course in the open sea, and these ships carried the knowledge of the compass to India, and from thence to the east coast of Africa. The Arabic terms zophron and aphron (for south and north) ([399]) which Vincent of Beauvais in his mirror of nature gives to the two ends of the magnetic needle, shew (as do the many Arabic names of stars which we still employ) the channel through which the nations of the West received much of their knowledge. In Christian Europe the use of the compass is first mentioned as a perfectly familiar subject in the politico-satirical poem called "La Bible," written by Guyot of Provence in 1190, and in the description of Palestine by Jacob of Vitry, Bishop of Ptolemais, between 1204 and 1215. Dante (Parad. xii. 29) alludes in a comparison to the "needle which points to the star."

The discovery of the mariner's compass was long ascribed to Flavio Gioja of Positano, a place not far from the beautiful Amalfi, which its widely extended maritime laws rendered so celebrated; perhaps he may have made (1302) some improvement in its construction. That the compass was used in European seas much earlier than the beginning of the fourteenth century is proved by a nautical treatise of Raymond Lully of Majorca, the highly ingenious and eccentric man whose doctrines inspired Giordano Bruno with enthusiasm when a boy, ([400]) and who was at once a philosophical systematiser, a practical chemist, a christian teacher, and a person skilled in navigation. He says in his book entitled "Fenix de las maravillas del orbe," written in 1286, that mariners made use in his time of "measuring instruments, of sea charts, and of magnetic needles." ([401]) The early voyages of the Catalans to the north coast of Scotland and to the west coast of tropical Africa, (Don Jayme Ferrer, in the month of August 1346, reached the mouth of the Rio de Ouro), and the discovery of the Azores (the Bracix Islands of Picigano's map of the world in 1367) by the Normans, remind us that the open western ocean was navigated long before Columbus. That navigation of the high seas which, under the Roman empire, had been ventured upon in the Indian Ocean between Ocelis and the coast of Malabar in reliance upon the regularity of the periodical direction of the winds, ([402]) was here performed under the guidance of the magnetic needle.

The application of astronomy to navigation was prepared by the influence which, from the thirteenth to the fifteenth century, was exerted, in Italy by Andalone del Nero and John Bianchini who corrected the Alphonsine astronomical tables, and in Germany by Nicolaus of Cusa, ([403]) Georg von Peuerbach, and Regiomontanus. Astrolabes capable of being used at sea for the determination of time, and of geographical latitudes by meridian altitudes, underwent gradual improvement from the instruments used by the pilots of Majorca described by Raymond Lully ([404]) in 1295, in his "Arte de Navegar," to that which Martin Behaim made in 1484 at Lisbon, and which was perhaps only a simplification of the meteoroscope of his friend Regiomontanus. When the Infante Henry (Duke of Viseo) the great encourager of navigation, and himself a navigator, founded a school of pilots at Sagres, Maestro Jayme of Majorca was named its director. Martin Behaim was desired by king John II. of Portugal to compute tables for the sun's declination, and to instruct pilots to "navigate by the altitudes of the sun and stars." Whether the log line, which makes it possible to estimate the length of the course passed over, whilst the direction is given by the compass, was known as early as the end of the fifteenth century, cannot be determined, but it is certain that Pigafetta, a companion of Magellan, speaks of the log (la catena a poppa) as of a long known means of measuring the distance passed over. ([405])

The influence of Arabian civilisation on Spanish and Portuguese navigation, through the astronomical schools of Cordova, Seville, and Granada, is not to be overlooked: the large instruments of Cairo and Bagdad were imitated on a small scale for maritime use. The names were also transferred; the "astrolabon" which Martin Behaim attached to the main mast belongs originally to Hipparchus. When Vasco de Gama landed on the east coast of Africa, he found the Indian pilots at Melinda acquainted with the use of astrolabes and cross staffs. ([406]) Thus, by intercommunication consequent on more extended intercourse between nations, as well as by original invention, and by the mutual aids to advancement furnished by mathematical and astronomical knowledge, every thing was gradually prepared for the great geographical achievements, which have distinguished the close of the fifteenth and the early portion of the sixteenth centuries, or the thirty years from 1492 to 1522, namely,—the discovery of tropical America, the rapid determination of its form, the passage round the southern point of Africa to India, and the first circumnavigation of the globe. Men's minds were also stimulated and rendered more acute to receive the immense accession of new phenomena, to work out the results of what was thus obtained, and by their comparison to render them available for the formation of higher and more general views of the physical Universe.

It will suffice to allude here to a few only of the principal elements of these higher views, which were capable of conducting men to a farther insight into the connection of the phenomena of the globe. In a careful study of the original works of the earliest historians of the Conquista, we often discover with astonishment in the Spanish writers of the sixteenth century the germ of important physical truths. At the sight of a continent in the wide waste of waters far removed from other lands, many of the important questions which occupy us in the present day presented themselves to the awakened curiosity both of the first voyagers and of those who collected their narrations;—questions respecting the unity of the human race, and its deviations from a common normal type;—the migrations of nations, and the relationship of languages which often shew greater differences in their radical words than in their flexions or grammatical forms;—the possibility of the migration of particular species of plants or animals;—the cause of the trade winds, and of the constant currents of the ocean;—the regular decrease of temperature on the declivities of the Cordilleras, and in successive strata of water in descending in the depths of the ocean;—and on the reciprocal operation upon each other of the different volcanoes forming chains, and their influence on the frequency of earthquakes as well as on the extent of the circles of commotion. The groundwork of what we now term physical geography, (abstracting from it mathematical considerations,) is found in the Jesuit Joseph Acosta's "Historia natural y moral de las Indias," as well as in the work by Gonzalo Hernandez de Oviedo, which appeared only twenty years after the death of Columbus. Never, since the commencement of civil society, was there an epoch in which the sphere of ideas as regards the external world and geographical relations was so suddenly and wonderfully enlarged, or in which the desire of observing nature under different latitudes and at different elevations above the level of the sea, and of multiplying the means by which her secrets might be interrogated, was more keenly felt.

It has, perhaps, as I have elsewhere remarked, ([407]) been erroneously supposed, that the value of these great discoveries, each of which in turn promoted others,—of these twofold conquests in the physical and in the intellectual world,—was not felt until its recognition in our own days, when the history of the intellectual cultivation of mankind is made a subject of philosophic study. Such a supposition is refuted by the writings of the cotemporaries of Columbus. The feelings of the most talented among them anticipated the influence which the events of the latter part of the fifteenth century would exert on mankind. Peter Martyr de Anghiera ([408]) says, in his letters written in 1493 and 1494,

"Every day brings to us new wonders from a new world, from those western antipodes which a certain Genoese (Christophorus quidam vir Ligur) has discovered. Sent by our monarchs, Ferdinand and Isabella, he could with difficulty obtain three ships, since what he said was regarded as fabulous. Our friend Pomponius Lætus" (one of the most distinguished promoters of classical literature, and persecuted at Rome on account of his religious opinions), "could hardly refrain from tears of joy, when I gave him the first tidings of an event so unhoped for." Anghiera, from whom these words are taken, was a highly intelligent and distinguished statesman at the court of Ferdinand the Catholic and Charles V., was once sent as ambassador to Egypt, and was a personal friend of Columbus, Amerigo Vespucci, Sebastian Cabot, and Cortes. His long life comprised the discovery of the westernmost of the Azores (Corvo), and the expeditions of Diaz, Columbus, Gama, and Magellan. Pope Leo X. "continued to a very late hour in the night" reading to his sister and the cardinals, Anghiera's Oceanica. Anghiera says, "henceforward I would not willingly leave Spain again, for I am here at the fountain-head of the tidings from the newly discovered lands, and I may hope, as the historian of such great events, to obtain for my name some fame with posterity. ([409]) Thus vividly did cotemporaries feel the splendour of events, of which the remembrance will survive through all ages.

Columbus, in sailing westward of the meridian of the Azores, through an entirely unexplored sea, and employing the newly-improved astrolabe for the determination of his position, sought the east of Asia by the western route, not as an adventurer, but according to a preconceived and steadfastly pursued plan. He had indeed on board, the sea-chart which the Florentine physician and astronomer, Toscanelli, had sent to him in 1477, and which fifty-three years after his death was still in the possession of Bartholomew de las Casas. According to the manuscript history of las Casas which I have examined, this was the Carta de Marear, (410) which the Admiral shewed, on the 25th of September, 1492, to Martin Alonso Pinzon, and on which several out-lying islands were drawn. But if Columbus had only followed the chart of his counsellor Toscanelli, he would have held a more northern course, and have kept along a parallel of latitude from Lisbon; instead of this, in the hope of reaching Zipangu (Japan) more quickly, he sailed for half the distance in the latitude of Gomera, one of the Canary Islands, and subsequently diminishing his latitude, found himself on the 7th of October 1492, in 25½°. Uneasy at not having yet discovered the coasts of Zipangu, which according to his reckoning he should have met with two huudred and sixteen nautical miles more to the East, he, after a long debate, gave way to the commander of the Caravel Pinta, Martin Alonso Pinzon, (one of the three rich and influential brothers who were hostile to Columbus), and steered towards the south-west. The course thus altered, led on the 12th of October, to the discovery of Guanahani.

We must here pause a while, in order to notice a very remarkable instance of the wonderful enchainment and connection, which links small and apparently trivial occurrences with great events affecting the world's destiny. Washington Irving has justly stated, that if Columbus, resisting the counsel of Martin Alonso Pinzon, had continued to sail on towards the west, he would have entered the warm current of the Gulf stream, have reached Florida, and thence perhaps have been carried to Cape Hatteras and Virginia; a circumstance of immeasurable importance, since it might have given the present United States of America a Roman Catholic Spanish population, instead of a later arriving Protestant English one. "It is," said Pinzon to the Admiral, "as if something whispered to my heart (el corazon me da) that we must change our course." He even maintained in the celebrated lawsuit (1513-1515), which he conducted against the heirs of Columbus, that on this account the discovery of America was due to him only. But Pinzon owed in fact this suggestion, or what "his heart whispered to him," as an old sailor from Moguer related in the same lawsuit, to the flight of a flock of parrots which he saw flying in the evening towards the southwest for the purpose, as he might suppose, of sleeping among trees or bushes on shore. Never had the flight of birds more important consequences. It may be said to have determined the first settlements on the new Continent, and its distribution between the Latin and Germanic races ([411]).

The march of great events, like the sequence of natural phenomena, is regulated by laws of which a few only are known to us. The fleet which King Emanuel of Portugal sent under the command of Pedro Alvarez Cabral to India, by the route discovered by Gama, was driven out of its course to the coast of Brazil, on the twenty-second of April, 1500. From the zeal which, from the time of the enterprise of Diaz (1487), the Portuguese shewed for sailing round the Cape of Good Hope, accidents similar to those which the currents of the ocean occasioned to the ships of Cabral, could hardly have failed to occur. Thus the African discoveries would have led to that of America south of the equator; and Robertson was justified in describing it as in the destiny of mankind, that before the end of the fifteenth century the new continent should be known to European navigators.

Amongst the characteristic qualities possessed by Christopher Columbus, we must especially distinguish the penetrating glance and keen sagacity with which, though without learned or scientific culture, and without acquired knowledge in physics or in natural history, he could seize and combine the various phenomena of the external world. On arriving "in a new world and under a new heaven," (412) he noticed carefully the form of the land, the physiognomy of the vegetation, the habits of the animals, the distribution of heat, and the variations of the earth's magnetism. The old navigator, whilst endeavouring to find the spices of India, and the rhubarb (ruibarba) which had already acquired so much celebrity through Arabian and Jewish physicians, and through the reports of Rubruquis and the Italian travellers, examined very closely the roots, fruits, and form of the leaves of the plants which fell under his observation. In this portion of our work, where we desire to recal the influence which the great epoch of nautical enterprizes and discoveries exercised on the enlargement of men's views of nature, our descriptions will become more animated by being attached to the individuality of a great man. In the journal of his voyage and in his accounts, which were published for the first time between 1825 and 1829, we find allusions to almost all the subjects to which scientific activity was afterwards directed in the latter half of the fifteenth and the whole of the sixteenth centuries.

It is sufficient to recal in a general manner, all that the geography of the western hemisphere gained from the period, when, at his country seat, Perça Naval, on the beautiful bay of Sagres, the Infante Dom Henry the Navigator sketched his first plan of discovery, to the epoch of the South Sea expeditions of Gaetano and Cabrillo. The daring enterprizes of the Portuguese, the Spaniards, and the English, testify how powerfully the desire for the great and boundless in geographical space had made itself felt, suddenly opening as it were a new sense. The advances in the art of navigation, and the application of astronomical methods to the correction of a ship's reckoning, favoured the efforts which gave to this age its peculiar character, and disclosed to men the true features of the globe which they inhabit. The discovery of the mainland of tropical America, which took place on the 1st of August, 1498, was seventeen months later than Cabot's arrival off the Labrador coast of North America. Columbus first saw the Terra firma of South America, not as has been hitherto believed on the mountainous coast of Paria, but in the Delta of the Orinoco east of Cano Macareo. (413) Sebastian Cabot (414) landed on the 24th of June, 1497, on the coast of Labrador between 56° and 58° of latitude. I have shewn above that this inhospitable coast had been visited five centuries earlier by the Icelander Leif Erikson.

Columbus on his third voyage set more value on the pearls of the islands of Margarita and Cubagua, than on the discovery of the Terra firma; as he was persuaded until his death, that, in his first voyage when at Cuba in November 1492, he had already touched a part of the continent of Asia. (415) From hence (as his son Don Fernando, and his friend the Cura de los Palacios, relate,) if he had

had sufficient provisions, his design would have been to have continued his navigation towards the west, and to have returned to Spain, [416] either by water, passing by Ceylon (Taprobane) and "rodeando toda la tierra de los Negros," or by land, by Jerusalem and Jaffa. Such were the projects which Columbus cherished in 1494, proposing to himself the circumnavigation of the globe, four years before Vasco de Gama, and twenty-seven years before Magellan and Sebastian de Elcano. The preparations for Cabot's second voyage, in which he penetrated among masses of ice as far as 67½° North latitude, seeking a North-West passage to Cathay (China), led him to think of a voyage to the North pole, (a lo del polo arctico), to be made at some future period. [417] The more it became gradually recognised, that the newly-discovered lands formed a connected continent stretching uninterruptedly from Labrador to the promontory of Paria,—and even as the celebrated lately-discovered map of Juan de la Cosa (1500) shewed, far beyond the equator into the Southern hemisphere,—the more ardent became the desire to find a passage to the westward, either in the North or in the South. Next to the rediscovery of the American continent, and the conviction of its extension in the direction of the meridian from Hudson's Bay to Cape Horn (discovered by Garcia Jofre de Loaysa,) [418] the knowledge of the South Sea or the Pacific Ocean, which bathes the Western coasts of America, was the most important cosmical occurrence in the great epoch which we are now describing.

Ten years before Balboa obtained the first sight of the South Sea, from the summit of the Sierra de Quarequa on the isthmus of Panama, Columbus in sailing along the coast of Veragua, had already received distinct accounts of a sea to the westward of that land, "which would conduct in less than nine days' voyage to the Chersonesus aurea of Ptolemy, and to the mouth of the Ganges." In the same Carta rarissima which contains the beautiful and highly poetic narration of a dream, the Admiral says that at the part near the Rio del Belen "the two opposite coasts of Veragua are situated relatively to each other like Tortosa near the Mediteranean and Fuenterrabia in Biscay, or like Venice and Pisa." This southern or western sea, the great Pacific ocean, was at that time still regarded as only a continuation of the Sinus Magnus (μεγας κολπος) of Ptolemy, beyond which lay the golden Chersonesus, whilst Cattigara and the land of the Sinæ (Thinæ) was supposed to form its eastern shore. The fanciful hypothesis of Hipparchus, according to which this eastern coast of the great Gulf, or Sinus Magnus, joined itself on to a part of the continent of Africa advancing far to the East, (419) (thus making the Indian ocean a closed inland sea,) was happily little regarded in the middle ages, notwithstanding their attachment to the opinions of Ptolemy; it would doubtless have exercised an unfavourable influence on the direction of the great nautical enterprizes of the age.

The discovery and navigation of the Pacific, mark an epoch so much the more important in reference to the recognition of great cosmical relations, as it was by their means, and scarcely therefore three centuries and a half ago, that not only the western coast of America and the eastern coast of Asia were first known, but also, what is of much greater importance, on account of the meteorological results which follow from it, that the prevailing highly erroneous views respecting the relative areas of land and water upon the surface of the globe, were first dispelled. The relative magnitude and distribution of these areas are most influential conditions in determining the quantity of moisture contained in the air, the variations of atmospheric pressure, the degree of vigour and luxuriance of vegetation, the more or less extensive distribution of particular kinds of animals, and many other great and general physical phenomena. The larger extent of fluid surface (in the proportion of 2⅘ to 1), does indeed restrict the habitable range for the settlements of man, and for the nourishment of the greater number of mammalia, birds, and reptiles; but it is nevertheless, under the present laws which govern organised beings, a beneficent arrangement and necessary condition for the preservation and well being of all the living inhabitants of continents.

When at the end of the fifteenth century there arose an earnest and pressing desire to find the shortest way to the Asiatic spice lands,—and when the idea of reaching the East, by sailing to the West, germinated almost simultaneously in the minds of two men of Italy, the navigator Columbus, and the physician and astronomer Paul Toscanelli,—([420]) it was generally believed, in conformity with the opinion put forward by Ptolemy in the Almagest, that the old continent from the western coast of the Iberian peninsula to the meridian of the easternmost Sinæ, occupied a space of 180°; or in other words, that it extended from East to West, over an entire half of the globe. Columbus, misled by a long series of erroneous inferences, extended this space to 240°, making the desired eastern coast of Asia advance as far as the meridian of San Diego in New California. Columbus hoped therefore that he would only have to sail over 120°, instead of the 231° which the rich trading city of Quinsay, for example, is actually situated to the westward of the extremity of the Iberian peninsula. Toscanelli, in his correspondence with the Admiral, diminished the breadth of the ocean in a manner still more singular and more favourable to his plans. He made the distance by sea from Portugal to China only 52° of longitude, leaving, according to the ancient saying of Esdras, six-sevenths of the earth dry. Columbus, in a letter which he addressed to Queen Isabella from Hayti immediately after the accomplishment of his third voyage, shewed himself the more inclined towards this view, because it was the same which had been defended by the man whom he regarded as the highest authority, Cardinal d'Ailly, in his "Imago Mundi." ([421])

Six years after Balboa sword in hand and advancing up to his knees in the waves had claimed possession of the entire South Sea for Castille, and two years after his head had fallen by the hand of the executioner in the revolt against the tyrannical Pedrarias Davila, ([422]) Magellan appeared in the Pacific (27 November 1520), and navigated the wide ocean for more than ten thousand geographical miles; by a singular fatality seeing only, before discovering the Marianas, (his Islas de los Ladrones or de las Velas Latinas), and the Philippines, two small uninhabited islands (the Desventuradas or Unfortunate islands), one of which, if we might trust his journal and ship's reckoning, would be to the East of the Low Islands, and the other a little to the South West of the Archipelago of Mendana. ([423]) Sebastian de Elcano, after the murder of Magellan in the island of Zebu, completed the first circumnavigation of the globe in the ship Victoria, and received for his armorial bearings a terrestrial globe, with the glorious inscription: "Primus circumdedisti me." He entered the harbour of San Lucar in September 1522; and before an entire year had elapsed, we find the

Emperor Charles urging, in a letter to Hernando Cortes, the discovery of a passage "which should shorten the distance to the spice lands by two-thirds." The expedition of Alvaro de Saavedra was sent from a harbour of the province of Zacatula on the west coast of Mexico, to the Moluccas; and in 1527, Hernando Cortes wrote, from the newly conquered Mexican capital of Tenochtilan, "to the kings of Zebu and Tidor in the Asiatic Archipelago." So rapid was the enlargement of the geographical horizon, and with it the desire for an extensive and animated intercourse with remote nations.

Subsequently the conqueror of New Spain went himself in search of discoveries in the Pacific, and of a north-eastern passage from thence to Europe. Men could not accustom themselves to the idea that the continent really extended uninterruptedly from such high southern to high northern latitudes. When the report came from the coast of California that the expedition of Cortes had perished, the wife of the great warrior Juana de Zuñiga, the beautiful daughter of the Conde de Aguilar, had two ships prepared in order to seek for more certain tidings. (424) In 1541 California was already known as an arid peninsula without wood, although this was again forgotten in the 17th century. We can discover in the accounts which we now possess of Balboa, Pedrarias Davila, and Hernando Cortes, that at that period men hoped to discover in the South Sea, as a part of the Indian ocean, groups of "islands rich in gold, precious stones, spices, and pearls." Excited fancy impelled men to great enterprizes; and the hardihood of these, whether successful or unfortunate, reacted on the imagination and nflamed it still more powerfully. Thus, at this extraordinary period of the Conquista, (a period when men's heads were dizzy with strenuous efforts, heroic achievements, deeds of violence, and discoveries by sea and land), notwithstanding the entire absence of political freedom, many circumstances conspired to favour individual development, and to cause some more highly gifted minds to attain to much that was noble. They err who regard the Conquistadores as led only by a thirst for gold, or even exclusively by religious fanaticism. Dangers always exalt the poetry of life; and moreover, the powerful age which we here seek to depict in regard to its influence on the development of cosmical ideas, gave to all enterprizes, as well as to the impressions of nature offered by distant voyages, the charm of novelty and surprise, which begins to be wanting to our present more learned age in the many regions of the earth which are now open to us. It was not only a hemisphere, but almost two-thirds of the surface of the globe, which was then still an unknown and unexplored world; as unseen as that half of the moon's disk which the laws of gravitation withdraw for ever from the view of the inhabitants of the earth. Our more deeply investigating age finds, in the increasing riches of ideas, a compensation for the lessening of that surprise, which the novelty of great and imposing natural phenomena once called forth; but this is a compensation not to the multitude, but to the small number of physicists acquainted with the state of science,—and to them it is ample. To them the increasing insight into the silent operation of the powers of nature;—whether in electro-magnetism, or in the polarisation of light, in the influence of diathermal substances, or in the physiological phenomena of living organised beings, offers a world of wonders gradually unveiling itself, and of which we have yet scarcely reached the threshold!

The Sandwich Islands, New Guinea, and some parts of New Holland, were all discovered in the first half of the 16th century. ([425]) These discoveries prepared the way for those of Cabrillo, Sebastian Vizcaino, Mendaña, ([426]) and Quiros, whose "Sagittaria" is Tahiti, and his "Archipelago del Espiritu Santo" the New Hebrides of Cook. Quiros was accompanied by the bold navigator who afterwards gave his name to Torres Straits. The Pacific no longer appeared as it had done to Magellan a desert waste; it was now enlivened by islands, which indeed for want of exact astronomical determinations of position, strayed to and fro on the map like floating lands. The Pacific long continued the exclusive theatre of the enterprizes of the Spaniards and Portuguese. The important South Indian Malayan Archipelago, obscurely described by Ptolemy, Cosmas, and Polo, began to shew itself with more definite outlines after Albuquerque had established himself in Malacca in 1511, and after the voyage of Anthony Abreu. It is the especial merit of the classical Portuguese historian Barros, a cotemporary of Magellan and of Camoens, to have apprehended the peculiarities of the physical and ethnical character of the Archipelago in so lively a manner, that he first proposed to distinguish Australian Polynesia as a fifth part of the globe. It was when the Dutch power acquired the ascendancy in the Molluccas, that this portion of the globe began to emerge from obscurity, and to become known to geographers; ([427]) and then also began the great epoch of Abel Tasman. We do not propose to ourselves to give the history of the several geographical discoveries, but merely to recal by a passing allusion the leading occurrences, by which, in a short space of time, and in close and connected succession, in obedience to the suddenly awakened desire to search out the wide, the unknown and the distant, two-thirds of the earth's surface were laid open.

Together with this enlarged and increasing geographical knowledge of land and sea, there arose also a more enlarged insight into the existence and the laws of the powers or forces of nature,—the distribution of heat over the surface of the earth,—the abundance of organic forms, and the limits of their distribution. The progress which different branches of science had made during the course of the middle ages, (which, as regards science, have been too little esteemed,) accelerated the just apprehension and thoughtful comparison of the unbounded wealth of physical phenomena, which was now presented at one time to observation. The impressions produced on men's minds were so much the more profound, and the more fitted to incite to the investigation of cosmical laws, as before the middle of the 16th century, the western nations of Europe had already explored the new continent, in the neighbourhood of the coasts at least, in the most different degrees of latitude; and because it was here that they first became acquainted with the true equatorial zone, where, moreover, the remarkable conformation of the earth's surface presented to their view in close approximation, at varying degrees of elevation, the most striking contrasts of vegetation and of climate. If I here find myself again induced to allude to the peculiar privileges of these regions, in the inspiring influence belonging to a land of lofty mountains in the equinoctial zone, I must plead once more as my justification that, to their inhabitants alone is it given, to behold at once all the stars of heaven, and almost all the families of forms of the vegetable world;—but to behold is not necessarily to observe, viz. to compare, and to combine.

Although in Columbus, as I think I have shewn in another work, notwithstanding the entire absence of any preliminary knowledge of natural history, the mere contact with great natural phenomena, developed in a remarkable and varied manner the perceptions and faculties required for accurate observation, yet we must by no means assume a similar development in those who composed the rude and warlike mass of the Conquistadores. That which Europe unquestionably owes to the discovery of America,—in the gradual enrichment of the physical knowledge of the constitution of the atmosphere, and its effects on human organization,—the distribution of climates on the declivities of the Cordilleras,—the elevation of the snow-line in different degrees of latitude in the two hemispheres,—the arrangement of volcanoes in chains,—the circumscribed area of the circle of commotion in earthquakes,—the laws of magnetism,—the direction of the currents of the ocean,—and the gradations of new forms of plants and animals,—it owes to a different and more peaceful class of travellers, and to a small number of distinguished men among municipal functionaries, ecclesiastics, and physicians. These men dwelling in old Indian towns, some of which are upwards of twelve thousand feet above the level of the sea, could observe with their own eyes, and could test and combine that which others had seen, with the superior advantage of long residence; and could collect, describe, and send to their European friends, the natural productions of the country. It is sufficient here to name Gomara, Oviedo, Acosta and Hernandez. Columbus brought home from his first voyage some natural productions,—fruits and skins of animals. In a letter written from Segovia (August 1494), Queen Isabella requests the Admiral to continue his collections, and particularly desires "all birds belonging to the shores and the woods of countries having a different climate and seasons." From the same west coast of Africa, from which, almost 2000 years earlier, Hanno brought "tanned skins of wild women," (the skins of the great Gorilla ape), to be suspended in a temple,—Martin Behaim's friend Cadamosto, brought to the Infante Henry the Navigator, black elephant's hair a palm and a half long. Hernandez, the surgeon of Philip II., and sent by that monarch to Mexico, to have all the most remarkable objects of the vegetable and animal kingdoms in that country represented by fine drawings, was able to augment his collections by copies of several very carefully executed pictures of specimens of natural history, which had been painted by command of a king of Tezcuco, Nezahualcoyotl, ([428]) half a century before the arrival of the Spaniards. Hernandez also availed himself of a collection of medicinal plants, which he found still growing in the ancient Mexican garden of Huaxtepec. Owing to its proximity to a newly established Spanish hospital, ([429]) this garden had not been laid waste by the Conquistadores. Almost at the same time, the fossil bones of Mastodons found on the plateaus of Mexico, New Granada, and Peru, which afterwards became of so much importance in reference to the theory of the successive elevation of different chains of mountains, were collected and described. The names of Giants' bones, and Giants' fields (Campos de Gigantes), shew how fanciful were the interpretations first attached to these remains.

During this active period, the enlargement of cosmical views was promoted by the immediate contact of numerous bodies of Europeans, not only with the free aspect and grand features of nature in the mountains and plains of America, but also (in consequence of the successful navigation of Vasco de Gama) with the eastern coast of Africa, and with India. As early as the commencement of the sixteenth century, a Portuguese physician, Garcia de Orta, to whom the Muse of Camoens has paid a patriotic tribute of praise, established, on the present site of Bombay, and under the auspices of the noble Martin Alfonso de Sousa, a botanic garden in which he cultivated the medicinal plants of the vicinity. The impulse to direct and independent observation was now every where awakened, whilst the cosmographic writings of the middle ages were rather compilations, reproducing the opinions of classical antiquity, than the results of personal observation. Two of the greatest men of the sixteenth century, Conrad Gesner and Andreas Cæsalpinus, honourably opened a new path in zoology and botany.

In order to afford a more lively idea of the early influence which the oceanic discoveries exercised on the enlargement of physical and astro-nautical knowledge, I will call attention at the close of this description to some bright points of light which we see already glimmering in the writings of Columbus. Their first feeble ray is the more deserving of careful regard because they contained the germ of general cosmical views. I pass over the proofs of the results here presented to my readers, because I have already given them in detail in an earlier work, entitled "Critical examination of the historic development of the geographical knowledge of the new world, and of nautical astronomy in the fifteenth and sixteenth centuries." In order, however, to avoid its being supposed that I have unduly mingled modern physical views with the remarks of Columbus, I will commence with the literal translation of a portion of a letter written by the Admiral in October 1498 from Hayti.

"Each time that I sail from Spain to the Indies, I find as soon as I arrive a hundred nautical miles to the west of the Azores, an extraordinary alteration in the movement of the heavenly bodies, in the temperature of the air, and in the character of the ocean. I have observed these alterations with particular care, and have recognised that the needle of the mariner's compass (agujas de marear), the declination of which had been to the north-east, now turned to the north-west; and when I had passed this line (raya), as if I had passed the ridge of a hill (como quien traspone una cuesta), I found the sea covered with such a mass of weed resembling small branches of pine trees with fruits like pistachio nuts, that we were led to expect there would not be sufficient water, and that the ships would run upon a shoal. Before we had arrived at this line no trace of such sea-weed was to be seen. Also at this boundary line (a hundred miles west of the Azores) the sea becomes at once still and calm, scarcely ever agitated by a breeze. As I came down from the Canary Islands to the parallel of Sierra Leone I had to sustain a terrible heat, but as soon as we had passed beyond the above-mentioned line (west of the meridian of the Azores) the climate altered, the air became temperate, and the freshness increased the farther we advanced."

This passage, which is elucidated by several others in

the writings of Columbus, contains views of physical geography, remarks on the influence of geographical longitude, on the declination of the magnetic needle, on the inflection of the isothermal lines between the west coast of the old and the east coast of the new Continent, on the situation of the great Sargasso bank in the basin of the Atlantic, and on the relations of this part of the ocean to the atmosphere above it. Erroneous observations ([430]) in the neighbourhood of the Azores on the position of the Pole star had misled Columbus as early as the period of his first voyage, from the deficiency of his mathematical knowledge, to entertain the belief of an irregularity in the spherical form of the earth. According to his view, the earth was protuberant in the western hemisphere, so that the ships gradually arrived nearer to the sky on approaching the line (raya) where the magnetic needle points to the true north; and this elevation he supposed to be the cause of the cooler temperature. The solemn reception of the Admiral at Barcelona took place in April 1493, and on the 4th of May of the same year Pope Alexander VI. signed the celebrated bull which "establishes for ever" the demarcation line ([431]) between the Spanish and Portuguese possessions at a hundred miles westward of the Azores. If we bear in mind that Columbus, immediately after his return from his first voyage of discovery, purposed to go himself to Rome, in order, as he said, "to report to the Pope all that he had discovered," and if we remember the importance which the cotemporaries of Columbus attached to the line of no variation, it may be admitted that there are grounds for a suggestion first put forward by myself, that at the moment of his highest court favour Columbus endeavoured to cause "a physical line of demarcation to be converted into a political one."

The influence which the discovery of America, and the great nautical enterprizes connected with it exercised so rapidly on all physical and astronomical knowledge, is most strikingly felt when we recal the first impressions of those who lived at the period, and the wide range of scientific endeavours of which the most important part belongs to the first half of the sixteenth century. Columbus has not only the incontestable merit of having first discovered a "line without magnetic variation," but also of having, by his considerations on the progressive increase of westerly declination in receding from that line, given the first impulse to the study of terrestrial magnetism in Europe. The circumstance, that almost every where the ends of a freely suspended magnet do not point exactly to the north and south geographical poles, might easily have been recognised, even with very imperfect instruments, in the Mediterranean, and in other places where the declination amounted in the twelfth century to more than eight or ten degrees. But it is not improbable that the Arabs, or the Crusaders who were in contact with Eastern nations from 1096 to 1270, in spreading the use of Chinese or Indian compasses, may also have called attention, even at that early period, to the circumstance of magnetic needles pointing in different parts of the world to the north-east or to the north-west, as to a long-known phenomenon. We know positively from the Chinese Penthsaoyan, which was written under the dynasty of the Song (432) between 1111 and 1117, that the manner of measuring the amount of westerly declination had been then long understood.

That which belongs to Columbus is not the first observation of the existence of the variation (which, for example, is noted in the map of Andrea Bianco in 1436), but the remark which he made on the 13th of September, 1492, that "2½° east of the Island of Corvo the magnetic variation changes, passing from N.E. to N.W."

This discovery of a "magnetic line without declination" marks a memorable era in nautical astronomy. It has been celebrated with just praise by Oviedo, Las Casas and Herrera. Those who with Livio Sanuto would attribute it to the famous navigator Sebastian Cabot, forget that the first voyage of the latter, made at the cost of some merchants of Bristol, and distinguished by its attaining the American continent, took place five years later than Columbus's first voyage of discovery. But not only has Columbus the merit of having discovered the part of the Atlantic in which at that period the geographic and magnetic meridians coincided; he also made at the same time the ingenious and thoughtful remark, that the magnetic variation might serve to determine the ship's position in respect to longitude. In the journal of the second voyage (April 1496) we find him really inferring his position from the observed declination. The difficulties which oppose this method of determining the longitude, (more especially in a part of the globe where the magnetic lines of declination are so much curved that they do not follow the direction of the meridian, but correspond even with the parallels of latitude for considerable distances), were at that period still unknown. Magnetical and astronomical methods were anxiously sought after, in order to determine, both on land and sea, the points intersected by the ideally constituted line of demarcation. Neither the state of science nor that of the imperfect instruments employed at sea in 1493, whether for measuring angles or time, were competent to the practical solution of so difficult a problem. Under these circumstances, Pope Alexander VI., in presumptuously dividing half the globe between two powerful states, rendered without knowing it an essential service to nautical astronomy and to the physical science of terrestrial magnetism. The great maritime powers were from that time continually solicited to entertain innumerable impracticable proposals. Sebastian Cabot, as we learn from his friend Richard Eden, still boasted on his death bed that there had been "divinely revealed to him an infallible method of finding the longitude." This revelation was no other than his firm belief that the magnetic declination changed rapidly and regularly with the meridian. The cosmographer Alonso de Santa Cruz, one of the instructors of Charles V., undertook the drawing up of the first general "Variation Chart", (433) although, indeed, from very imperfect observations, as early as 1530, or a century and a half before Halley.

The "movement" of the magnetic lines, the first recognition of which is usually ascribed to Gassendi, was not even yet conjectured by William Gilbert; but at an earlier period, Acosta, "from the information of Portuguese navigators," assumed four lines of no declination upon the surface of the globe. (434) Hardly had the inclinometer, or dipping needle, been invented in England by Robert Norman, in 1576, than Gilbert boasted that, by means of this instrument, he could determine the position of a ship in a dark and starless night (aere caliginoso). (435) From my own observations in the Pacific, I shewed soon after my return to Europe that, in certain parts of the earth, and under particular local circumstanees, for example on the coasts of Peru in the season of constant fogs (garua), the *latitude* might be determined from the inclination of the magnetic needle with sufficient accuracy for the purposes of navigation. I have dwelt so long on these details, with the view of shewing that all the points with which we are now occupied, in reference to an important cosmical subject (with the exception of the measurement of the intensity of the magnetic force, and of the horary variations of the declination), were already spoken of in the 16th century. In the remarkable map of America appended to the Roman edition of the Geography of Ptolemy in 1508, we find to the north of Gruentlant (Greenland) a part of Asia represented, and "the magnetic pole" marked as an insular mountain. Martin Cortez, in the Breve Compendio de la Sphera (1545), and Livio Sanuto, in the Geographia di Tolomeo (1588), place it more to the south. Sanuto entertained a prejudice which, strange to say, has existed even in later times, that a man who should be so fortunate as to reach the magnetic pole (il calamitico), would experience there "alcun miracoloso stupendo effetto."

In the department of the distribution of temperature and meteorology, attention was already directed, at the end of the 15th and beginning of the 16th centuries, to the decrease of temperature (436) with increasing western longitude, (the inflection of the isothermal lines); to the law of rotation of the winds (437) generalized by Francis Bacon; to the diminution of atmospheric moisture and of the quantity of rain, caused by the destruction of forests; (438) and to the decrease of temperature with increasing elevation above the level of the sea, and the lower limit of perpetual snow. That this limit is "a function of the geographical latitude" was first recognised by Petrus Martyr Anghiera in 1510. Alonso de Hojeda and Amerigo Vespucci had seen the snowy mountains of Santa Marta (tierras nevadas de Citarma) as early as 1500; Rodrigo Bastidas and Juan de la Cosa examined them more closely in 1501; but it was not until the accounts of the expedition of Colmenares, which the pilot Juan Vespucci, nephew of Amerigo, communicated to his patron and friend Anghiera, that the "tropical snow region" seen on the mountainous shore of the Caribbean sea acquired a great, and it might be said a cosmical, signification. The lower limit of perpetual snow was now brought into connection with the general relations of the decrease of temperature and the diversity of climates. Herodotus, in discussing the causes of the rising of the Nile (ii. 22), had positively denied the existence of snowy mountains south of the tropic of Cancer. Alexander's expeditions, indeed, conducted the Greeks to the Nevados of the Hindoo Coosh (ορη αγαννιφα); but these are situated between 34° and 36° of north latitude. The only notice with which I am acquainted of "snow in the equatorial zone," prior to the discovery of America and the year 1500, is one which has been very little attended to by men of science, and which is contained in the celebrated inscription of Adulis, which Niebuhr considers to be later than Juba and than Augustus. The recognition of the dependence of the lower limit of perpetual snow on the latitude of the place, (439) and the first insight into the law of the decrease of temperature in an ascending vertical line, and the consequent gradual lowering, from the equator towards the poles, of a stratum of air of equal coolness, mark no unimportant era in the history of our physical knowledge.

If this knowledge was favoured by observations which were accidental and wholly unscientific in their origin, the age which we are describing lost on the other hand, by an unfortunate combination of circumstances, a great advantage which it might have received from a purely scientific impulse. The greatest physicist of the 15th century, who combined distinguished mathematical knowledge with the most admirable and profound insight into nature, Leonardo da Vinci, was the cotemporary of Columbus, and died three years after him. This great artist had occupied himself in meteorology, as well as in hydraulics and optics. His influence on the age in which he lived was exercised through the great works of painting which he created, and by his eloquent discourse, but not by his writings. If the physical views of Leonardo da Vinci had not remained buried in his manuscripts, the field of observation which the new world offered would have been already cultivated scientifically in many of its parts before the great epoch of Galileo, Pascal, and Huygens. Like Francis Bacon, and a full century before him, he regarded induction as the only sure method in natural science; "dobbiamo comminciare dall' esperienza, e per mezzo di questa scoprirne la ragione." (440)

As, notwithstanding the want of measuring instruments, climatic relations in the tropical mountainous regions, the distribution of temperature, the extremes of atmospheric dryness and humidity, and the frequency of electric explosions, were often spoken of in the commentaries on the first land journeys; so also the mariners very early embraced just views in regard to the direction and rapidity of currents, which, like rivers of variable breadth, traverse the Atlantic Ocean. The proper "equatorial current," the movement of the waters between the tropics, was first described by Columbus. "The waters move con los cielos (or like the vault of heaven) from east to west." Even the direction of separate floating masses of sea-weed confirmed this belief. ([441]) A light pan of wrought iron, which he found in the hands of the natives of the island of Guadaloupe, led Columbus to conjecture that it might be of European origin, belonging to a shipwrecked vessel which the equatorial current might have brought from the Iberian to the American coasts. In his geognostical fancies he regarded the existence of the series of the smaller West India Islands, as well as the peculiar form of the larger islands, (the coincidence of the direction of their coast with the parallels of latitude,) as caused by the long-continued action of the movement of the sea within the tropics from east to west.

When on his fourth and last voyage the Admiral discovered the north and south direction of the coast of America, from Cape Gracias a Dios to the Laguna de Chiriqui, he felt the action of the strong current which sets to the N. and N.N.W., and results from the impinging of the equatorial current against the opposing line of coast. Anghiera survived Columbus long enough to be aware of the deflection of the waters of the Atlantic in its whole course, to recognise the rotation round the Gulf of Mexico, and the propagation of this movement to the Tierra de los Bacallaos (Newfoundland), and the mouth of the St. Lawrence. I have shewn circumstantially in another place, how much the expedition of Ponce de Leon, in 1512, contributed to the formation of more accurate opinions, and have noticed that, in a memoir written by Sir Humphry Gilbert between 1567 and 1576, the movement of the waters of the Atlantic, from the Cape of Good Hope to the Banks of Newfoundland, was treated according to views which agree almost entirely with those of my excellent deceased friend, Major Rennell.

The knowledge of the oceanic currents was accompanied by that of the great banks of sea-weed (Fucus natans), the "oceanic meadows" which offer the remarkable spectacle of the accumulation of a "social plant" over a surface almost seven times greater than that of France. The "great Fucus bank," the proper "Mar de Sargasso," extends between 19° and 34° of north latitude. Its principal axis is about 7° west of the Island of Corvo. The "lesser Fucus bank" is situated in the space between the Bermudas and the Bahamas. Winds and partial currents affect in different years the position and extent of these Atlantic sea-weed meadows, for the first description of which we are indebted to Columbus. No other sea in either hemisphere shews an assemblage of social plants, on a similar scale of magnitude. [442]

But the important epoch of the great geographical discoveries, besides suddenly laying open an unknown hemisphere of the terrestrial globe, also enlarged the view of the regions of space, or to speak more distinctly, of the visible celestial vault. As man, to quote a fine expression of Garcilaso de la Vega, "in wandering to distant lands, sees earth and stars change together, [443]" so the advance to the equator, on both sides of Africa, and in the western hemisphere beyond the southern extremity of America, offered to the navigators and land travellers of the period of which we are treating, the magnificent spectacle of the southern constellations, longer and more frequently than could have been the case in the time of Hiram or of the Ptolemies, or under the Roman Empire, or in the course of the commerce of the Arabians in the Red Sea, and in the Indian Ocean between the Straits of Bab-el-Mandeb and the western peninsula of India. Amerigo Vespucci, in his letters, Vicente Yañez Pinzon, Pigafetta who accompanied Magellan and Elcano, as well as Andrea Corsali in his voyage to Cochin, in Eastern India, in the beginning of the 16th century, have given us a record of the vivid impressions produced by the earliest contemplation of the southern heavens beyond the feet of the Centaur, and the fine constellation of the Ship. Amerigo, who had more literary acquirement than the others, but who was also more inclined to a vain-glorious display, praises not unpleasingly the brightness, the picturesque beauty, and the novel aspect of the constellations which circle round the southern pole, of which the more immediate vicinity is poor in stars. He affirms in his letter to Pierfrancesco de Medici, that on his third voyage he occupied himself carefully with observing the southern constellations, measuring the polar distance of the principal amongst them, and making drawings of them. What he communicates on the subject does not indeed lead us greatly to regret the loss of his measurements.

I find the first description of the enigmatical black patches, (Coalbags) given by Anghiera in 1510. They had been remarked as early as 1499 by the companions of Vicente Yañez Pinzon, on the expedition which went from Palos and took possession of the Brazilian Cape St

Augustine. (444) The Canopo fosco (Canopus niger) of Amerigo, is probably one of these "coal bags." The acute Acosta compares it to the darkened portion of the moon's disk in partial eclipses, and appears to ascribe it to a void in space, or to the absence of stars. Rigaud has shewn how the mention of the "coal bags," of which Acosta expressly says that they are visible in Peru but not in Europe, and that they move like other stars round the South Pole, has been mistaken by a celebrated astronomer for the first notice of spots in the sun. (445) The knowledge of the two Magellanic clouds has been erroneously ascribed to Pigafetta; I find that Anghiera, from the observations of Portuguese navigators, mentions these clouds eight years before the completion of Magellan's circumnavigation of the globe: he compares their mild brightness with that of the Milky Way. The larger of these two clouds, however, appears not to have escaped the clear sight of the Arabians. It was very probably the White Ox "el Bakar" of their southern sky; the "white patch," of which the astronomer Abdurrahman Sofi says that it cannot be seen in Bagdad, or in the North of Arabia, but is seen in the Tehama, and in the parallel of the Straits of Bab-el-Mandeb. Under the Lagidæ and subsequently, Greeks and Romans had passed over those regions without noticing, or at least without mentioning in any writing which has come down to us, this luminous cloud, which yet, in the latitude of between 11° and 12° N., rose in the time of Ptolemy 3° above the horizon, and in that of Abdurrahmann (1000 A.D.), more than 4°. (446) The meridian altitude of the middle of the Nubecula Major may be now about 5° at Aden. It usually happens that mariners first distinctly recognise the Magellanic clouds in much more southerly latitudes, viz.

near the equator, or even south of it; but the reason of this is to be ascribed to atmospheric differences, and to the presence of vapours near the horizon reflecting white light. In the interior of southern Arabia, the azure of the celestial vault, and the great dryness of the atmosphere, must have favoured the recognition of the Magellanic clouds. The probability that such was the case is shewn by examples of the visibility of comets' tails in clear daylight between the tropics, and even in more southern latitudes.

The arrangement of the stars near the southern pole into new constellations belongs to the 17th century. What the Dutch navigators, Petrus Theodori of Embden, and Frederic Houtman, who (1596—1599) was a prisoner to the king of Bantam and Atschin, in Java and Sumatra, had observed with imperfect instruments, was laid down in the celestial charts of Hondius Bleaw (Jansonius Cæsius) and Bayer.

The more unequal distribution of the masses of light gives to that zone of the southern heavens, between the parallels of 50° and 80°, which is so rich in crowded nebulæ and clusters of stars, a peculiar, and one might almost say a picturesque character; a charm arising from the grouping of the stars of the first and second magnitude, and from the intervention of regions which, to the naked eye, appear dark and desert. These singular contrasts,—the Milky Way, which at several parts of its course shews a greatly increased brilliancy,—the insulated, revolving, rounded Magellanic clouds,—and the "coal bags," of which the largest is so near to a fine constellation,—increase the variety of this natural picture, and rivet the attention of susceptible spectators to particular regions in the southern celestial hemisphere. Religious associations have given to one of these regions,—that of the Southern Cross,—a peculiar interest to Christian navigators, travellers, and missionaries, in the tropical and southern seas, and in both the Indies. The four principal stars of which the Cross is composed were regarded, in the Almagest, and in the age of Hadrian and Antoninus Pius, as part of the constellation of the Centaur. (447) The form of the Southern Cross is so striking, and so remarkably individualised and detached,—as is the case of the Greater and Lesser Bear, the Scorpion, Cassiopea, the Eagle, and the Dolphin,—that it is almost surprising that those four stars should not have been earlier separated from the large ancient constellation of the Centaur; it is, indeed, the more surprising, because the Persian Kazwini and other Mahometan astronomers were at pains to make out crosses from stars in the Dolphin and Dragon. Whether the courtly flattery of the Alexandrian learned men, who transformed Canopus into a "Ptolemæon," also applied the stars of our present Southern Cross to the glorification of Augustus, by forming them into a "Cæsaris thronon" (448) which was never visible in Italy, remains somewhat uncertain. In the time of Claudius Ptolemæus, the fine star at the foot of the Southern Cross had still an altitude of 6° 10′ at its meridian passage at Alexandria; whilst, at the present day, it culminates several degrees below the horizon of that place. At this time (1847), in order to see α Crucis at an altitude of 6° 10′, and taking refraction into account, we must be 10° to the south of Alexandria, or in 21° 43′ of N. lat. The Christian anchorites in the Thebais may still have seen the cross at an altitude of 10° in the fourth century. I doubt, however, whether it received its name from them; for Dante, in the celebrated passage of the Purgatorio—

> "Io mi volsi a man destra, e posi mente
> All' altro polo, e vidi quattro stelle
> Non viste mai fuor ch' alla prima gente:"

and Amerigo Vespucci,—who, at the aspect of the southern firmament in his third voyage, first recalled these lines, and even boasted that "he now beheld in his own person the four stars never before seen save by the first human pair,"—were still unacquainted with the denomination of "Southern Cross." Vespucci says simply, that the four stars form a rhomboidal figure (una mandorla); and this remark belongs to the year 1501. As sea voyages round the Cape of Good Hope and in the Pacific Ocean, by the routes which Gama and Magellan had opened, multiplied, and as Christian missionaries pressed forward into the newly discovered tropical lands of America, the fame of this constellation increased more and more. I find it first mentioned as a "wondrous cross (croce maravigliosa), more glorious than all the constellations of the entire heavens," by the Florentine Andrea Corsali (1517), and afterwards, in 1520, by Pigafetta. The Florentine extols Dante's "prophetic spirit,"—as if the great poet had not possessed as much erudition as creative genius,—as if he had not seen Arabian celestial globes, and held communication with many oriental travellers from Pisa (449). That in the Spanish settlements in tropical America, the first settlers were accustomed to infer the hour of the night from the inclined or perpendicular position of the Southern Cross, as is still done, was already remarked by Acosta in his "Historia natural y moral de las Indias." (450)

By the precession of the equinoxes the aspect of the starry heavens from every point of the earth's surface is constantly changing. The earlier inhabitants of our high northern latitudes might see magnificent southern constellations rise to their view, which, now long unseen, will not reappear for thousands of years. In the time of Columbus, Canopus was already fully 1° 20′ below the horizon at Toledo (lat. 39° 54′); it is now about the same quantity above the horizon at Cadiz. For Berlin and the northern latitudes the stars of the Southern Cross, as well as α and β Centauri, are receding more and more; whilst the Magellanic clouds slowly approach our latitudes. Canopus has had its greatest northerly approximation during the thousand years which have closed, and is now moving (though, on account of its proximity to the south pole of the ecliptic, with extreme slowness) progressively to the south. The Southern Cross began to be invisible in 52½° north latitude, 2900 years before the Christian era. According to Galle it might previously have reached, in that latitude, an altitude of more than 10°; and when it vanished from the horizon of the countries adjoining the Baltic, the great Pyramid of Cheops had already been standing in Egypt for five centuries. The pastoral nation of the Hyksos made their invasion 700 years later. Former times seem to draw sensibly nearer to us, when we connect their measurement with memorable occurrences.

The extension of a knowledge of the celestial spaces,—a knowledge, however, limited to their outward aspect,—was accompanied by advances in nautical astronomy; that is to say, in the improvement of all the methods of determining a ship's place, or its geographical latitude and longitude.

All that in the course of time has contributed to favour these advances in the art of navigation;—the compass, and the more correct knowledge of the magnetic declination,—the measurement of a ship's way by the more exact apparatus of the log,—the use of chronometers and of lunar distances,—the better construction of vessels,—the substitution of another propelling force for the force of the wind,—and in all respects, the skilful application of astronomy to a ship's reckoning,—must be regarded as powerful means of throwing open all parts of the earth's surface, of accelerating the animating intercourse of nations with each other, and of advancing the investigation of cosmical relations. Taking this as our point of view, we would here recal the fact that as early as the middle of the 13th century "nautical instruments were in use for determining time by the altitudes of stars" in the vessels of the Catalans and of the Island of Majorca; and that the astrolabe described by Raymond Lully, in his Arte de Navegar, is almost two centuries older than that of Martin Behaim. The importance of astronomical methods was so vividly recognised in Portugal, that about the year 1484 Behaim was named president of a Junta de Mathematicos, "who were to compute tables of the sun's declination," and, as Barros says, (451) to teach pilots the "maneira de navegar per altura do sol." The navigation "by the meridian altitudes of the sun" was already at that period clearly distinguished from the navigation by determinations of longitude, or "por la altura del este-oeste." (452)

The desirability of fixing the locality of the Papal line of demarcation, for the sake of settling the boundary between the claims of the Spanish and Portuguese crowns in the newly discovered Brazils and in the South Indian Islands, augmented the anxiety for the discovery of practical methods for finding the longitude. It was felt how rarely the ancient imperfect Hipparchian method by lunar eclipses could be applied, and the use of lunar distances was already recommended, in 1514, by the Nuremberg astronomer Johann Werner, and soon afterwards by Orontius Finæus and Gemma Frisius. Unfortunately this method long continued impracticable, until, after many vain attempts with the instruments of Peter Apianus (Bienewitz) and Alonso de Santa Cruz, the mirror sextant was invented in 1700 by Newton, and brought into use among mariners by Hadley in 1731.

The influence of the Arabian astronomers was also operative, in and through Spain, on the progress of nautical astronomy. Many modes were, indeed, tried for determining the longitude, which did not succeed; but the failure was less often attributed, at the time, to the imperfection of the observation, than to errors of the press in the astronomical ephemerides of Regiomontanus. The Portuguese even suspected the results of the astronomical data of the Spaniards, whose tables were supposed to have been falsified from political motives. (453) The suddenly awakened sense of the want of those means which nautical astronomy, theoretically at least, promised, shews itself in a particularly vivid manner in the narratives of Columbus, Amerigo Vespucci, Pigafetta, and Andres de San Martin the celebrated pilot of Magellan's expedition, who was in possession of Ruy Falero's method of finding the longitude. Oppositions of planets, occultations of stars, differences of altitude between the Moon and Jupiter, and changes of the Moon's declination, were all tried with more or less success. We have observations of conjunction by Columbus, in the night of the 13th of January, 1493, from Haiti. The necessity of giving to each great expedition a well-instructed astronomer, in addition to the naval officers, was so generally felt, that Queen Isabella wrote to Columbus on the 5th of September, 1493, that "although he had shewn in his enterprises that he knew more than any other mortal man (que ninguno de los nacidos), yet she advised him to take with him Fray Antonio de Marchena, as a learned and skilful man in the knowledge of the stars." Columbus says, in the description of his fourth voyage: "there is but one infallible method of keeping a ship's reckoning, namely, the astronomical one. Those who understand it may be content. What it yields is like a 'vision profetica.' (454) Our ignorant pilots, when they have lost sight of the coast for many days, know not where they are; they would not be able to find again the lands which I have discovered. To navigate requires 'compas y arte,' the compass, and the knowledge or art of the astronomer."

I have given these characteristic details, because they bring more sensibly before us the manner in which nautical astronomy, the powerful instrument of rendering navigation secure and certain and thereby facilitating access to all regions of the globe, received its first development in the epoch of which we are treating; and how, in the general movement of men's minds, there was an early recognition of the possibility of methods, which had to await for their extensive practical application the improvement of timekeepers and of instruments for measuring angles, as well as correct solar and lunar tables. If the character of an age be "the manifestation of the human mind in a definite epoch of time," the age of Columbus, and of the great nautical discoveries, whilst augmenting in an unexpected manner the objects of knowledge and contemplation, also opened to succeeding centuries a new and higher range of attainment. It is the peculiarity of great discoveries at once to extend the field of our conquests, and our prospect into new regions which yet remain to be conquered. Weak spirits in every age believe complacently that mankind have reached the highest point of their intellectual progress; forgetful that through the intimate mutual relation of all natural phænomena, in proportion as we advance, the field to be travelled over obtains a wider extension,—that it is bounded by an horizon which recedes continually before the march of the explorer.

Where, in the history of nations, can we point to an epoch similar to that in which events so fruitful in consequences, as the discovery and first colonisation of America, the navigation to India by the Cape of Good Hope, and Magellan's first circumnavigation of the globe, coincided with the highest and most flourishing period of art, with the attainment of intellectual and religious liberty, and with the sudden enlargement of the knowledge of the heavens and of the earth? Such an epoch owes but a very small portion of its grandeur to the distance from which we regard it, or to the circumstance that it comes before us only in historical remembrance, unobscured by the disturbing actuality of the present. But here too, as in all terrestrial things, the period of greatest brilliancy is closely associated with events which call forth emotions of the deepest sorrow. The progress of cosmical knowledge was purchased by all the violence and all the horrors which conquerors, the so-called extenders of civilisation, spread over the earth. Yet it would be an indiscreet and rash boldness which, in the interrupted history of the development of humanity, should venture to decide dogmatically on the balance of good or ill. It is not for men to pronounce judgment on events which, slowly prepared in the womb of time, belong but partially to the age in which we place them.

The first discovery of the middle and southern parts of the United States of America by the Scandinavians almost coincides in point of time with the appearance and mysterious arrival of Manco Capac in the highlands of Peru; it preceded by almost 200 years the arrival of the Aztecs in the valley of Mexico. The foundation of the principal city, Tenochtitlan, dates fully 325 years later. If these colonizations by Northmen had been more permanent in their results,—if they had been fostered and protected by a powerful and politically united mother country,—the advancing Germanic race would have still found many wandering tribes of hunters, (455) where the Spanish conquerors found settled agriculturists.

The period of the conquista, the end of the 15th and beginning of the 16th centuries, is marked by a wonderful coincidence of great events in the political and moral life of the nations of Europe. In the same month in which Hernan Cortes, after the battle of Otumba, advanced to besiege Mexico, Martin Luther burnt the papal bull at Wittenberg, and laid the foundation of the Reformation, which promised to the mind of man freedom and progress in almost untried paths. (456) Somewhat earlier, those long buried glorious monuments of ancient Grecian art, the Laocoon, the Torso, the Belvedere Apollo, and the Medicean Venus had been disclosed. Michael Angelo, Leonardo da Vinci, Titian, and Raphael flourished in Italy, and Holbein and Albert Durer in our German country. In the year in which Columbus died, fourteen years after the discovery of the new continent, the order of the universe was discovered, though not publicly announced, by Copernicus.

The consideration of the importance of the discovery of America, and of the first European settlements therein, touches on other fields of thought besides those to which these pages are especially devoted; it would include all those intellectual and moral influences, which the sudden enlargement of the entire mass of ideas exercised on the improvement of the social state. We recal only by a passing allusion, how, since that great era, a new activity of thought and feeling, courageous wishes, and hopes hard to relinquish, have gradually pervaded all classes of civil society;—how the scantiness of the population of one hemisphere of the globe, especially on the coasts opposite to Europe, favoured the settlement of colonies, which by their extent and position have been transformed into independent states, unrestricted in the choice of free forms of government,—and how, lastly, the religious Reformation, the precursor of great political revolutions, passed through the different phases of its development in a region which became the refuge of all religious opinions, and of the most different views in Divine things. The boldness of the Genoese navigator is the first link in the immeasurable chain of these fate-fraught events; and it was accident, and not fraud or strife, (457) which deprived the continent of America of his name. The new world, brought during the last half century continually nearer to Europe by commercial intercourse, and by the improvement of navigation, has exercised an important influence on the political institutions, (458) and on the ideas and tendencies of those nations who dwell on the eastern shore of the constantly narrowing valley of the Atlantic Ocean.

EPOCHS IN THE HISTORY OF THE CONTEMPLATION OF THE UNIVERSE.—DISCOVERIES IN THE CELESTIAL SPACES.

VII.

Great Discoveries in Space by the application of the Telescope.—The great Epoch of Astronomy and Mathematics from Galileo and Kepler to Newton and Leibnitz.—Laws of the Planetary Motions, and general Theory of Gravitation.

In attempting to recount the most distinctly marked periods and gradations of the development of cosmical contemplation, we have in the last section endeavoured to depict the epoch, in which one hemisphere of the globe first became known to the cultivated nations inhabiting the other. The epoch of the most extensive discoveries upon the surface of our planet was immediately succeeded by man's first taking possession of a considerable part of the celestial spaces by the telescope. The application of a newly formed organ, of an instrument of space-penetrating power, called forth a new world of ideas. Now began a brilliant age of astronomy and mathematics; and in the latter the long series of profound investigators, leading to the "all-transforming" Leonard Euler, the year of whose birth (1707) is so near the year of Jacob Bernouilli's death.

A few names may suffice to recal the giant strides with which the human mind advanced in the 17th century, less from any outward incitements than from its own independent energies, and especially in the development of mathematical thought. The laws that regulate the fall of bodies, and the planetary motions, were recognised; the pressure of the atmosphere, the propagation of light, and its refraction and polarisation, were investigated. Mathematico-physical science was created, and established on firm foundations. The invention of the infinitesimal calculus marks the close of the century; and, reinforced by its aid, the human intellect has been enabled, in the succeeding hundred and fifty years, to attempt successfully the solution of problems presented by the perturbations of the heavenly bodies, by the polarisation and interference of the waves of light, by radiant heat, by the electro-magnetic re-entering currents, by vibrating chords and surfaces, by the capillary attraction of tubes of small diameter, and by so many other natural phænomena.

In this world of thought the work proceeds uninterruptedly, and its different portions lend to each other mutual support. No earlier fruitful germ is stifled. We see increase, simultaneously, the abundance of materials, the strict accuracy of methods, and the perfection of instruments. I propose to limit myself principally to the consideration of the 17th century, the age of Kepler, Galileo, and Bacon, of Tycho Brahe, Descartes, and Huygens, of Fermat, Newton, and Leibnitz. What they have done is so generally known, that slight indications will suffice to point out through what part of their achievements they have more especially contributed to the enlargement of cosmical views.

We have already shewn (459) how, by the discovery of telescopic vision, there was lent to the eye,—the organ of the sensuous contemplation of the visible universe,—a power of which we are yet far from having reached the limit, but of which the first feeble commencement (magnifying hardly as much as 32 times in linear dimension), ([460]) sufficed to penetrate into cosmical depths before unknown. The exact knowledge of many heavenly bodies belonging to our solar system, the unchanging laws according to which they revolve in their orbits, and the perfected insight into the true structure of the universe, are the characteristics of the epoch which we here attempt to describe. The results which this age produced have defined the leading outlines of the picture of nature or sketch of the Cosmos, and have added an intelligent recognition of the contents of the celestial spaces,—at least in the well-understood arrangement of one planetary group,—to the earlier explored contents of terrestrial space. Seeking to fix attention on general views, I here name only the most important objects of the astronomical labours of the 17th century; and would point to their influence in inciting at once to great and unexpected mathematical discoveries, and to a more comprehensive and grander contemplation of the material universe.

I have already remarked, that the age of Columbus, Gama, and Magellan, the age of nautical discoveries, coincided with other great and deeply influential events, with the awakening of religious liberty of thought, with the development of art, and with the promulgation of the Copernican system of the universe. Nicholas Copernicus (in two still existing letters he calls himself Kopernik) had already attained his 21st year, and had observed with the astronomer Albert Brudzewski, at Cracow, when Columbus discovered America. Hardly a year after the death of the great discoverer, Copernicus having returned to Cracow from a six years' residence at Padua, Bologna, and Rome, we find him occupied with an entire revolution in the astronomical view of the universe.

By the favour of his uncle, Lucas Waisselrode von Allen, (461) Bishop of Ermland, he was named, in 1510, Canon at Frauenburg, where he was engaged for thirty-three years in the completion of his work "De Revolutionibus Orbium Cœlestium." The first printed copy was brought to him when in immediate preparation for death, and when his strength of body and mind were failing: he saw it and touched it; but temporal things were no farther heeded, and he died, not, as Gassendi says, a few hours, (462) but some days afterwards, on the 24th of May, 1543. Two years previously, an important part of his doctrine had been made known in print, by a letter from one of his most zealous pupils and adherents, Joachim Rhæticus, to Johann Schoner, Professor at Nuremberg. Yet it was not the promulgation of the Copernican theory, the renewed doctrine of the solar orb forming the centre of our system, which led, somewhat more than half a century after its first appearance, to the brilliant discoveries in space which mark the beginning of the 17th century:—these discoveries were the result of an invention accidentally made,—that of the Telescope. Through them the doctrine of Copernicus was perfected and enlarged. His fundamental views, confirmed and extended by the results of physical astronomy (by the newly discovered system of the satellites of Jupiter, and by the phases of Venus),—pointed out to theoretical astronomy the paths which must conduct to the sure attainment of her aims, and incited to the solution of problems which required that the analytical calculus should be carried to still higher degrees of perfection. As George Peuerbach and Regiomontanus (Johann Müller, of Königsberg, in Franconia), exerted a beneficial influence on Copernicus and his scholars, Rhæticus, Reinhold, and Möstlin, so also did these (though divided from them by a longer interval of time) exert a similar influence on the labours of Kepler, Galileo, and Newton. This is the connecting link which, in the enchainment of ideas, unites the 16th and 17th centuries, and requires that, in describing the enlarged astronomical views of the later of these two periods, we should allude to the incitements which descended to it from the former.

An erroneous, and unhappily still recently prevailing opinion, (463) regards Copernicus as having, through timidity and fear of priestly persecution, represented the earth's planetary movement, and the sun's position in the centre of the whole planetary system, as a mere "hypothesis," which fulfilled the astronomical object of subjecting the orbits of the heavenly bodies to convenient calculation, "but which need not be regarded as true, or even as probable." These singular words (464) are indeed found in the anonymous preface placed at the commencement of Copernicus's work, and entitled "De Hypothesibus hujus operis;" but they do not belong to Copernicus, and are in direct contradiction to his dedication to the Pope, Paul III. The author of this preliminary notice was, as Gassendi says most distinctly in his life of Copernicus, a mathematician named Andreas Osiander, then living at Nuremberg, who, conjointly with Schoner, superintended the printing of the book "De Revolutionibus," and who, although he does not make express mention of any religious scruples, would appear to have thought it advisable to term the new views an hypothesis, and not, like Copernicus, a demonstrated truth. The founder of our present system of the universe (the most important parts of that system, the grandest traits in the picture of the universe, unquestionably belong to him) was no less distinguished by the courage and confidence with which he propounded it, than by his knowledge. He was in a high degree deserving of the fine eulogium of Kepler, who, speaking of him in the introduction to the Rudolphine Tables, says, "vir fuit maximo ingenio, et quod in hoc exercitio (in combating prejudices) magni momenti est, *animo liber*." Copernicus, in his dedication to the Pope, does not hesitate to term the generally received opinion of the immobility and central position of the earth an "absurd acroama," and to expose the stupidity of those who adhere to so erroneous a belief. "If," said he, "any empty babbler (ματαιολογοι), ignorant of mathematical knowledge, should yet rashly pronounce sentence upon his work, by wresting for that purpose some passage from Holy Scripture (propter aliquem locum scripturæ male ad suum propositum detortum), he should despise so presumptuous an assault. It was, indeed, generally known that the celebrated Lactantius (who could not, it is true, be reckoned among mathematicians), had spoken very childishly (pueriliter) of the form of the earth, deriding those who hold it to be spherical. On mathematical subjects one must write for mathematicians only. In order to shew that, deeply penetrated with the truth of his results, he had no cause to fear any condemnation, he addressed himself, from a remote corner of the world, to the supreme visible head of the Church, that he might protect him from the tooth of slander; adding, that the Church would, moreover, be advantaged by his investigations on the length of the year and the movements of the moon." In regard to this last remark it may be noticed, that astrology, and amendments in the Calendar, were long chiefly efficacious in obtaining for astronomy the protection of secular or ecclesiastical power; as chemistry and botany were long regarded solely as subservient to medicinal knowledge.

The free and powerful language employed by Copernicus, the evident outpouring of deep internal conviction, sufficiently refutes the assertion, that the system which bears his immortal name was proposed as an hypothesis convenient to calculating astronomers, but which might very well be without foundation. "By no other arrangement," he exclaims, with inspired enthusiasm, "have I been able to discover so admirable a symmetry of the universe, so harmonious a combination of orbits, than by placing the light of the world (lucernam mundi), the sun, as on a kingly throne, in the midst of the beautiful temple of nature, guiding from thence the entire family of circum-revolving planets (circumagentem gubernans astrorum familiam)." ([465]) Even the idea of universal gravitation or attraction (appetentia quædam naturalis partibus indita) towards the centre of the world (centrum mundi), the sun, inferred from the force of gravity in spherical bodies, appears to have floated before the mind of this great man, as is shewn by a remarkable passage ([466]) in the 9th chapter of the 1st book of the "Revolutions."

In passing in review the different stages of the development of cosmical contemplations, we discover from the earliest times more or less obscure anticipations of the attraction of masses, and of centrifugal forces. Jacobi, in his investigations on the mathematical knowledge of the Greeks, (which are unfortunately still in manuscript), dwells with justice on "the deep consideration of Nature by Anaxagoras, from whom we hear, not without astonishment, that the moon (467) if its force of rotation ceased would fall to the earth as a stone discharged from a sling." I have already, in my first volume, when treating of the fall of aerolites, (468) noticed similar expressions of the Clazomenian, and of Diogenes of Apollonia, respecting the "cessation or interruption of the force of rotation." Of the attracting force which the centre of the earth exerts on all heavy masses removed from it, Plato had a clearer idea than Aristotle; who was, indeed, like Hipparchus, acquainted with the acceleration of bodies in falling, but who did not correctly apprehend its cause. In Plato, and according to Democritus, attraction is limited to bodies which have affinity with each other; or in other words, to the tending together of homogeneous elementary substances. (469) But at a later period, probably in the 6th century, the Alexandrian John Philoponus, a pupil of Ammonius Hermeæ, ascribes the movements of cosmical bodies to a primitive impulse, and combines with this idea that of the fall of bodies, or the tendency of all substances, heavy or light, to come to the ground. (470) But the idea which Copernicus divined, and which Kepler enunciated more clearly in his fine work "de Stella Martis," even applying it (471) to the ebb and flood of the Ocean, we find invested with new life, and rendered more fruitful (1666 and 1674) by the sagacity of the ingenious Robert Hooke. The Newtonian theory of gravitation came next, and presented the grand means of transforming the whole of physical astronomy into a system of celestial mechanics. (472)

Copernicus, as we perceive not only from his dedication to the Pope, but also from several passages in the book itself, was tolerably well accquainted with the representations which the ancients formed to themselves of the structure of the Universe. In the period before Hipparchus, he however only names Hicetas of Syracuse, (whom he always calls Nicetas), Philolaus the Pythagorean, the Timæus of Plato, Ecphantus, Heraclides of Pontus, and the great geometer Apollonius of Perga. Of the two mathematicians who came nearest to his system, Aristarchus of Samos, and Seleucus the Babylonian, ([473]) he only names the first without farther notice, and does not mention the second at all. It has often been said that Copernicus was not accquainted with the opinion of Aristarchus of Samos, relative to the central position of the Sun and the planetary character of the Earth, because the "Arenarius," and all the works of Archimedes, were only published a year after his death, a full century after the invention of the art of printing; but in saying this, it is forgotten that, in the dedication to Pope Paul III., Copernicus quotes a long passage on Philolaus, Ecphantus, and Heraclides of Pontus, from Plutarch's work "on the opinions of Philosophers" (iii. 13), and that he might have read in the same work (ii. 24), that Aristarchus of Samos regarded the Sun as one of the fixed stars. Among all the opinions of the Ancients, the greatest influence on the direction and gradual development of the views of Copernicus, would appear, from Gassendi's statements, to have been exercised by a passage in the encyclopædic work of Martianus Mineus Capella of Madaura, written in a semi-barbarous language, and by the System of the World of Apollonius of Perga. According to the system described by Martianus Mineus, which has been confidently ascribed ([474]) sometimes to the Egyptians, and sometimes to the Chaldeans, the Earth rests immoveably in the centre, and the Sun revolves round it as a planet, while Mercury and Venus accompany, and revolve round the Sun as his satellites. Such a view of the structure of the Universe might tend to prepare the way for that of the Sun's central force. There is nothing either in the Almagest, or in the writings of the Ancients generally, or in the work of Copernicus "de Revolutionibus," to justify Gassendi's decided assertion as to the perfect similarity of the System of Tycho Brahe with that of Apollonius of Perga. After Böckh's complete investigation, nothing more need be said respecting the confusion of the System of Copernicus with that of the Pythagorean Philolaus, in which the non-rotating Earth (the Antichton or opposite earth is not itself a planet, but only the opposite hemisphere of our planet,) moves, as well as the sun, round the "hearth of the world," the central fire or flame of life of the entire planetary system.

The scientific revolution commenced by Copernicus had the rare good fortune (setting aside a brief retrograde movement in Tycho Brahe's hypothesis), of proceeding uninterruptedly forward to its object,—the discovery of the true structure of the universe. The rich supply of exact observations which were furnished by Tycho Brahe himself, the zealous opponent of Copernicus, laid the foundation of the discovery of those unchanging laws of the planetary movements, which prepared for Kepler imperishable fame, and which, when interpreted by Newton, and shewn by him to be theoretically necessary, were transferred to the bright domain of thought, and became the "intelligent recognition of nature." It has been ingeniously said, (475) though perhaps with too feeble an appreciation of the free, great, and independent spirit which conceived the theory of gravitation, "Kepler wrote a book of laws, Newton the spirit of the laws."

The figurative poetic myths of the Pythagorean and Platonic pictures of the universe, (476) variable as the imagination from which they had their birth, still found a partial reflex in Kepler; they warmed and cheered his often saddened spirits, but they did not divert him from the earnest path which he steadfastly pursued, and of which he reached the goal, (477) 12 years before his death, on the memorable night of the 15th of May, 1618. Copernicus had afforded a sufficient explanation of the apparent revolution of the heaven of the fixed stars, by the diurnal rotation of the Earth around her axis; and by the annual movement round the sun, had given an equally perfect solution of the most striking movements of the planets (their retrogressions and stationary appearances),—and had thus found the true cause of what is called the "second inequality of the planets." The first inequality, the non-uniform movement of the planets in their orbits, he left unexplained. True to the ancient Pythagorean principle of the inherent perfection of circular movements, Copernicus, in his structure of the universe, needed to add to the "excentric" circles having unoccupied centres, some of the epicycles of Apollonius of Perga. Bold as was the path struck out, men could not free themselves at once from all earlier views.

The equal distance at which the fixed stars continue from each other, whilst the whole heavenly vault moves from East to West, had led to the representation of a firmament,—a solid crystal sphere,—in which Anaximenes, (who was perhaps not much later than Pythagoras), imagined the stars to be fastened as if nailed. (478) Geminus the Rhodian, a cotemporary of Cicero's, doubted the constellations being all in the same plane; some, he thought, were higher and some lower. This manner of representing the heaven of the fixed stars was transferred to the planets; and thus arose the theory of the excentric intercalated spheres of Eudoxus, Menaechmus, and Aristotle who invented retrograding spheres. After a century, the acute mind of Apollonius caused the theory of epicycles,—a construction which adapted itself more easily to the representation and calculation of the motions of the planets,—to supersede the solid spheres. Whether, as Ideler believes, it was not until after the establishment of the Alexandrian Museum, that philosophers began to regard "a free movement of the planets in space as possible,"—whether previously to that period the intercalated transparent spheres, (27 according to Eudoxus, 55 according to Aristotle), as well as the epicycles which passed from Hipparchus and Ptolemy to the middle ages, were generally regarded, not as actual solid substances having material thickness, but simply as ideal abstractions,—I refrain here from any attempt to decide historically, greatly as I incline to the latter view.

It is more certain, that in the middle of the 16th century, when the theory of the 77 homocentric spheres of the learned Polyhistor, Girolamo Fracastoro, was received with applause, and when, subsequently, the opponents of Copernicus sought for every means of supporting the system of Ptolemy,—the representation of the existence of solid spheres, circles and epicycles, which had been particularly favoured by the fathers of the Church, was still extremely prevalent. Tycho Brahe expressly boasts, that by his considerations on the paths of comets, he first demonstrated the impossibility of solid spheres, and thus shattered the whole artificial fabric. He filled the free celestial spaces with air, and even believed that the "resisting medium," made to vibrate by the revolving heavenly bodies, might produce sounds. The unpoetic Rothmann thought it incumbent upon him to refute this renewal of the Pythagorean myth of the music of the spheres.

The great discovery of Kepler, that all the planets move round the sun in ellipses, and that the sun is placed in one of the foci of these ellipses, finally freed the original Copernican system from the excentric circles, and from all epicycles. ([479]) The planetary fabric of the universe now appeared objectively, and as it were architecturally, in its simple grandeur; but the play and connection of indwelling, impelling and maintaining forces, were first unveiled by Isaac Newton. In the history of the gradual development of human knowledge, we have already often remarked the appearance, within short intervals of time, of important though seemingly accidental discoveries, and of great minds clustered as it were together; and we see this phenomenon repeated in the most striking manner in the first ten years of the 17th century. Tycho Brahe, the founder of modern practical astronomy, Kepler, Galileo, and Francis Bacon, were cotemporaries. All, except Tycho, were cotemporaneous in their maturer years with the labours of Descartes and Fermat. The fundamental traits of Bacon's Instauratio Magna appeared in the English language as early as 1605, fifteen years before the Novum Organon. The invention of the telescope, and the greatest discoveries in physical astronomy, (Jupiter's satellites, the solar spots, the phases of Venus, and the wonderful form of Saturn), fall between the years 1609 and 1612. Kepler's speculations on the elliptic orbit of Mars (480) were began in 1601, and gave occasion to the "Astronomia nova seu Physica cœlestis" completed eight years later. "By the study of the orbit of the planet Mars," writes Kepler, "we must arrive at the knowledge of the mysteries of astronomy, or we must remain ever ignorant of them. By resolutely continued labour I have succeeded in subjecting the inequalities of the motion of Mars to a natural law." The generalization of the same thought conducted Kepler to the great truths and cosmical conjectures which he presented ten years later in his "Harmonices Mundi" (libri quinque). "I believe," he writes, in a letter to the Danish astronomer Longomontanus, "that astronomy and physics are so closely connected, that neither can be perfected without the other." The results of his investigations on the structure of the eye and the theory of vision appeared in the "Paralipomena ad Vitellionem," in 1604, and the "Dioptrica," (481) in 1611. Thus rapid, in regard both to the most important objects in the phænomena of the celestial spaces, and to the mode of apprehending these objects through the invention of new organs, was the extension of knowledge in the short interval of the first ten or twelve years of the century, which opened with Galileo and Kepler, and closed with Newton and Leibnitz.

The accidental discovery of the space-penetrating power of the telescope was first made in Holland, probably as early as the close of 1608. According to the latest documentary investigations, (482) this great invention may be claimed by Hans Lippershey, a native of Wesel, and spectacle-maker at Middelburg,—Jacob Adriansz, also called Metius, who is said to have made burning-glasses of ice,—

and Zacharias Jansen. The first named of these three parties is always called Laprey in the important letter of the Dutch ambassador Boreel to the physician Borelli, the author of the memoir "De vero telescopii inventore." (1655.) If the priority were to be determined by the precise times when the offers were made to the States General, it would belong to Hans Lippershey, who offered to the Government, on the 2d of October, 1608, three instruments "with which one can see to a distance." The offer of Metius is dated the 17th of October of the same year; but he says expressly in his petition, that "through meditation and industry he had constructed such instruments for two years." Zacharias Jansen (who, like Lippershey, was a spectacle-maker at Middelburg), together with his father Hans Jansen, invented the compound microscope, the eye-piece of which is a concave lens, towards the end of the 16th century (probably about 1590), but discovered the telescope only in 1610, as the ambassador Boreel testifies. Jansen and his friends directed the telescope towards remote terrestrial, but not towards celestial objects. The inappreciable importance and magnitude of the influence exerted by the microscope in communicating a more profound knowledge of all organic objects in respect to the conformation and movements of their parts, and by the telescope in the sudden opening of regions of cosmical space before unknown, required this detailed reference to the history of their discovery.

When the news of the recent Dutch invention, or of the discovery of telescopic vision, reached Venice, Galileo was accidentally present; he at once divined what were the essential conditions of the construction, and immediately completed a telescope at Padua for his own use. He directed it first to the mountains of the moon, and shewed the method of measuring their heights; attributing, like Leonardo da Vinci and Möstlin, the ashy coloured light of the moon to the light of the sun reflected back upon her from the earth. He examined with small magnifying powers the group of the Pleiades, the cluster of stars in Cancer, the Milky Way, and the group of stars in the head of Orion. Then followed in quick succession the great discoveries of the four satellites of Jupiter,—the two "handles" of Saturn, or his surrounding ring imperfectly seen so that its true character was not at once recognised,—the solar spots,—and the crescent form of Venus.

The satellites or moons of Jupiter, (the first of all the secondary planets of which the telescope disclosed the existence), were discovered, as it would appear, almost simultaneously, and quite independently, on the 29th of December, 1609, by Simon Marius, at Ansbach; and on the 7th of January, 1610, by Galileo, at Padua. In the publication of this discovery, Galileo, by the Nuncius Sidereus (1610), preceded the Mundus Jovialis of Simon Marius (1614). (484) Simon Marius wished to call Jupiter's satellites Sidera Brandenburgica; Galileo proposed Sidera Cosmica or Medicea, of which names the last was most approved at Florence. The collective name was not, however, sufficient to meet the love of flattery; and the satellites, instead of being designated as they are by us, by numbers, having been called by Simon Marius, Io, Europa, Ganymede, and Callisto,—Galileo substituted for these mythological personages the names of the members of the family of the Medicean ruling house, Catharina, Maria, Cosimo the elder, and Cosimo the younger.

The knowledge of Jupiter's satellites and of the phases of Venus was most influential in confirming and extending the Copernican system. The little world composed of the planet Jupiter and his satellites (Mundus Jovialis) offered to the intellectual eye a perfect image of the great solar and planetary system. It was recognised that the satellites of Jupiter obeyed the laws discovered by Kepler, and, in the first place, that the squares of their periods of revolution were in the ratio of the cubes of their mean distances from the central planet. This led Kepler, in the Harmonice Mundi, to exclaim with the confidence and courage which belongs to intellectual freedom, addressing himself to those whose voices bore sway beyond the Alps:—"Eighty years (485) have elapsed, during which the Copernican doctrine of the motion of the earth and the immobility of the sun has been taught unhindered, because it was held permissible to dispute concerning natural things, and to throw light upon the works of God; and now, when new documents have been discovered for the proof of this doctrine, documents which were unknown to the (ecclesiastical) judges, the promulgation of the true system of the fabric of the universe is by you prohibited!" This prohibition or ban,—a consequence of the ancient feud between ecclesiastical authorities and natural science,—had been already experienced by Kepler even in Protestant Germany. (486)

The discovery of Jupiter's satellites marks a memorable epoch in the history of astronomy, and in the permanent establishment of the principles upon which it is founded. (487)

The occultations of the satellites, or their entrance into the shadow of Jupiter, led to the knowledge of the velocity of light (1675), and through this, in 1727, to the explanation of the "aberration-ellipse" of the fixed stars, in which the orbit of the earth, in her annual revolution round the sun, is, as it were, reflected on the celestial vault. These discoveries of Römer's and Bradley's have justly been termed the "key-stone of the Copernican system;" the visible demonstration of the earth's movement of translation.

The importance of the occultations of Jupiter's satellites for geographical determinations of longitude on land was early perceived by Galileo (Sept. 1612). He proposed this method of determining longitudes, first to the Court of Spain (1616), and subsequently to the States General of Holland; he proposed it, indeed, as a method available at sea, (488) apparently little aware of the insuperable difficulties which oppose its practical application on the unstable ocean. He wished either to go himself, or to send his son Vicenzio, to Spain, with a hundred telescopes which he should prepare; requiring for recompense "una Croce di S. Jago," and an annual pension of 4000 crowns; a small sum, he says, as at first, in Cardinal Borgia's house, he had been led to expect 6000 ducats a year.

The discovery of Jupiter's satellites was soon after followed by the observation of Saturn as a triple star,—"planeta tergeminus." As early as November 1610, Galileo wrote to Kepler that "Saturn consists of three heavenly bodies in contact with each other." In this observation there was the germ of the discovery of Saturn's ring. Hevelius described, in 1656, the variations in the form of Saturn, the unequal opening of the "handles," and their occasional entire disappearance. But the merit of having explained scientifically all the phænomena of the ring of Saturn taken as one, belongs to Huygens (1655), who, according to the mistrustful manner of the time, and like Galileo, concealed his discovery in an anagram, consisting in this case of 88 letters. It was Dominic Cassini who first saw the black stripes in the ring (1684), and recognised its division into at least two concentric rings. I have here brought together the information gained in the course of a century respecting the most wonderful and least anticipated of all the forms of celestial bodies with which we are yet acquainted; a form which has led to ingenious conjectures respecting the original mode of formation of the planets and satellites.

The spots on the sun were first observed through telescopes by John Fabricius of East Friesland, and by Galileo either at Padua or at Venice. In the publication of the discovery, Fabricius (June, 1611) was certainly a year in advance of Galileo (first letter to the burgomaster Marcus Welser, May 4, 1612.) The first observations of Fabricius appear, by Arago's careful researches, (489) to have been made in March 1611, or, according to Sir David Brewster, even at the close of the preceding year; while Christopher Scheiner does not himself refer his observations to an earlier period than April 1611, and probably did not begin to occupy himself in earnest with the solar spots until the month of October of the same year. Respecting Galileo we have only obscure and discordant information. He was acquainted with the solar spots in April 1611, for he shewed them publicly at Rome, in the garden of the Cardinal Bandini, on the Quirinal, in April and May of that year. Harriot, to whom Baron Zach attributes the discovery of the solar spots (16th Jan. 1610 !) did indeed see three of them on the 8th of December, 1610, and marked their position in a register of observations; but he was not aware that they were solar spots, as Flamstead, on the 23d of December, 1690, and Tobias Mayer, on the 25th of September, 1756, did not recognise Uranus as a planet when seen in their telescopes. Harriot first recognised them as solar spots Dec. 1, 1611, five months after Fabricius had published his discovery. Galileo remarked thus early, that the solar spots, "of which many are larger than the Mediterranean, and even than Africa and Asia," occupy a distinct zone in the sun's disk. He noticed that the same spots sometimes returned, and was persuaded that they belonged to the sun itself. The differences in their dimensions at the centre of the disk, and when near disappearing at the margin, particularly arrested his attention; but I do not find, in the remarkable second letter to Marcus Welser (Aug. 14, 1612), anything that could be interpreted to indicate that he had observed the inequality of the ashy coloured border at the two sides of the black nucleus when approaching the limb of the sun (Alexander Wilson's fine remark in 1773 !) The Canon Tarde, in 1620, and Malapertus, in 1633, ascribed all obscurations of the sun to small revolving cosmical bodies which intercepted his light, and to which the names of Borbonia and Austriaca Sidera were given. [490] Fabricius recognised, like Galileo, that the spots belong to the sun itself; [491] he also saw that spots which he had observed disappeared and returned again; and these phænomena taught him the rotation of the sun, which Kepler had conjectured before the discovery of the spots.

The most exact determinations of the period of rotation were made by the diligent Scheiner (1630). Since the strongest light which man has yet been able to produce, Drummond's incandescent lime, appears of an inky black when projected upon the sun's disk, we need not wonder that Galileo, who doubtless first described the great solar faculæ, should have considered the light of the nuclei of the solar spots to be more intense than that of the full moon, or of the atmosphere near the solar disk. (492) Fancies respecting the many envelopes of air, cloud, and light surrounding the black earth-like nucleus of the sun, may be found in the writings of Cardinal Nicolaus of Cuss, in the middle of the 15th century. (493)

The cycle of admirable discoveries which scarcely occupied two years, and in which the immortal name of the Florentine shines foremost, was completed by the observation of the phases of Venus. As early as 1610 Galileo saw the sickle or crescent-form of the planet, and, according to a practice already alluded to, concealed the important discovery in an anagram, which Kepler recals in the preface to his Dioptrica. He says also, in a letter to Benedetto Castelli (Dec. 30, 1610), that he thinks he has recognised changes in the enlightened disk of Mars, notwithstanding the small power of his telescope. The discovery of the moon-like crescent shape of Venus was the triumph of the Copernican system. The necessity of the existence of these phases could certainly not have escaped the founder of that system; he discusses in detail, in the tenth chapter of his first book, the doubts which the later adherents of the Platonic opinions had raised against the Ptolemaic system on account of the moon's phases. But in the development of his own system he makes no particular remark respecting the phases of Venus, as in Thomas Smith's Optics he is stated to have done.

These enlargements of cosmical knowledge, (the description of which cannot be kept entirely free from the unhappy contests respecting claims of priority in discovery), like all that belongs to physical astronomy, excited more general interest than might otherwise have been the case, from the invention of the telescope (1608) having occurred at a period when popular attention had been roused by three great and surprising events in the regions of space: I allude to the sudden appearance and extinction of three new stars; one in Cassiopea in 1572, one in Cygnus in 1600, and one in the foot of Ophiuchus in 1604. All these surpassed in brightness stars of the first magnitude; and that which Kepler observed in Cygnus continued to shine in the vault of heaven for twenty-one years, through the whole period of Galileo's discoveries. Almost three centuries and a half have since elapsed, and no new star of the first or second magnitude has subsequently appeared; for the remarkable cosmical event witnessed by Sir John Herschel in the southern hemisphere in 1837, ([494]) was a great increase of luminous intensity in a long known star of the second magnitude (η Argûs), which had not until then been seen to be of variable brightness. The writings of Kepler, and the sensation produced at the present time by the appearance of comets visible to the naked eye, enable us to comprehend how powerfully the three new stars which appeared between 1572 and 1604 arrested curiosity—how much they increased the interest felt in astronomical discoveries, and what a stimulus they afforded to imaginative combinations. Striking terrestrial natural events, such as earthquakes in countries where they are rarely felt, the outbreak of volcanoes after long periods of repose, the rushing sound of aerolites which traverse our atmosphere and become suddenly heated in it, awaken for a time a lively interest in problems which appear even more mysterious to persons in general than to dogmatising philosophers.

In the foregoing remarks on the influence exerted by the direct visible contemplation of particular heavenly bodies, I have named Kepler more particularly, for the sake of recalling how, in this great, richly gifted, and extraordinary man, the love for imaginative combinations was united with a remarkable talent for observation, a grave and severe method of induction, a courageous and almost unexampled perseverance in calculation, and a depth of mathematical thought which, displayed in his Stereometria doliorum, exercised a happy influence on Fermat, and through him on the invention of the infinitesimal calculus. ([495]) The possessor of such a mind ([496]) was pre-eminently suited, by the richness and mobility of his ideas, and even by the boldness of the cosmological speculations which he hazarded, to promote and animate the movement which carried the 17th century uninterruptedly forward towards the attainment of its exalted object, the enlarged contemplation of the universe. The many comets visible to the naked eye from 1577 to the appearance of Halley's comet in 1607 (eight in number), and the apparition, almost within the same period, of the three new stars already spoken of, led to speculations in which these heavenly bodies were viewed as originating from, or being formed out of, a cosmical vapour filling the regions of space. Kepler, like Tycho Brahe, believed the new stars to have been condensed from this vapour, and redissolved into it again. (497) In his "new and strange discourse on long-haired stars," he represented comets also (to which, before the actual investigation of the elliptic orbits of the planets, he attributed a rectilinear not a closed or re-entering path), as formed from the "celestial air." He even added, in accordance with the old fancies of spontaneous generation, that comets were formed "like the herbs which grow without seed from the earth, and as fishes are produced from salt water by generatio spontanea."

More happy in his other cosmical anticipations, Kepler adventured the following propositions:—That the fixed stars are all suns like our own, surrounded by planetary systems; that our sun is enveloped in an atmosphere which shews itself as a white corona in total solar eclipses; that the situation of our sun in the great island of the universe to which it belongs is in the centre of the crowded ring of stars which forms the Milky Way; (498) that the sun rotates round its axis as do the planets and the fixed stars (this was before the discovery of the solar spots); that satellites, like those which Galileo had discovered revolving round Jupiter, would be discovered round Saturn (and round Mars); and that in the much too large interval (499) between Mars and Jupiter, where we are now acquainted with seven asteroids, (and also between Venus and Mercury), there moved planets, which their small size rendered invisible to the naked eye. Anticipatory annunciations of this nature—felicitous conjectures, which have been for the most part realised by subsequent discoveries—excited general interest; while none of Kepler's cotemporaries, not even Galileo, paid any just tribute of praise to the discovery of the three laws which, since Newton and the promulgation of the theory of gravitation, have immortalised the name of Kepler. (500) Cosmical speculations, even such as are not founded on observation, but only on faint analogies, then, as is still often the case, arrested attention more than the most important results of "calculating astronomy."

Having thus described the important discoveries which, in so small a cycle of years, enlarged the knowledge of the regions of space, I have still to recal the advances in physical astronomy which marked the second half of the great century of which we are treating. The improvement of telescopes occasioned the discovery of the satellites of Saturn. Huygens, with an object-glass polished by himself, first discovered one of them (the sixth), on the 25th of March, 1655, forty-five years after the discovery of Jupiter's satellites. From a prejudice which Huygens shared with several astronomers of the period, that the number of satellites or secondary planets could not exceed that of the larger or primary planets, (501) he did not seek to discover any more of the satellites of Saturn. Four of them, Sidera Lodovicea, were discovered by Dominic Cassini: the seventh, or outermost, which has great alternations of brightness, in 1671; the fifth in 1672; and the third and fourth in 1684, with an object-glass of Campani's having a focal length of 100—136 feet. The two innermost, or the first and second satellites, were discovered more than a century later (1788 and 1789), by William Herschel with his colossal telescope. The second satellite offers the remarkable phænomenon of performing its revolution round the principal planet in less than one of our days.

Soon after Huygens' discovery of a satellite of Saturn,

Childrey (1658—1661) discovered the Zodiacal Light, of which, however, the true relations in space were first determined by Dominic Cassini in 1683. Cassini regarded it not as a part of the solar atmosphere, but, like Schubert, Laplace, and Poisson, as a detached separately revolving nebulous ring. ([502]) Next to the demonstration of the existence of secondary planets or satellites, and of the detached and concentrically divided ring of Saturn, the discovery of the probable existence of the nebulous ring of the Zodiacal Light unquestionably constitutes one of the grandest enlargements in our view of the planetary system, which at first appeared so simple. In our own days the closely interwoven orbits of the small planets between Mars and Jupiter, the comets of short period which remain within our system (the first of which was shewn to be such by Encke), and the showers of shooting stars occurring on particular days (if we may regard these bodies as small cosmical masses moving with planetary velocity), have enriched the view of our solar system with new and wonderfully varied objects of contemplation.

In the first part of the period of which we are treating, in the age of Kepler and Galileo, great additions were also made to the view of the contents of space, or of the distribution of the material creation beyond the outermost planetary orbit, and beyond the path of any comet. In the same period (1572—1604) in which three new stars of the first magnitude appeared in Cassiopea, Cygnus, and Ophiuchus, David Fabricius, Protestant minister of Ostell in East Friesland (the father of the discoverer of the solar spots), in 1596, and Johann Bayer, at Augsburg, in 1603, remarked in the neck of Cetus a star which disappeared again, the varying brightness of which, however, as Arago has shewn in an important memoir on the history of astronomical discovery, (503) was first recognised by Johannes Phocylides Holwarda, professor at Franeker, in 1638 and 1639. Other phænomena of the same class were observed in the latter half of the 17th century; stars of periodically variable brilliancy were discovered in the head of Medusa, in Hydra, and in Cygnus. In the memoir of Arago in 1842 above referred to, it is very ingeniously shewn, how exact observations of the change of light of Algol might lead directly to the determination of the velocity of the light of that star.

The use of the telescope now stimulated astronomers to the observation of another class of phænomena, some of which could not escape the notice even of the unassisted eye. Simon Marius described the nebula in Andromeda in 1612, and in 1656 Huygens drew a sketch of the nebula in the sword of Orion. These two nebulæ may serve as types of different states of condensation, more or less advanced, of the nebulous cosmical matter. Marius, in comparing the nebula of Andromeda with the light of a taper seen through a semi-transparent substance, indicates very appropriately the difference between it and the groups or clusters of stars examined by Galileo in the Pleiades and in Cancer. As early as the commencement of the 16th century, Spanish and Portuguese navigators, though without the advantage of telescopic vision, had observed and admired the two Magellanic luminous clouds which revolve round the southern pole, and of which one, as we have already remarked, was known as the "white patch," or "white ox," of the Persian astronomer Abdurrahman Sofi, in the middle of the

10th century. Galileo, in the Nuncius Siderius, employs the appellations "stellæ nebulosæ," and "nebulosæ," to denote clusters of stars, which, as he expresses it, like "areolæ sparsim per æthera subfulgent." As he bestowed no particular attention on the nebula of Andromeda, which is visible to the naked eye but has not yet shewn any stars even under the highest magnifying powers, he regarded all nebulous appearances, all his nebulosæ, as being like the Milky Way, masses of light formed of closely crowded stars. He did not distinguish between nebula and star, as Huygens did in the case of the nebula of Orion. Such were the first commencements of the great works on nebulæ, which have so honourably occupied the first astronomers of our age in both hemispheres.

Although the 17th century owed its chief splendour, at its commencement, to the sudden enlargement by Galileo and Kepler of the knowledge of the celestial spaces, and, at its close, to Newton and Leibnitz's advances in pure mathematical knowledge, yet it was not without a beneficial influence on the greater part of the physical problems in which we are engaged at the present day. In order not to depart from the character of this history of the contemplation of the universe, I merely mention the works which exercised a direct and essential influence on general or cosmical views of nature. In reference to Light, Heat, and Magnetism, we must name first Huygens, Galileo, and Gilbert. When Huygens was occupied with the double refraction of light in crystals of Iceland spar, *i. e.* with the separation of the pencils of light into two parts, he also discovered, in 1678, that kind of polarisation of light which bears his name. More than a century elapsed before the discovery of this insulated phænomenon (which was not published until 1690, within five years of Huygens' death) was followed by the great discoveries of Malus, Arago, Fresnel, Brewster [504] and Biot. Malus, in 1808, discovered polarisation by reflection from polished surfaces; and Arago, in 1811, discovered coloured polarisation. A world of wonders —of variously modified waves of light gifted with new properties, was now opened. A ray of light which reaches our eyes from the regions of space, from a heavenly body many millions of miles distant, when received in Arago's polariscope, tells as it were of itself whether it is reflected or refracted, whether it emanates from a solid, a fluid, or a gaseous body, [505] and even announces its degree of intensity. Advancing in this path, which takes us back through Huygens to the 17th century, we are instructed respecting the constitution of the solar orb and its envelopes,—the reflected or the proper light of the tails of comets and of the Zodiacal Light,—the optical properties of our atmosphere, and the position of the four neutral points of polarisation, [506] which Arago, Babinet, and Brewster discovered. Thus man makes for himself, as it were, new organs, which, when skilfully used, open to him new views of nature.

We should next name, by the side of the polarisation of light, the most striking of all the phænomena of optics—the phænomenon of "interferences," faint indications of which were also observed in the 17th century, though without any understanding of their causal conditions, by Grimaldi, in 1665, and by Hooke. [507] Our own time is indebted for the discovery of these conditions, and the clear recognition of the laws according to which rays of light (unpolarised), when they proceed from one and the same source, but with a different length of path, destroy each other and produce darkness, to the acute and successful penetration of Thomas Young. The laws of the interference of polarised light were discovered in 1816, by Arago and Fresnel. The theory of undulations, advanced by Huygens and Hooke, and defended by Euler, at last found a firm basis.

But if the latter half of the 17th century was distinguished by an important enlargement of optical knowledge, in the attainment of an insight into the nature of double refraction, it has been invested with a far higher splendour by Newton's experimental researches, and by Olaus Römer's discovery (in 1675) of the measurable velocity of light; a discovery which enabled Bradley, half a century later (in 1728), to regard the variation which he found in the apparent place of the stars as a consequence of the movement of the earth in her orbit combined with the propagation of light. Newton's Optics appeared in 1704, not being published in English for personal reasons until two years after Hooke's death; but this magnificent work may be regarded as belonging to the 17th century, for we are assured that, even previously to the years 1666 and 1667, its great author was in possession (508) of the essential points of his optical discoveries, of his theory of gravitation, and of the method of fluxions.

In order not to break the links of the common bond which unites the general "primitive phænomena of matter," I place here, immediately after the above brief notice of Huygens, Grimaldi, and Newton, considerations on terrestrial magnetism and atmospheric temperature,—so far at least as the foundations of these studies were established in the century which it is the object of this section to describe.

The most ingenious and important work on electric and magnetic forces, William Gilbert's Physiologia nova de Magnete, of which I have already several times had occasion to speak, (509) was published in 1600. Gilbert, whose sagacious mind was so highly admired by Galileo, (510) anticipated by his conjectures much of our present knowledge. He regarded magnetism and electricity as two emanations of one fundamental force pervading all matter, and therefore treated of both at once. Such obscure anticipations, founded on analogies of the attracting power of the Heraclean magnetic stone on iron, and of amber (when animated, as Pliny says, with a soul by warmth and friction) on dry straws, have been common to all periods, and even to the most different races; for they were shared by the followers of the Ionic philosophy of nature, and by Chinese physicists. (511) William Gilbert regarded the earth itself as a magnet, and the lines of equal declination and inclination as having their inflections determined by distribution of mass, or by the form of continents and the extent of the deep intervening oceanic basins. It is difficult to reconcile the periodic variation which characterises the three elementary forms of the magnetic phænomena (the isoclinal, isogonic, and isodynamic lines) with this rigid system of distribution of force and mass, unless we imagine the attractive force of the material particles modified by similarly periodical variations in the interior of the globe.

In Gilbert's theory, as in gravitation, the quantity of material particles only is estimated, without regard to the specific heterogeneity of substances. This circumstance gave to his work, in the period of Galileo and Kepler, a character of cosmical grandeur. By the unexpected discovery of "rotation-magnetism" by Arago (1825), it has been practically proved that all kinds of matter are susceptible of magnetism; and Faraday's latest researches on diamagnetic substances have, under particular conditions of "axial or equatorial direction," and of solid, fluid, or gaseous inactive conditions of the bodies, confirmed this important result. Gilbert had so clear an idea of the imparting of the telluric magnetic force, that he already ascribed the magnetic state of iron bars in the crosses on old church towers or steeples to this circumstance. (512)

In the 17th century, by the increasing activity of navigation to the higher latitudes, and by the improvement of magnetic instruments, to which, since 1576, the dipping needle or inclinatorium, constructed by Robert Norman of Ratcliffe, had been added, a general knowledge of the progressive motion of a part of the magnetic curves—*i. e.* of the lines of no variation—was first obtained. The position of the magnetic equator (or line of no inclination), which was long believed to be identical with the geographical equator, was not examined. Observations of inclination were made only in a few of the principal cities of western and southern Europe: the intensity of the earth's magnetic force, which varies both with place and with time, was indeed attempted to be measured by Graham in London, in 1723, by the oscillations of a magnetic needle; but after the failure of Borda's endeavour on his last voyage to the Canaries in 1776, it was Lamanon who, in 1785, in the expedition of La Perouse, first succeeded in comparing the intensity in different regions of the earth.

Edmund Halley, availing himself of a great mass of existing observations of declination, of very unequal value (by

Baffin, Hudson, James Hall, and Schouten), sketched, in 1683, his theory of four magnetic poles or points of attraction, and of the periodical movement of the magnetic lines of no variation. In order to test this theory, and to render it more perfect by the aid of new and more exact observations, he was permitted by the English Government to make (1698—1702) three voyages in the Atlantic Ocean, in a ship of which he was given the command. On one of these voyages he proceeded as far as 52° south latitude. This undertaking forms an epoch in the history of terrestrial magnetism. A general "variation chart," or a chart on which the points at which the navigator had found the same amount of declination were connected by curved lines, was its result. Never before, I believe, did any Government equip a naval expedition for an object, which, whilst its attainment promised considerable advantages for practical navigation, yet so properly deserved to be entitled scientific or physico-mathematical.

As no phænomenon can be examined by an attentive investigator without being considered in its relation to others, Halley, as soon as he returned from his voyages, hazarded the conjecture that the Aurora Borealis is a magnetic phænomenon. I have remarked, in the picture of nature contained in the first volume of this work, that Faraday's brilliant discovery of the evolution of light by magnetism has raised this hypothesis, enounced in 1714, to the rank of an experimental certainty.

But if the laws of terrestrial magnetism are to be thoroughly sought out,—that is to say, if they are to be investigated in the great cycle of the periodical movement in geographical space of the three classes of magnetic curves,—

it is not sufficient that the diurnal, regular, or disturbed march of the needle should be observed at the magnetic stations, which, since 1828, have begun to cover a considerable portion of the earth's surface, both in northern and southern latitudes; ([513]) it would also be requisite to send four times in each century an expedition of three ships, which should have to examine, as nearly as possible at the same time, the state of magnetism over all the accessible parts of the globe which are covered by the ocean. The magnetic equator, or the line where the inclination is 0, must not merely be inferred from the geographical positions of its nodes (or intersections with the geographical equator), but the course of the ship should be made to vary continually, in accordance with the observations of inclination, so as never to quit the line forming the magnetic equator at that time. Land expeditions should be combined with the undertaking, in order, where masses of land cannot be entirely traversed, to determine exactly at what points of the coast the magnetic lines (and especially the lines of no variation) enter. The two isolated "closed systems" or ovals, in eastern Asia, and in the Pacific in the meridian of the Marquesas, ([514]) may, in their movements and gradual changes of form, be deserving of particular attention. Since the memorable antarctic expedition of Sir James Clark Ross (1839—1843), provided with excellent instruments, has thrown a great light over the high latitudes of the southern hemisphere, and determined empirically the place of the magnetic south pole, and since my honoured friend Friedrich Gauss has succeeded in establishing the first general theory of terrestrial magnetism, we need not abandon the hope that the many wants of science and of navigation will some day be satisfied by the execution of this plan so often desired by me. May the year 1850 deserve to be marked as the first normal epoch in which the materials of a "magnetic map of the world" shall be assembled; and may permanent scientific institutions impose on themselves the duty of reminding, every quarter of a century, a Government favourable to the prosperity and progress of navigation, of the importance of an undertaking the great cosmical value of which is attached to long-continued repetitions! ([514 bis].)

The invention of instruments for measuring temperature (Galileo's thermoscopes ([515]) of 1593 and 1602 were dependent concurrently on changes of temperature and on variations in the pressure of the external air) first gave rise to the idea of investigating the modifications of the atmosphere by a series of connected and successive observations. We learn from the Diario of the Academia del Cimento,—which, during the short continuance of its activity, exercised so happy an influence on the disposition for experiments and researches on a systematic plan,—that, as early as 1641, observations of temperature were made five times a day at many stations, ([516]) with spirit thermometers similar to our own; at Florence, at the Convent degli Angeli, in the plains of Lombardy, in the mountains near Pistoia, and even in the elevated plain of Innspruck. The Grand Duke Ferdinand II. charged the monks of many convents in his states with this task. ([517]) The temperatures of mineral springs were also determined, giving occasion to many questions respecting the temperature of the earth. As all telluric natural phænomena, *i. e.* all the alterations which terrestrial matter undergoes, are connected with modifications of heat, light, and electricity, either in repose or moving in currents,—and as the phænomena of temperature operating by expansion are most accessible to visible perception and cognizance, it follows that, as I have elsewhere observed, the invention and improvement of thermometric instruments marks an important epoch in the progress of the general knowledge of nature. The field of application of the thermometer, and the conclusions founded on its indications, are commensurate with the domain of those forces or powers of nature which exert their dominion alike in the aerial ocean, on the dry land, in the superimposed aqueous strata of the sea, and in inorganic substances, as well as in the chemical and vital processes of organic tissues.

More than a century previous to Scheele's extensive labours, the action of radiant heat was also investigated by the Florentine members of the Academia del Cimento, by remarkable experiments made with concave mirrors, towards which, non-luminous heated bodies, and masses of ice of 500lbs. in weight, radiated actually and apparently. [518] Mariotte, at the close of the 17th century, investigated the relations of radiant heat in its passage through glass plates. I have here recalled these detached experiments, because, since that period, the doctrine of the "radiation of heat" has thrown considerable light on the cooling of the ground, the origin of dew, and many general climatic modifications, and through Melloni's admirable sagacity, has even conducted to the contrasted diathermism of rock salt and alum.

With investigations on the variations of atmospheric temperature, coincident with changes of latitude, season and elevation, were soon associated others respecting the variations of pressure, and of the quantity of vapour in the atmosphere; as well as respecting the often observed periodical succession of the winds, or the "law of rotation" of the wind. Galileo's just views of atmospheric pressure conducted Torricelli, a year after the death of his great teacher, to the construction of the barometer. That the column of mercury in the Torricellian tube stood higher at the foot of a tower or of a hill, than on its summit, would appear to have been first remarked at Pisa, by Claudio Beriguardi; (519) and was observed five years later in France by Perrier, who, at the request of his brother-in-law, Pascal, ascended the Puy de Dôme, a mountain 840 French feet higher than Vesuvius. The idea of employing the barometer for the measurement of heights now presented itself readily; it may possibly have been first awakened in Pascal's mind by a letter from Descartes. (520) It is unnecessary to explain at length all that the barometer employed as a hypsometric instrument for the determination of differences of elevation upon the surface of the earth, and as a meterological instrument for investigating the influence of currents of air, has contributed to the extension of physical geography and meteorological knowledge. The foundations of the theory of the currents of the atmosphere were laid before the close of the 17th century. Bacon in 1644, in his celebrated "Historia naturalis et experimentalis de ventis," (521) had the merit of considering the direction of winds in connection with temperature and aqueous precipitations; but unmathematically denying the truth of the Copernican system, he reasoned on the possibility "that our atmosphere may turn daily round the Earth like the heavens, and may thus occasion the East wind."

Hooke's comprehensive genius acted here also as the restorer of light and order. (522) He recognised the influence of the Earth's rotation, as well as the existence of upper and lower currents of warm and cold air, passing from the equator to the poles, and returning from the poles to the equator. Galileo, in his last Dialogo, had indeed also considered the trade winds as a result of the Earth's rotation; but he ascribed the remaining behind of the particles of air within the tropics to a vapourless purity of the air in those regions. (523) Hooke's juster view was not revived until the 18th century, when it was again put forward by Halley, and explained more circumstantially and satisfactorily in regard to the operation of the velocity of rotation proper to each parallel of latitude. Halley had been previously led by his long sojourn in the torrid zone to publish an excellent work on the geographical extension of the trade winds and monsoons. It is surprising that in his magnetic expeditions he makes no mention of the "law of the winds"—so important for the whole of meteorology,—as its general features had been recognised by Bacon, and by Johannes Christian Sturm of Hippolstein, who, according to Brewster, (524) was the true discoverer of the differential thermometer.

In the brilliant period of the foundation of "mathematical natural philosophy," attempts to investigate the moisture of the atmosphere in its connection with variations of temperature, and with the direction of the wind, were not wanting. The Academia del Cimento conceived the happy idea of determining the quantity of vapour by evaporation and precipitation. The oldest Florentine hygrometer was accordingly a condensation hygrometer, an apparatus in which the quantity of precipitated water which ran off was determined by its weight. (525) To this condensation hygrometer, which, aided by the ideas of Le Roy, has gradually led in our own days to the exact psychrometric methods of Dalton, Daniell, and Auguste, there were added, according to the example previously set by Leonardo da Vinci, (526) the absorption hygrometers made of animal or vegetable substances, of Santori (1625), Torricelli (1626), and Molineux. Catgut, and the beards of a wild oat, were used almost at the same time. Instruments of this kind, founded on the absorption of the aqueous vapour contained in the atmosphere by organic substances, were provided with indexes and counterpoises, and were very similar in construction to Saussure's and Deluc's hair and whalebone hygrometers; but the instruments of the 17th century were deficient in the determination of fixed dry and wet points, so necessary for the comparison and understanding of the results. This desideratum was at last supplied by Regnault, but without reference to the variation which might be occasioned by time in the susceptibility of the hygrometric substances employed. Pictet, (527) however, found that the hair of a Guanche mummy from Teneriffe, which might be a thousand years old, employed in a Saussure's hygrometer, still possessed a satisfactory degree of sensibility.

Electric action was recognised by William Gilbert as the operation of a natural force or power allied to magnetism. The book in which this view was first enounced, and even in which the terms "electric force," "electric emanations," and "electric attraction" (528) were first employed, is the work to which I have already so often referred, published in 1600, and entitled "Physiology of Magnets, and of the Earth as a great Magnet" (de magno magnete tellure). "The faculty of attracting, when rubbed, light substances, whatever may be their nature, does not," says Gilbert, "belong exclusively to amber, which is a condensed earth-juice thrown up by the waves of the sea, and in which flying insects, ants, and worms, are inclosed as in perpetual tombs, (æternis sepulchris). The attracting power belongs to a whole class of very different substances; such as glass, sulphur, sealing wax and all resins, rock crystal, and all kinds of precious stones, alum and rock salt." The strength of the electricity excited was measured by Gilbert by means of an iron needle (not very small), moving freely on a point (versorium electricum): very similar to the apparatus employed by Haüy and by Brewster, in trying the electricity excited in different minerals by warmth and friction.

Gilbert says farther on, that "friction is found to produce more effect in dry than in damp air, and that rubbing with silk is most advantageous. The terrestrial globe is held together as by an electric force (?) (Globus telluris per se electrice congregatur et cohæret); for the electric action tends to produce the cohesion of matter (motus electricus est motus coacervationis materiæ)." In these obscure axioms is expressed the view of a telluric electricity,—the manifestation of a force like magnetism belonging to matter as such. Nothing was yet said of repulsion, or of the difference between insulators and conductors.

The ingenious discoverer of the air-pump, Otto-von

Guerike, was the first who observed more than mere phenomena of attraction. In his experiments, made with a rubbed cake of sulphur, he recognised phenomena of repulsion, which afterwards led to a knowledge of the laws of the sphere of action and of the distribution of electricity. He heard the first sound and saw the first light in artificially elicited electricity. In an experiment made by Newton in 1675, the first traces of the "electric charge" in a rubbed plate of glass were seen. ([529]) We have here sought out only the first germs of the science of electricity, which, in its great and singularly retarded development, has not only become one of the most important parts of meteorology, but also, since we have learned that magnetism is one of the manifold forms in which electricity discloses itself, has cleared up to us so much belonging to the internal operation of terrestrial powers or forces.

Although Wall in 1708, Stephen Gray in 1734, and Nollet, conjectured the identity of friction electricity and of lightning, yet the experimental certainty was first attained about the middle of the 18th century by the successful endeavours of the illustrious Benjamin Franklin. From this epoch the electric process passed from the domain of speculative physics to that of the cosmical contemplation of nature—from the chamber of the student to the open field. The doctrine of electricity, like that of optics and of magnetism, has had long periods of exceedingly slow development, until in these three branches the labours of Franklin and Volta, Thomas Young and Malus, Oersted and Faraday, aroused their cotemporaries to an admirable activity. The progress of human knowledge is generally connected with such alternations of slumber and of suddenly awakened activity.

But if, as we have already remarked, by the invention of appropriate although still very imperfect physical instruments, and by the sagacity of Galileo, Torricelli, and the members of the Academia del Cimento, the relations of temperature, the variations of the atmospheric pressure, and the quantity of vapour in the air, became objects of immediate research; on the other hand, all that regards the chemical composition of the atmosphere remained wrapped in obscurity. The foundations of "pneumatic chemistry" were indeed laid by Johann Baptist, Van Helmont and Jean Rey, in the first half,—and by Hooke, Mayow, Boyle, and the dogmatising Becher in the latter half of the 17th century; but however striking was the correct apprehension of particular and important phenomena, yet the insight into their connection was wanting. The old belief in the elementary simplicity of the air which acts in combustion, in the oxydation of metals, and in respiration, formed an obstacle difficult to be overcome.

The inflammable or light-extinguishing kinds of gas occurring in caves and mines (the "spiritus letales" of Pliny), and the escape of these gases in the shape of bubbles in marshes and mineral springs, had already arrested the attention of the Erfurt Benedictine monk Basilius Valentinus, who probably belonged to the close of the 15th century, and of Libavius, an admirer of Paracelsus, in 1612. Comparisons were drawn between what was accidentally remarked in alchemistic laboratories, and what was seen to have been prepared in the great laboratories of nature, especially in the interior of the earth. Mining operations in beds rich in ore, (particularly such as contained pyrites which become heated by oxydation and contact electricity), led to anticipations of the chemical relations between metals, acids, and the air which gained access from without. Paracelsus, whose fancies belong to the epoch of the first conquests in America, already remarked the disengagement of gas when iron was dissolved in sulphuric acid. Van Helmont, who first made use of the word "gas," distinguishes gases from atmospheric air, and also, on account of their non-condensability, from vapours. He regards the clouds as vapours, which, when the sky is very clear, are changed into gas "by cold and by the influence of the heavenly bodies." Gas, he says, can only become water when it has previously been retransformed into vapour. These views of meteorological processes belonged to the first half of the 17th century. Van Helmont was not yet acquainted with the simple means of receiving and separating his "Gas sylvestre," (under which name he included all uninflammable gases different from pure atmospheric air, and incapable of supporting flame and respiration); yet he made a light burn in a vessel having its mouth in water, and remarked that as the flame went out, the water entered, and the "volume of air" diminished. Van Helmont also sought to demonstrate by determinations of weight, (which we find already in Cardanus), that all the solid parts of plants are formed from water.

The mediæval alchemistic opinions of the composition of metals, and of their combustion in air whereby their brilliancy was destroyed, incited to the examination of what took place during the process, and of the changes undergone by the metals themselves, and by the air in contact with them. Cardanus had already become aware in 1553 of the increase of weight that takes place during the oxidation of lead, and, quite in the spirit of the phlogistic hypothesis, had ascribed it to the escape of a "celestial fiery substance" causing levity; but it was not until eighty years afterwards, that Jean Rey, an exceedingly skilful experimenter at Bergerac, who had examined with great accuracy the increase of weight during the calcination of lead, tin and antimony, enounced the important result that the increase of weight was to be attributed to the accession of air to the metallic calx, saying, "Je responds et soustiens glorieusement que ce surcroît de poids vient de l'air qui dans le vase a ésté espessi." (530)

Men had now entered on the path which was to conduct to the chemistry of our days, and through it to the knowledge of a great cosmical phenomenon, the connection between the oxygen of the atmosphere and the life of plants. But the combination of ideas which next presented itself to distinguished men was of a singularly complicated nature. Towards the end of the 17th century there arose,—obscurely with Hooke in his Micrographia (1665), and more distinctly with Mayow (1669,) and Willis (1671),—a belief in the existence of nitro-aerial particles, (spiritus nitro-aereus, pabulum nitrosum),—identical with those which are fixed in saltpetre,—contained in the air and constituting the necessary condition of combustion. "It was stated that the extinction of flame in a close space does not take place from the air being over-saturated with vapours proceeding from the burning body, but that this extinction is a consequence of the entire absorption of the nitro-aerial particles ("spiritus nitro-aereus") which the air at first contained." The suddenly increased glow when melting saltpetre (emitting oxygen) is strewed upon the coals, and the exudation of saltpetre on clay walls in contact with the atmosphere, appear to have conduced to this opinion. According to Mayow, the respiration in animals, of which the production of animal heat, and the conversion of black into red blood are the result, the processes of combustion and the calcination of metals, are all dependent on these nitro-aerial particles of the atmosphere: in the antiphlogistic chemistry, they play nearly the part of oxygen. The cautiously doubting Robert Boyle recognised that the presence of a certain constituent of atmospheric air is necessary to the process of combustion; but he remained uncertain as to its nitrous nature.

Oxygen was to Hooke and Mayow an ideal object or a fiction of the imagination. The acute chemist and vegetable physiologist Hales, in 1727, first saw oxygen escape as gas in large quantities from the lead which he calcined under an intense heat. He saw the gas escape, but without examining its nature or remarking the vividness of the flame occasioned by it. Hales did not divine the importance of the substance which he had produced. The vivid evolution of light in bodies burning in oxygen gas, and its properties, were discovered, as many believe quite independently, (531) —by Priestley in 1772-1774, by Scheele in 1774-1775, and by Lavoisier and Trudaine in 1775.

The commencements of pneumatic chemistry have been touched upon in these pages in their historic connection, because, like the feeble beginnings of electric science, they prepared the way for the enlarged views, which the succeeding century has been able to form of the constitution of the atmosphere and of its meteorological variations. The idea of specifically-distinct gases was never perfectly clear to those who in the seventeenth century produced those gases. Men began again to attribute the difference between atmospheric air and the irrespirable, light-extinguishing, or inflammable gases, exclusively to the admixture of certain vapours. Black and Cavendish first shewed in 1766 that carbonic acid (fixed air) and hydrogen (combustible air) are specifically distinct aeriform fluids. So long had the ancient belief in the elementary simplicity of the atmosphere impeded the progress of knowledge. The final investigation of the chemical composition of the atmosphere, by a most accurate determination of the quantitative relations of its constituent parts by Boussingault and Dumas, is one of the brilliant points of modern meteorology.

The extension of physical and chemical knowledge, which has been here described in a fragmentary manner, could not remain without influence on the early progress of Geology. A great part of the geological questions with the solution of which our age is occupied, were stirred by a man of the most comprehensive knowledge, the great Danish anatomist Nicolaus Steno (Stenson) in the service of the Grand Duke of Tuscany, by an English physician Martin Lister, and by "Newton's worthy rival," ([532]) Robert Hooke. Steno's merits in respect to the superposition of rocks have been developed by me more fully in another work. ([533]) Previously to this period, and towards the end of the fifteenth century, Leonardo da Vinci, probably in laying out the canals in Lombardy which cut through alluvium and tertiary strata,—Fracastoro in 1517, on the occasion of seeing rocky strata containing fossil fish accidentally uncovered at Monte Bolca near Verona,—and Bernard Palissy in his investigations respecting fountains,—had recognised the traces of a former oceanic world of animal life. Leonardo, as if with the presentiment of a more philosophical division of animal forms, terms the shells "animali che hanno l'ossa di fuori." Steno, in his work on the substances contained in rocks, (de Solido intra Solidum naturaliter contento) (1669), distinguishes "rocky strata (primitive?), hardened before the existence of plants and animals, and, therefore, never containing organic remains, from sedimentary strata (turbidi maris sedimenta sibi invicem imposita), which alternate with each other and cover those other strata first spoken of. All deposited strata containing fossils were originally horizontal. Their inclination has arisen partly from the outbreak of subterranean vapours which the central heat (ignis in medio terræ) produces, and partly by the giving way of lower supporting strata. ([534]) The valleys are the result of the falling in, consequent on the removal of support."

Steno's theory of the formation of valleys is that of Deluc, whereas Leonardo da Vinci, ([535]) like Cuvier, considers the valleys as formed by the action of running water. In the geological character of the ground in Tuscany, Steno thought he recognised revolutions which must be attributed to six great natural epochs, (sex sunt distinctæ Etruriæ facies, ex præsenti facie Etruriæ collectæ): at six recurring periods the sea had broken in, and after continuing for a long time to cover the interior of the country, had withdrawn again within its ancient limits. Steno did not, however, regard all petrifactions as belonging to the sea; he distinguishes between pelagic and fresh-water petrifactions. Scilla, in 1670, gave drawings of the petrifactions or fossils of Calabria and Malta: our great zoologist and anatomist Johannes Müller has recognised among the latter the oldest drawing of the teeth of the gigantic Hydrarchus of Alabama (the Zeuglodon Cetoides of Owen), a mammal of the great order of the Cetaceæ: [536] the crown of these teeth is formed like those of seals.

Lister, as early as 1678, made the important statement, that each kind of rock is characterised by its own fossils, and that "the species of Murex, Tellina and Trochus, which are found in the quarries of Northamptonshire, do, indeed, resemble those of the present sea, but when closely examined are found to differ from them." "They are," he said, "specifically different." [537] In the then imperfect state of descriptive morphology, strict proofs of the justness of these grand anticipations or conjectures could not indeed be given. We here point out an early dawning and soon extinguished light, anterior to the great paleontological labours of Cuvier and Alexander Brongniart which have given a new form to the geology of the sedimentary formations. [538] Lister, attentive to the regular succession of strata in England, was the first who felt the want of geological maps. Although these phenomena in their connexion with ancient inundations (single or repeated) attracted interest and attention, and, mingling together belief and knowledge, produced in England the "systems" of Ray, Woodward, Burnet, and Whiston, yet, from the entire want of mineralogical distinction of the constituent parts of compound rocks, all that relates to the crystalline and massive eruptive rocks and their transformations remained unstudied. Notwithstanding the assumption of a central heat in the globe, earthquakes, thermal springs, and volcanic eruptions, were not regarded as the results of the reaction of the planet against its external crust, but were ascribed to such small local causes, as, for example, the spontaneous combustion of beds of pyrites. Even experiments made in sport by Lemery in the year 1700 exerted a long-continued influence on volcanic theories, although these might have been raised to more general views by the imaginative Protogæa of Leibnitz (1680).

The Protogæa, which is sometimes more poetic than the many metrical attempts of the same philosopher which have recently been brought to light, ([539]) teaches the scorification of the cavernous, glowing, and once self-luminous crust of the earth;—the gradual cooling of the heat-radiating surface enveloped in vapours;—the condensation, and precipitation into water, of the gradually cooled atmosphere of vapour;—the lowering of the sea by the sinking of its waters into internal hollows in the earth;—and finally the falling in of these caves or hollows causing the inclination of the strata. The physical part of these wild fancies offers some traits which, to the adherents of our modern and every way more advanced geological science, will not appear altogether deserving of rejection. Such are, the transference of heat in the interior of the globe, and the cooling by radiation from the surface; the existence of an atmosphere of vapour; the pressure exerted by these vapours upon the strata during their consolidation; and the double origin of the masses as either fused and solidified or precipitated from the waters. The typical character and mineral differences of rocks, *i. e.* the associations of certain substances, chiefly crystalline, recurring in the most distant regions of the earth, are as little spoken of in the Protogæa as in Hooke's geognostical views. In the last named writer, also, physical speculations on the operation of subterranean forces in earthquakes, in the sudden elevation of the bottom of the sea and of coast districts, and in the formation of mountains and islands, predominate. The nature of the organic remains of the ancient world even led Hooke to form a conjecture that the Temperate Zone must once have enjoyed the temperature of a tropical climate.

We have still to speak of the greatest of all geognostical phenomena, the Mathematical Figure of the Earth, in which we recognise as in a mirror the primitive condition of fluidity of the rotating mass, and its solidification into the present form of the terrestrial spheroid. The figure of the earth was sketched theoretically in its general outlines at the end of the seventeenth century, although the numerical ratio of the polar and equatorial diameters was not assigned with accuracy. Picard's measurement of a degree, executed with measuring instruments which he had himself improved (1670), is the more deserving of regard, because it first induced Newton to resume with renewed zeal the theory of gravitation, which he had already discovered in 1666 and had subsequently neglected: it offered to that profound and successful investigator, the means of demonstrating the manner in which the attraction of the earth maintained in her orbit the moon impelled onward by the centrifugal force. The much earlier recognised fact of the flattening of the poles of Jupiter (540) had, it is supposed, led Newton to reflect on the cause of such a departure from sphericity. The experiments on the length of the seconds' pendulum made at Cayenne by Richer in 1673, and on the west coast of Africa by Varin, had been preceded by others less decisive (541) made in London, Lyons, and Bologna, including a difference of 7° of latitude. The decrease of gravity from the pole to the equator, which had long been denied even by Picard, was now generally admitted. Newton recognised the compression of the earth at the poles as a result of its rotation: he even ventured, upon the assumption of homogeneity of mass, to assign the amount of the compression. It remained for the comparison of degrees measured in the eighteenth and nineteenth centuries under the equator, near the North Pole, and in the temperate zones of both hemispheres, to furnish a more correct deduction of the mean compression, or the true figure of the earth. As has already been remarked in the Picture of Nature in the first volume of the present work, (542) the existence of the compression announces of itself what may be termed the most ancient geognostical event, viz. the state of general fluidity of the planet, and its progressive solidification.

We commenced the description of the great epoch of Galileo and Kepler, Newton and Leibnitz, with the discoveries made in the celestial spaces by the aid of the newly invented teloscope; we terminate it with the figure of the earth as then recognised from theoretical considerations. "Newton attained to the explanation of the system of the Universe, because he succeeded in discovering the force (543) of whose operation the Keplerian laws are the necessary consequences, and which could not but correspond to the phenomena, since those laws corresponded to and foretold them." The discovery of such a force, the existence of which Newton has developed in his immortal work, the Principia, (which may be regarded as a general theory of Nature), was almost simultaneous with that of the Infinitesimal Calculus, which opened the way to new mathematical discoveries. The work of the intellect shews itself in its most exalted grandeur, where, instead of requiring the aid of outward material means, it receives its light exclusively from the pure abstraction of the mathematical development of thought. There dwells a powerful charm, deeply felt and acknowledged in all antiquity, in the contemplation of mathematical truths; in the eternal relations of time and space, as they disclose themselves in harmonics, numbers, and lines. (544) The improvement of an intellectual instrument of research—analysis, has powerfully promoted and advanced that mutual fructification of ideas, which is no less important than their abundant production. It has opened to us new regions of measureless extent in the physical contemplation of the Universe both in its terrestrial and celestial spheres, in the tidal fluctuations of the Ocean, as well as in the periodic perturbations of the planets.

VIII.

RETROSPECT OF THE PRINCIPAL EPOCHS IN THE CONTEMPLATION OF THE UNIVERSE.

Retrospective View of the Epochs or Periods which have been successively considered.—Influence of External Events on the Development of the Recognition of the Universe as a Whole.—Wide and varied Scope and close mutual Connexion of the Scientific Endeavours of modern times.—The History of the Physical Sciences gradually becomes coincident with that of the Cosmos.

I APPROACH the termination of a comprehensive and hazardous undertaking. More than two thousand years have been passed in review, from the earliest state of intellectual cultivation among the nations who dwelt round the basin of the Mediterranean and in the fertile river districts of Western Asia, to a period the views and feelings of which pass by almost imperceptible shades into those of our own age. I have sought to present the history of the gradually developed knowledge and recognition of the Universe as a whole, in seven distinctly marked sections, or as it were in a series of as many distinct pictures. Whether any measure of success has attended this attempt to maintain in their due subordination the mass of accumulated materials, to seize the character of the leading epochs, and to mark the paths in which ideas and civilisation have been conducted onwards, cannot be determined by him who, with a just mistrust of his remaining powers, knows only that the type of so great an undertaking has floated in clear, though general, outlines before his mental eye.

In the early part of the section occupied by the epoch of the Arabians, in beginning to describe the powerful influence exerted by the blending of a foreign element with European civilisation, I determined the period from which the history of the Cosmos becomes coincident with that of the physical sciences. According to my conception, an historical view of the gradual extension of natural knowledge, both in its terrestrial and celestial spheres, is connected with definite epochs, or with certain events which have exerted a powerful intellectual influence within definite geographical limits, and which impart to those epochs their peculiar character and colouring. Such were the enterprises which conducted the Greeks into the Euxine, and led them to anticipate the existence of another sea shore beyond the Phasis,—the expeditions to the tropical lands which furnished gold and incense;—the passage through the Western Straits into the Atlantic Ocean, and the opening of that great maritime highway of nations on which were discovered at widely separated intervals of time, Cerne and the Hesperides, the Northern Tin and Amber Islands, the Volcanic Azores, and the New Continent of Columbus south of the ancient Scandinavian settlements. The movements which proceeded from the basin of the Mediterranean, and from the northern extremity of the neighbouring Arabian Gulf, and the voyages to the Euxine and to Ophir, are followed in my historic description by the military expeditions of the Macedonian conqueror, and his attempt to fuse together the nations of the West and of the East,—by the operation of the Indian maritime commerce, and of the Alexandrian Institute under the Lagidæ;—by the Roman Universal Empire under the Cæsars;—and by the epoch of the Arabians, from whose attachment to the study of nature and of her powers, and especially to astronomical and mathematical knowledge, and to practical chemistry, great benefits were derived. The series of external events which suddenly enlarged the intellectual horizon, stimulating men to the research of physical laws, and animating them to the endeavour to rise to the ultimate apprehension of the Universe as a Whole, closed, according to my view, with those geographical discoveries,—the greatest ever achieved,—which placed the nations of the Old Continent in possession of an entire terrestrial hemisphere till then concealed. From thenceforward, as we have already remarked, the human intellect produces great results, no longer from the incitement of external events, but through the operation of its own internal power; and this simultaneously in all directions. Nevertheless, amongst the instruments which men formed for themselves, constituting as it were new organs augmenting their powers of sensuous perception, there was one which acted like a great and sudden event. By the space-penetrating power of the telescope, a considerable portion of the heavens was explored as it were at once; the number of known celestial bodies was increased, and their form and orbits began to be determined. Mankind now first entered on the possession of the "celestial sphere" of the Cosmos. It appeared possible to found a seventh section of the history of the physical contemplation of the Universe, on the importance of these occurrences, and on the unity of the endeavours which the employment of the telescope called forth. If we compare with the discovery of this optical instrument, another great discovery belonging to a very recent period,—that of the voltaic pile,—and the influence which it has exercised on the ingenious electro-chemical theory,—on the production of the metals of the earths and alkalies,—and on the long sought discovery of electro-magnetism;—we arrive at a series of phenomena called forth at will, which in many directions enter deeply into the knowledge of the dominion of the powers of nature; but which may rather seem to form a section in the history of the Physical Sciences, than to belong directly to the history of the contemplation of the Cosmos. The multiplied connections which link together the different branches of our modern science, render it more difficult to distinguish and circumscribe them. We have even seen, most recently, electro-magnetism acting upon the direction of the polarised ray of light, and producing modifications like chemical mixtures. Where, through the mental labours of the age, the progressive development of knowledge is so rapid, it is no less dangerous to attempt to lay a daring hand on the intellectual process, and to paint that which is incessantly advancing as if the goal were already attained, than it is for one sensible of his own limited powers, to venture to pronounce on the relative importance of the honourable efforts of those still living or recently departed.

In the historical considerations, describing the earlier germs of our natural knowledge, I have, in almost all cases, indicated the latest degree of development to which they have attained. The third and last portion of my work is designed to furnish towards the elucidation of the general picture of nature, contained in the first volume, those results of observation on which the present state of our scientific opinions is principally founded. Much, which according to other views than mine of the composition of a book of nature may have appeared wanting, will there find its place. Excited by the brilliancy of new discoveries, and fed with hopes of which the delusiveness is often not discovered till late, every age dreams that it has approached near to the culminating point of the knowledge and comprehension of nature. I doubt whether upon serious reflection such a belief will really appear to enhance the enjoyment of the present. A more animating conviction, and one more suitable to the idea of the destinies of our race, is, that the possessions yet achieved are but a very inconsiderable portion of those which, in the advance of activity and of general cultivation, mankind in their freedom will attain in succeeding ages. In the unfailing connection and course of events, every successful investigation becomes a step to the attainment of something beyond.

That which has especially promoted the progress of knowledge in the 19th century, and has formed the chief character of the age, is the general and highly useful endeavour, not to limit our regards to that which has been just achieved, but to test rigidly by weight and measure all earlier as well as more recent acquisitions; to distinguish between mere inferences from analogies, and certain knowledge; and to subject to the same severe critical method all departments of knowledge, physical astronomy, the study of the telluric powers or forces of nature, geology, and the study of antiquity. The generality of this method of criticism has especially contributed to shew on each occasion the boundaries of the several sciences, and to discover the weakness of certain systems, in which unfounded opinions or conjectures assume the place of facts, and symbolising myths present themselves as grave theories. Vagueness of language, and the transference of the nomenclature of one science to another, have conducted to erroneous views and delusive analogies. The progress of zoology was long endangered by its being believed that, in the lower classes of animals, all the vital actions must be attached to organs similar in form to those of the highest classes; and the knowledge of the development of vegetation in what have been called the Cryptogamic Cormophytes (mosses, liverworts, ferns and lycopodiums), or in the still lower Thallophytes (sea weeds, lichens and fungi) has been still more obscured, by the expectation of finding everywhere analogies to the sexual propagation of the animal kingdom. (545) If art and poetry, dwelling within the magic circle of the imagination, belong rather to the inner powers of the mind,—the extension of knowledge, on the other hand, rests by preference on contact with the external world; and this contact becomes closer and more varied as the intercourse between different nations increases. The creation of new organs or instruments of observation augments the intellectual, and often also the physical powers of man. More rapid than light, the closed electric current now carries thought and will to the remotest distance. Forces, whose silent operation in elementary nature, as well as in the delicate cells of organic tissues, still escapes the cognizance of our senses, will one day become known to us; and called into the service of man, and awakened by him to a higher degree of activity, will be included in a series of indefinite extent, through the medium of which, the subjection of the different domains of nature, and the more vivid understanding of the Universe as a Whole, are brought continually nearer.

NOTES.

(1) p. 4.—Kosmos, Bd. i. S. 50 (English edition, Vol. i. p. 43).

(2) p. 5.—See my Relation historique du Voyage aux Régions équin. T. i. p. 208.

(3) p. 5.—Dante, Purg. i. 25—28:

"Goder pareva il ciel di lor fiammelle:
O settentrional vedovo sito,
Poi che privato se' di mirar quelle."

(4) p. 6.—Schiller's sämmtliche Werke, 1826, Bd. xviii. S. 231, 473, 480, and 486; Gervinus, neuere Gesch. der poet. National-Litteratur der Deutchen, 1840, Bd. i. S. 135; Adolph Becker im Charikles, Th. i. S. 219. Compare therewith Edward Müller über Sophokleische Naturanschauung, und die tiefe Naturempfindung der Griechen, 1842, S. 10 und 26.

(5) p. 7.—Schnaase, Geschichte der bildenden Künste bei den Alten, Bd. ii. 1843, S. 128—138.

(6) p. 8.—Plut. de El. apud Delphos, c. 9. Compare on a passage of Apollonius Dyscolus of Alexandria (Mirab. Hist. c. 40), Otfried Müller's last work, Gesch. der griech. Litteratur, Bd. i. 1845, S. 31.

(7) p. 8.—Hesiodi Opera et Dies, v. 502, 561; Göttling, in Hes. Carm. 1831, p. xix.; Ulrici, Gesch. der hellenischen Dichtkunst, Th. i. 1835, S. 337; Bernhardy, Grundriss der griech. Litteratur, Th. ii. S. 176; Gottfried Hermann (Opuscula, Vol. vi. p. 239) remarks, that Hesiod's picturesque description of winter has all the indications of great antiquity.

(8) p. 8.—Hes. Theog. v. 233—264. May not the name of the Nereid Mära (Od. xi. 326; Il. xviii. 48) express the phosphoric flashing of the sur-

face of the sea, as the same name, Μαιρα, expresses the sparkling dog-star Sirius?

([9]) p. 8.—Compare Jacobs, Leben und Kunst der Alten, i. Abth. i. S. vii.

([10]) p. 9.—Ilias, viii. 555—559; iv. 452—455; xi. 115—199. Compare also the accumulated but animated descriptions taken from the animal world which precede the review of the army, ii. 458—475.

([11]) p. 10.—Od. xix. 431—445; vi. 290; ix. 115—199. Compare the "verdant overshadowing grove" near Calypso's cave, "where an immortal might linger with admiration, and gaze with cordial delight," v. 55—73; the breakers at the Pheacian Islands, v. 400—442; and the gardens of Alcinous, vii. 113—130. On the vernal dithyrambus of Pindar, see Böckh, Pindari Opera, T. ii. P. ii. p. 575—579.

([12]) p. 11.—Œd. Kolon, v. 668—719. Amongst descriptions of scenery disclosing a deep feeling for nature, I would instance those of Cithæron, in the Bacchæ of Euripides, v. 1045, when the messenger emerges from the valley of Asopos (see Leake, Northern Greece, Vol. ii. p. 370); of the sunrise in the Delphic valley, in the Ion of Euripides, v. 82; and the picture, in gloomy colours, of the aspect of the sacred Delos, "surrounded by hovering sea-gulls, and scourged by the stormy waves" in Callimachus, in the Hymn. on Delos, v. 11.

([13]) p. 11.—According to Strabo (Lib. viii. p. 366, Casaub.), where he accuses the tragedian of giving to Elis a boundary geographically incorrect. This fine passage of Euripides is from the Cresphontes. The description of the excellence of the country of Messenia is closely connected with the exposition of political circumstances (the division of the territory among the Heraclides). Here, therefore, as Böckh has well remarked, the description of nature is connected with human affairs.

([14]) p. 13.—Meleagri Reliquiæ, ed. Manso, p. 5. Compare Jacobs, Leben und Kunst der Alten, Bd. i. Abth. i. S. xv.; Abth. ii. S. 150—190. Zenobetti, in the middle of the eighteenth century, supposed himself the first discoverer of Meleager's poem on the Spring (Mel. Gadareni in Ver Idyllion, 1759, p. 5). See Brunckii Anal. T. iii. p. 105. There are two fine sylvan poems by Marianos in the Anthol. Græca, ii. 511 and 512. Meleager's poetry is strongly contrasted with the praises of spring in the Eclogues of Himerius, a sophist and teacher of rhetoric in Athens under Julian. The style of Himerius is generally ornate and cold, but in particular parts, and especially in his form of description, he sometimes comes very near the modern manner of contemplating the universe. Himerii Sophistæ Eclogæ et Declamationes, ed. Wernsdorf, 1790 (Oratio iii. 3—6, and xxi. 5). The magnificent situation of Constantinople could not inspire the sophists (Orat. vii. 5—7, and xvi. 3—8). The passages of Nonnus referred to in the text are found in Dionys. ed. Petri Cunæi, 1610, Lib. ii. p. 70; vi. p. 199; xxiii. p. 16 and 619; xxvi. p. 694. Compare also Ouwaroff, Nonnos von Panopolis, der Dichter, 1817, S. 3, 16 und 21.

(15) p. 13.—Æliani Var. Hist. et Fragm. Lib. iii. cap. i. p. 139, Kuhn. Compare A. Buttmann, Quæst. de Dicæarcho, Naumb. 1832, p. 32, and Geogr. gr. min. ed. Gail. Vol. ii. p. 140—145. We find in the tragic poet Chæremon, a remarkable love of nature, and especially a fondness for flowers, which Sir William Jones has noticed as resembling that of the Indian poets: see Welcker, griechische Tragödien, Abth. iii. S. 1088.

(16) p. 14.—Longi Pastoralia (Daphnis et Chloe, ed. Seiler, 1843), Lib. i. 9; iii. 12; and iv. 1—3; p. 92, 125, and 137. See Villemain sur les romans grecs, in his Mélanges de Littérature, T. ii. p. 435—448, where Longus is compared with Bernardin de St.-Pierre.

(17) p. 14.—Pseudo-Aristot. de Mundo, c. 3, 14—20, p. 392, Bekker.

(18) p. 14.—See Stahr's Aristoteles bei den Römern, 1834, S. 173—177; and Osann, Beiträge zur griech. und röm. Litteraturgeschichte, Bd. i. 1835, S. 165—192. Stahr conjectures (S. 172), as does Heumann, that the present Greek is an altered version of the Latin text of Appuleius. The latter says distinctly (De Mundo, p. 250, Bip.), "that in the composition of his work he has kept in view Aristotle and Theophrastus."

(19) p. 14.—Osann, Beiträge zur griech. und röm. Litteraturgeschichte, Bd. i. S. 194—266.

(20) p. 14.—Cicero de Natura Deorum, ii. 37; a passage, in which Sextus Empiricus (Adversus Physicos, Lib. ix. 22, p. 554, Fabr.) adduces an expression of Aristotle's to the same effect, deserves the more attention, because he has alluded a short time before (ix. 20) to another lost work, on divination and dreams.

(21) p. 15.—"Aristoteles flumen orationis aureum fundens" (Cic. Acad. Quæst. ii. cap. 38.) (Compare Stahr, in Aristotelia, Th. ii. S. 161; and in Aristoteles bei den Römern, S. 53.)

(22) p. 16.—Menandri Rhetoris Comment. de Encomiis, ex rec. Heeren, 1785, § i. cap. 5, p. 38 and 39. The severe critic terms the didactic poem on Nature a "frigid" (ψυχρότερον) composition, in which the forces of nature are brought forward divested of their personality: Apollo is light, Hera the whole of the phænomena of the atmosphere, and Jove is heat. Plutarch also

ridicules the so-called poems of nature, which have only the mere external form of poetry (De Aud. Poet. p. 27, Steph.) The Stagirite (De Poet. c. i.) considers Empedocles rather a physiologist than a poet, having nothing in common with Homer, except the measure in which his verses are written.

(23) p. 16.—"It may seem strange to endeavour to connect poetry, which rejoices always in variety, form, and colour, with those ideas which are most simple and abstruse; but it is not the less correct. Poetry, science, philosophy, and history, are not in themselves, and essentially, divided from each other; they are united, either where man's particular stage of progress places him in a state of unity, or where the true poetic mood restores him to such a state (Wilhelm von Humboldt, gesammelte Werke, Bd. i. S. 98—102. Compare also Bernhardy, röm. Litteratur, S. 215—218, and Friedrich Schlegel's sammtliche Werke, Bd. i. S. 108—110. Cicero (ad Quint. fratrem, ii. 11) indeed ascribes to Lucretius, who Virgil, Ovid, and Quintilian, have praised so highly, more art than creative talent (ingenium).

(24) p. 17.—Lucret. Lib. v. V. 930—1455.

(25) p. 17.—Plato, Phædr. p. 230; Cicero de Leg. i. 5, 15, ii. 2, 1—3, ii. 3, 6 (compare Wagner, Comment Perp. in Cic. de Leg. 1804, p. 6); Cic. de Oratore, i. 7, 28 (p. 15 Ellendt).

(26) p. 17.—See the excellent work of Rudolph Abeken, Rector of the Gymnasium at Osnabrück, published in 1835, under the title of Cicero in seinen Briefen, S. 431—434. The valuable addition relative to Cicero's birthplace is by H. Abeken, the learned nephew of the author, who was formerly chaplain to the Prussian embassy at Rome, and is now taking part in the important Egyptian expedition of Lepsius. Respecting the place of Cicero's birth, see also Valery, Voy. hist. en Italie, T. iii. p. 421.

(27) p. 18.—Cic. Ep. ad Atticum, xii. 9 and 15.

(28) p. 19.—The passages from Virgil adduced by Malte-Brun (Annales des Voyages, T. iii. 1808, p. 235—266) as being actual local descriptions, merely shew that the poet was acquainted with the productions of different countries: that he knew the saffron of Mount Tmolus, the incense of the Sabeans, the true names of several small rivers, and even the mephitic vapours which rise from a cavern in the Apennines near Amsanctus.

(29) p. 19.—Virg. Georg. i. 356—392, iii. 349—380; Æn. iii. 191—211, iv. 246—251, iv. 522—528, xii. 684—689.

(30) p. 20.—Kosmos, Bd. i. S. 252 and 453 (English edit. Vol. i. p. 230 and 434). As separate pictures of natural scenes, compare Ovid, Met. i. 568—576, iii. 155—164, iii. 407—412, vii. 180—188, xv. 296—306;

Trist. Lib. i. El. 3, 60, Lib. iii. El. 4, 49, El. 12, 15. Ex Ponto, Lib. iii. Ep. 7—9. Ross has remarked, as being one of the rarely occurring instances of individual pictures relating to a determinate locality, the pleasing description of a fountain on Mount Hymettus, beginning, "Est prope purpureos colles florentis Hymetti" (Ovid de Arte Am. iii. 687). The poet is describing the fountain of Kallia, celebrated in antiquity, and consecrated to Aphrodite, which issues forth on the western side of Hymettus, which is otherwise very deficient in waters (see Ross, Letter to Professor Vuros, in the griech. medicin. Zeitschrift, June 1837.)

(31) p. 20.—Tibullus, ed. Voss, 1811, Eleg. Lib. i. 6, 21—34; Lib. ii. 1, 37—66.

(32) p. 20.—Lucan, Phars. iii. 400—452 (Vol. i. p. 374—384, Weber.)

(33) p. 20.—Kosmos, Bd. i. S. 298 (English edit. Vol. i. p. 273).

(34) p, 21,—Idem. S. 455 (English edit. p. 436). The poem of Lucilius, entitled Ætna, is very probably part of a longer poem on the remarkable natural objects of the island of Sicily, and is ascribed by Wernsdorf to Cornelius Severus. I would refer to some passages deserving of particular attention: to the praises of general knowledge of nature considered as "the fruits of the mind," v. 270—280; the lava currents, v. 360--370 and 474—515; the eruptions of water at the foot of the volcano (?) v. 395; the formation of pumice, v. 425 (p. xvi.—xx. 32, 42, 46, 50, and 55, ed. Jacob. 1826).

(35) p. 21.—Decii Magni Ausonii Mosella, v. 189—199 (p. 15 and 44, Böcking.) Consult also v. 85—150 (p. 9—12), the notice of the fish of the Moselle, which is not unimportant as regards natural history, and has been made use of by Valenciennes; and a pendant to Oppian (Bernhardy, griech. Litt. Th. ii. S. 1049). The Orthinogonia and Theriaca of Æmilius Macer of Verona, which were imitated from the works of the Colophonian Nicander, and which have not come down to us, also belonged to the same dry didactic class of poems treating of natural productions. A natural description of the south coast of Gaul, contained in a poem by Claudius Rutilius Numatianus, a statesman under Honorius, is more attractive than the Mosella of Ausonius. Rutilius, driven from Rome by the irruption of the Barbarians, is returning to his estates in Gaul. Unfortunately we possess only a fragment of the second book of the poem which gives a narrative of his travels; and this leaves off at the quarries of Carrara. Vide Rutilii Claudii Numatiani de Reditu suo (e Roma in Galliam Narbonensem) libri duo, rec. A.W. Zumpt, 1840. p. xv. 31,

and 219 (with a fine map by Kiepert); Wernsdorf, Poetæ Lat. Min. T. v. P. i. p. 125.

([36]) p. 22.—Tac. Ann. ii. 23, 24; Hist. v. 6. The only fragment which we possess of the heroic poem in which Pedo Albinovanus, the friend of Ovid, sung the exploits of Germanicus, which was preserved by the rhetor Seneca (Suasor. i. p. 11, Bipont.), also describes the unfortunate navigation on the Amisia (Ped. Albinov. Elegiæ, Amst. 1703, p. 172). Seneca considers this description of the stormy sea more picturesque than any thing which the Roman poets had produced; remarking, however, "Latini declamatores in oceani descriptione non nimis viguerunt; nam aut tumide scripserunt aut curiose."

([37]) p. 22.—Curt. in Alex. Magno, vi. 16 (see Droysen, Gesch. Alexanders des Grossen, 1833, S. 265). In Lucius Annæus Seneca (Quæst. Natur. Lib. iii. c. 27—30, p. 677—686, ed. Lips. 1741), we find a remarkable description of the destruction of mankind, once pure, but subsequently defiled by sin, by an almost universal deluge. "Cum fatalis dies diluvii venerit,.........bis peracto exitio generis humana exstinctisque pariter feris in quarum homines ingenia transierant." Compare the description of chaotic terrestrial revolutions in the Bhagavata-Purana, Book iii. c. 17 (Burnouf, T. i. p. 441).

([38]) p. 23.—Plin. Epist. ii. 17, v. 6, ix. 7; Plin. Hist. Nat. xii. 6; Hirt, Gesch. der Baukunst bei den Alten, Bd. ii. S. 241, 291, and 376. The villa Laurentina of the younger Pliny was situated near the present Torre di Paterno, in the coast valley of La Palombara, east of Ostia (see Viaggio da Ostia a la Villa di Plinio, 1802, p. 9; and Le Laurentin, par Haudelcourt, 1838, p. 62.) A deep feeling for nature breaks forth in the few lines written by Pliny from Laurentinum to Minutius Fundanus: "Mecum tantum et cum libellis loquor. Rectam sinceramque vitam! dulce otium honestumque! O mare, o littus, verum secretumque (πουσειον)! quam multa invenitis, quam multa dictatis!" (i. 9.) Hirt was persuaded that the beginning in Italy, in the 15th and 16th centuries, of the artificial style of gardening, which has long been termed the French style, and contrasted with the freer landscape gardening of the English, is to be attributed to the desire of imitating what the younger Pliny had described in his letters (Geschichte der Baukunst bei den Alten, Th. ii. S. 366).

([39]) p. 24.—Plin. Epist. iii. 19; viii. 16.

([40]) p. 24.—Suet. in Julio Cæsare, cap. 56. The lost poem of Cæsar (Iter.) described the journey to Spain, when he led his army to his last military exploit from Rome to Cordova, by land, in twenty-four days, according to Suetonius, or in twenty-seven days according to Strabo and Appian; the remains of Pompey's party, defeated in Africa, having assembled in Spain.

(41) p. 24.—Sil. Ital. Punica, Lib. iii. V. 477.

(42) p. 24.—Idem. Lib. iv. V. 348; Lib. viii. V. 399.

(43) p. 25.—See, on elegiac poetry, Nicol. Bach, in the allg. Schul-Zeitung, 1829, Abth. ii. No. 134, S. 1097.

(44) p. 26.—Minucii Felicis Octavius, ex rec. Gron. Roterod. 1743, cap. 2 and 3 (p. 12—28), cap. 16—18 (p. 151—171).

(45) p. 26.—On the Death of Naucratius, about the year 357, see Basilii Magni Opp. omnia, ed. Par. 1730, T. iii. p. xlv. The Jewish Essenes, two centuries before the Christian era, led an anchoritic life on the western shores of the Dead Sea, "in intercourse with nature." Pliny says of them (v. 15), "mira gens, socia palmarum." The Therapeutes dwelt originally more in conventual communities, in a pleasant district near Lake Mœris (Neander, allg. Geschichte der christl. Religion und Kirche, Bd. i. Abth. i. 1842, S. 73 and 103.)

(46) p. 28.—Basilii M. Epist. xiv. p. 93, Ep. ccxxiii. p. 339. On the beautiful letter to Gregory of Nazianzum, and on the poetic tone of mind of Saint Basil, see Villemain de l'Eloquence chrétienne dans le quatrième Siécle, in his Mélanges historiques et littéraires, T. iii. p. 320—325. The Iris, on the banks of which the family of the great Basil had ancient possessions in land, rises in Armenia, flows through Pontus, and, after mingling with the waters of the Lycus, pours itself into the Black Sea.

(47) p. 28.—Gregorius of Nazianzum was not, however, so much charmed with the description of the hermitage on the banks of the Iris, but that he preferred Arianzus, in the Tiberina Regio, though termed, with dissatisfaction, by his friend an impure *βαραθρον*. See Basilii Ep. ii. p. 70; and the Vita Sancti Bas., p. xlvi., and lix. in the edition of 1730.

(48) p. 28.—Basilii Homil. in Hexæm. vi., and iv. 6 (Bas. Opp. omnia, ed. Gul. Garnier, 1839, T. i. p. 54 and 70). Compare therewith the expression of profound melancholy in the beautiful poem of Gregory of Nazianzum, entitled, "On the Nature of Man." (Gregor. Naz. Opp. omnia, ed. Par. 1611, T. ii. Carm. xiii. p. 85).

(49) p. 29.—The quotation from Gregory of Nyssa given in the text, consists of separate fragments closely translated. They will be found in S. Gregorii Nysseni Opp. ed. Par. 1615, T. i. p. 49 C, p. 589 D, p. 210 C, p. 780 C; T. ii. p. 860 B, p. 619 B, p. 619 D, p. 324 D. "Be thou gentle towards the emotions of melancholy," says Thalassius, in aphoristic sayings, which

were admired by his contemporaries. (Biblioth. Patrum, ed. Par. 1624, T. ii. p. 1180 C.)

(50) p. 29.—See Joannis Chrysostomi Opp. omnia, Par. 1838 (8vo.) T. ix. p. 687 A, T. ii. p. 821 A, and 851 E, T. i. p. 79. Compare also Joannis Philoponi, in cap i. Geneseos de creatione Mundi, libri septem, Viennæ Austr. 1630, p. 192, 236, and 272; and also Georgii Pisidæ Mundi opificium, ed. 1596, v. 367—375, 560, 933, and 1248. The works of Basil and of Gregory of Nazianzum early arrested my attention after I began to collect descriptions of nature; but I am indebted for all the excellent (German) translations from Gregory of Nyssa, Chrysostom, and Thalassius, to my old and always kind colleague and friend, M. Hase, Member of the Institute, and Conservator of the Bibliothèque du Roi, at Paris.

(51) p. 30.—On the Concilium Turonense, under Pope Alexander III., see Ziegelbauer, Hist. Rei litter. ordinis S. Benedicti, T. ii. p. 248, ed. 1754; on the Council at Paris of 1209, and the Bull of Gregory IX. of the year 1231, see Jourdain, Recherches crit. sur les traductions d'Aristote, 1819, p. 204—206. Heavy penances were attached to the reading of the physical books of Aristotle. In the Concilium Lateranense of 1139 (Sacror. Concil. nova collectio, ed. Ven. 1776, T. xxi. p. 528), monks were forbidden to exercise the art of medicine. Consult also the learned and interesting writing of the young Wolfgang von Göthe, entitled, "der Mensch und die elementarische Natur," 1844, S. 10.

(52) p. 32.—Fried. Schlegel, über nordische Dichtkunst, in his sammtlichen Werken, Bd. x. S. 71 and 90. I may cite farther, from the very early time of Charlemagne, the poetic description of the Thiergarten at Aix, enclosing both woods and meadows, which is given in the life of the great emperor, written by Angilbertus, Abbot of St. Riques. (See Pertz, Monum. Vol. i. p. 393—403).

(53) p. 33.—See, in Gervinus's Geschichte der deutschen Litt., Bd. i. S. 354—381, the comparison of the two epics, the poem of the Niebelungen, (describing the vengeance of Chriemhild, the wife of Siegfried), and that of Gudrun, the daughter of King Hetel.

(54) p. 34.—On the romantic description of the grotto of the lovers, in the Tristan of Gottfried of Strasburg, see Gervinus, in the work above referred to, Bd. i. S. 450.

(55) p. 35.—Vridankes Bescheidenheit, by Wilhelm Grimm, 1834, S. 50, and 128. All that refers to the German Volks-epos and the Minnesingers (from p. 33 to p. 36) is taken from a letter of Wilhelm Grimm to myself

(Oct. 1845). In a very old Anglo-Saxon poem on the names of the Runes, which was first published by Hickes, there is the following pleasing description of the birch tree:—"Beorc is beautiful in its branches: its leafy top rustles sweetly, moved to and fro by the air." The greeting of the light of day is simple and noble:—"The messenger of the Lord, dear to man, the glorious light of God, bringing gladness and confidence to rich and poor, beneficent to all!" See also Wilhelm Grimm, über deutsche Runen, 1821, S. 94, 225, and 234.

([56]) p. 36.—Jacob Grimm, in Reinhart Fuchs, 1834, S. ccxciv. (Compare also Christian Lassen, in his indischer Alterthumskunde, Bd. i. 1843, S. 296.)

([57]) p. 37.—On "the non-genuineness of the Ossianic songs, and of Macpherson's Ossian in particular," see a memoir by the ingenious translatress of the Volkspoesie of Servia (die Unächtheit der Lieder Ossian's und des Macpherson'schen Ossian's insbesondere, von Talvj, 1840). The first publication of Ossian by Macpherson was in 1760. The Fingalian songs are, indeed, heard in the Scottish Highlands, as well as in Ireland, but they have been carried to Scotland from Ireland, according to O'Reilly and Drummond.

([58]) p. 37.—Lassen, ind. Alterthumskunde, Bd. i. S. 412—415.

([59]) p. 38.—Respecting the Indian forest-hermits, Vanaprestiæ (Sylvicolæ) and Sramâni (a name which has been altered into Sarmani and Garmani), see Lassen, "de nominibus quibus veteribus appellantur Indorum philosophi," in the Rhein. Museum für Philologie, 1833, S. 178—180. Wilhelm Grimm thinks he recognises something of Indian colouring in the description of the magic forest in the "Song of Alexander," composed more than 1200 years ago by a priest, named Lambrecht, in immediate imitation of a French original. The hero comes to a wood, where maidens, adorned with supernatural charms, spring from large flowers, and he remains with them so long that both flowers and maidens fade away. (Compare Gervinus, Bd. i. S. 282, and Massmann's Denkmaler, Bd. i. S. 16.) These are the same as the maidens of Edrisi's oriental magic Island of Vacvac, called, in the Latin version of Masudi, Chothbeddin puellas vasvakienses. (Humboldt, Examen crit. de la Géographie, T. i. p. 53.)

([60]) p. 39.—Kalidasa lived at the court of Vikramaditya, about 56 years before our era. It is highly probable that the age of the two great heroic poems, Ramayana and Mahabharata, is much earlier than that of the appearance of Buddha, or much earlier than the middle of the sixth century before our era. (Burnouf, Bhagavata-Purana, T. i. p. cxi. and cxviii.; Lassen, ind. Alterthumskunde, Bd. i. S. 356 and 492.) George Forster, by the translation of Sacontala, *i. e.* by his tasteful presentation in a German garb of an English version by Sir William Jones (1791), contributed greatly to the enthusiasm for Indian poetry, which then first shewed itself in Germany. I take pleasure in recalling two fine distichs of Göthe's, which appeared in 1792:—

"Willst du die Blüthe des frühen, die Früchte des späteren Jahres,
Willst du was reizt und entzückt, willst du, was sättigt und nährt,
Willst du den Himmel, die Erde mit einem Namen begreifen;
Nenn' ich Sakontala, Dich, und so ist alles gesagt."

The most recent German translation of this Indian drama is that of Otto Böhtlingk (Bonn, 1842), from the important original text found by Brockhaus.

(61) p. 40.—Humboldt, on steppes and deserts (ueber Steppen und Wüsten), in the Ansichten der Natur, 2te Ausgabe, 1826, Bd. i. S. 33—37.

(62) p. 40.—In order to render more complete the small portion of the text which belongs to Indian literature, and to enable me to point out, as in Greek and Roman literature, the several works referred to, I will here introduce some manuscript notices, kindly communicated to me by a distinguished and philosophical scholar thoroughly versed in Indian poetry, Herr Theodor Goldstücker:—

"Among all the influences which have affected the intellectual development of the Indian nation, the first and most important appears to me to have been that exercised by the rich aspect of nature in the country inhabited by them. A profound love of nature has been at all times a fundamental character of the Indian mind. In reference to the manner in which this feeling has manifested itself, three successive epochs may be pointed out, each of which has a determinate character, of which the foundations were deeply laid in the mode of life and tendencies of the people. A few examples may thus be sufficient to indicate the activity of the Indian imagination. The Vedas mark the first epoch of the expression of a vivid feeling for nature: we would refer in the Rigveda to the sublime and simple descriptions of the dawn of day (Rigveda-Sanhitâ, ed. Rosen, 1838, Hymn. xlvi. p. 88; Hymn. xlviii. p. 92; Hymn. xcii. p. 184; Hymn. cxiii. p. 233: see also Höfer, Ind. Gedichte, 1841, Lese i. S. 3,) and of the "golden-handed sun," (Rigveda-Sanhitâ, Hymn. xxii. p. 31; Hymn. xxxv. p. 65). The veneration of nature, connected here, as in other nations, with an early stage of their religious belief, has in the Vedas a peculiarly determinate direction, being always conceived in the most intimate connection with the external and internal life of man. The second epoch is very different: in it a popular mythology was formed, having for its object to mould the contents of the Vedas into a shape more easily comprehensible by an age already far removed in character from that which had given them birth, and to interweave them with historical events to which a mythical character is given. To this second epoch belong the two great heroic poems, the Ramayana and the Mahabharata; the latter had also the additional object of rendering the Brahmins the most influential of the four ancient Indian castes. The Ramayana is the older and more beautiful poem of the two: it is more rich in natural feeling, and has kept more strictly on poetic ground, not having been constrained to take up elements alien and almost hostile to poetry. In both poems, nature no longer constitutes, as in the Vedas, the entire picture, but only a portion of it. There are two points which essentially distinguish the conception of nature at the period of the heroic poems from that which the Vedas present, independently of the wide difference between the language of adoration and that of narrative. One of these points is the localising of the description. According to Wilhelm von Schlegel, the first book of the Ramayana, or Balakanda, and the second book, or Ayodhyakanda, are examples: see also Lassen, Ind. Alterthumskunde, Bd. i. S. 482, on the differences between these two epics. Narrative, whether historical, legendary, or fabulous, leads to the specification of particular localities, rather than to general descriptions. These early epic poets, whether Valmiki, who sings the exploits of Rama, or the authors of the Mahabharata, named collectively, by tradition, Vyasa, all show themselves transported, and as it were overpowered, by emotions connected with external nature. Rama's journey from Ayodhya to Dschanaka's capital; his life in the forest; his expedition to Lanka (Ceylon), where dwelt the savage Ravana, the robber of his bride, Sita; and the hermit life of the Panduides; all furnish to the poet the opportunity of following the bent of the Indian mind, and of blending, with the relation of heroic deeds, the rich imagery of tropical nature. (Ramayana, ed. Schlegel, lib. i. cap. 26, v. 13—15: lib. ii. cap. 56, v. 6—11: compare Nalus, ed. Bopp, 1832, Ges. xii. V. 1—10.) The other point in which the second epoch differs from that of the Vedas in regard to external nature, is closely connected with the first, and consists in the greater richness of materials employed, comprehending the whole of nature,—the heavens and the earth, with the world of plants and of animals in all their luxuriance and variety, and viewed in their influence on the mind and feelings of men. In the third epoch of poetic literature (if we except the Puranas, which have a particular object,) external nature exereises an undivided sovereignty, but the descriptive portion is based on more scientific and more local observation. Among the great poems belonging to this epoch is the Bhatti-kâvya (or Bhatti's poem), which, like the Ramayana, has for its subject the exploits and adventures of Rama, and in which fine descriptions of a forest life during banishment, of the sea and of its beautiful shores, and of the breaking of the day in Ceylon (Lanka), occur successively. (Bhatti-kâvya, ed. Calc. P. i. canto vii. p. 432; canto x. p. 715; canto xi. p. 814. Compare also Schütz, Prof. zu Bielefeld, fünf Gesänge des Bhatti-kâvya, 1837, S. 1—18.) I would also refer to an agreeable description of the different periods of the day in Magha's Sisupalabdha, and to the Naischada-tscharita of Sri Harscha. In the last-named poem, however, in the story of Nalus and Damayanti, the expression of the feeling for external nature passes into a vague exaggeration, which contrasts with the noble simplicity of the Ramayana, where Visvamitra leads his pupil to the shores of the Sona. (Sisupaladha, ed. Calc. p. 298 and 372; compare Schütz, fünf Ges. des Bhatti-kâvya, S, 25—28; Naischada-tscharita, ed. Calc. P. 1, v. 77—129; Ramayana, ed. Schlegel, lib. 1, cap. 35, v. 15—18.) Kalidasa, the celebrated author of Sacontala, represents, with a master's hand, the influence which the aspect of nature exercises on the minds and feelings of lovers. The forest scene pourtrayed by him in the drama of Vikrama and Urvasi is one of the finest poetic creations of any period. (Vikramorvasi, ed. Calc. 1830, p. 71; see the English translation in Wilson's Select Specimens of the Theatre of the Hindus, Calc. 1827, Vol. ii. p. 63.) In the poem of "The Seasons," I would particularly refer to the rainy season and to that of spring (Ritusanhara, ed. Bohlen, 1840, p. 11—18, and 37—45, S. 80—88, and S. 107—114, of Bohlen's translation). In the "Cloud Messenger," also by Kalidasa, the influence of external nature on human feeling is also the leading subject of the composition. This poem (the Meghaduta, or Cloud Messenger, which has been edited by Gildemeister and translated both by Wilson and by Chézy) describes the grief of an exile on the mountain Ramagiri, longing for the presence of his beloved from whom he is separated: he entreats a passing cloud to convey to her tidings of his sorrows; he describes to the cloud the path which it must pursue, and paints the landscape as reflected in a mind agitated with deep emotion. Among the treasures which the Indian poetry of the third period owes to the influence of nature on the national mind, the Gitagovinda of Dschayadeva deserves the highest praise. (Rückert, in the Zeitschrift für die Kunde des Morgenlandes, Bd. i. 1837, S. 129—173; Gitagovinda Jayadevæ poetæ indici drama lyricum, ed. Chr. Lassen, 1836.) We possess a masterly metrical translation of this poem by Rückert, which is one of the most pleasing and at the same time one of the most difficult in the whole of Indian literature. The translation renders the spirit of the original with admirable fidelity, and presents a conception of nature the intimate truth of which animates every part of this great composition.

(63) p. 41.—Journal of the Royal Geogr. Soc. of London, Vol. x. 1841, p. 2—3; Rückert, Makamen Hariri's, S. 261.

(64) p. 41.—Göthe im Commentar zum west-östlichen Divan: Bd. vi. 1828, S. 73, 78, and 111, of his works.

(65) p. 42.—Vide le Livre des Rois, publié par Jules Mohl, T. i. 1838, p. 487.

(66) p. 42.—Jos. von Hammer, Gesch. der schönen Redekunste Persiens, 1818, S. 96 (Ewhadeddin Enweri, who lived in the 12th century, in whose poem on the Schedschai some have discovered a remarkable allusion to the mutual attraction of the heavenly bodies; S. 183 (Dschelaleddin Rumi, the mystic); S. 259 (Dschelaleddin Ahdad); S. 403 (Feisi, who came forward at the court of Akbar as a defender of the religion of Brahma, and in whose Ghazuls there breathes an Indian tenderness of feeling).

(67) p. 42.—"Night comes on when the ink-bottle of heaven is overturned," is the tasteless expression of Chodschah Abdullah Wassaf, a poet, who has, however, the merit of having been the first to describe the great astronomical observatory of Meragha, with its lofty gnomon. Hilali, of Asterabad, makes the disk of the moon glow with heat, and calls the evening dew "the sweat of the moon." (Jos. von Hammer, S. 247 and 371.)

(68) p. 42.—Tuirja or Turan are names of which the derivation is still undiscovered. Burnouf (Yacna, T. i. p. 427—430) has acutely called attention in reference to them to the Bactrian Satrapy of Turina or Turiva mentioned in Strabo (xi. 11, 3, pag. 517, lat.): Du Theil and Groskard, however, Th. ii. S. 410) propose to read Tapyria.

(69) p. 42.—Ueber ein finnisches Epos, Jacob Grimm, 1845, S. 5.

(70) p. 46.—I have followed in the Psalms the excellent translation of Moses Mendelsohn (see his Gesammelte Schriften, Bd. vi. S. 220, 238, and 280). Noble after-echoes of the ancient Hebrew poetry are found in the 11th century in the hymns of the Spanish synagogue poet, Salomo ben Jehudah Gabirol: they also contain a poetic paraphrase of the pseudo-Aristotelian book, De Mundo. Vide die religiöse Poesie der Juden in Spanien, by Michael Sachs, 1845, S. 7, 217, and 229. Sketches drawn from nature, and

full of vigour and grandeur, are found in the writings of Mose ben Jakob ben Esra (S. 69, 77, and 285).

(71) p. 47.—I have taken the passages in the book of Job from the translation and exposition of Umbreit (1824), S. xxix.—xlii. and 290—314. (Consult generally Gesenius, Geschichte der hebr. Sprache und Schrift, S. 33; and Jobi antiquissimi carminis hebr. natura atque virtutes, ed. Ilgen, p. 28.) The longest and most characteristic description of an animal which we meet with in the book of Job, is that of the crocodile (xl. 25—xli. 26), and yet it contains one of the evidences of the writer having been himself a native of Palestine, Umbreit, S. xli. & 308. As the river-horse of the Nile and the crocodile were formerly found throughout the whole Delta of the Nile, it is not surprising that the knowledge of animals of such strange and peculiar form should have spread into the neighbouring country of Palestine.

(72) p. 48.—Göthe im Commentar zum west-östlichen Divan, S. 8.

(73) p. 48.—Antar, a Bedouin romance, translated from the Arabic by Terrick Hamilton, Vol. i. p. xxvi.; Hammer, in the Wiener Jahrbüchern der Litteratur, Bd. vi. 1819, S. 229; Rosenmüller, in the Charakteren der vornehmsten Dichter aller Nationem, Bd. v. (1798) S. 251.

(74) p. 49.—Antara cum schol. Sunsenii, ed. Menil. 1816, v. 15.

(75) p. 49.—Amrulkeisi Moallakat, ed. E. G. Henstenberg, 1823; Hamasa, ed. Freytag, P. i. 1828, lib. vii. p. 785. See also in the pleasing work, entitled, "Amrilkais, the Poet and King," translated by Fr. Rückert, 1843, pp. 29 and 62, where southern showers are twice described with exceeding truth to nature. The royal poet visited the court of the Emperor Justinian several years before the birth of Mahommed, for the purpose of obtaining assistance against his enemies. See Le Diwan d'Amro 'lkaịs, accompagné d'une traduction par le Baron MacGuckin de Slane, 1837, p. 111.

(76) p. 49.—Nabeghah Dhobyani, in Silvestre de Sacy's Chrestom. arabe, 1806, T. iii. p. 47. On the early Arabian literature generally, see Weil's Poet. Litteratur der Araber vor Mohammed, 1837, S. 15 and 90, as well as Freytag's Darstellung der arabischen Verskunst, 1830, S. 372—392. We may soon expect a truly fine and complete version of the Arabian poetry connected with nature in the writings of Hamasa from our great poet Friedrich Rückert.

(77) p. 49.—Hamasæ Carmina, ed. Freytag, P. i. 1828, p. 788. "Here finishes," it is said in page 796, "the chapter on travel and sleepiness."

(78) p. 51.—Dante, Purgatorio, canto i. v. 115:

"L' alba vinceva l' ora mattutina

Che fugia innanzi, sì che di lontano
Conobbi il tremolar de la marina"......

([79]) p. 51.—Purg. canto v., v. 109—127:

" Ben sai come nell' aer si raccoglie
Quell' umido vapor, che in acqua riede,
Tosto che sale, dove 'l freddo il coglie"......

([80]) p. 51.—Purg. canto xxviii. v. 1—24.

([81]) p. 51.—Parad. canto xxx. v. 61—69:

" E vidi lume in forma di riviera
Fulvido di fulgore intra duo rive
Dipinte di mirabil primavera.

Di tal fiumana uscian faville vive,
E d' ogni parte si mettean ne' fiori,
Quasi rubin, che oro circonscrive.

Poi como inebriate dagli odori,
Riprofondavan se nel miro gurge,
E s' una entrava, un' altra n' uscia fuori."

I do not refer to the Canzones of the Vita Nuova, because the comparisons and images which they contain do not belong to the purely natural range of terrestrial phænomena.

([82]) p. 51.—I would recal Boiardo's sonnet commencing,

" Ombrosa selva, che il mio duolo ascolti,"

and the fine stanzas of Vittoria Colonna, which begin,

" Quando miro la terra ornata e bella,
Di mille vaghi ed odorati fiori."

A beautiful and very characteristic natural description of the country seat of Fracastoro on the hill of Incassi (Mons Caphius), near Verona, is given by that distinguished doctor in medicine, mathematician, and poet, in his "Naugerius de poetica dialogus" (Hieron. Fracastorii Opp. 1591, P. i. p. 321—326). See also in a didactic poem, lib. ii. v. 208—219 (Opp. p. 636), the pleasing passage on the culture of the lemon in Italy. I miss with astonishment any expression of feeling connected with the aspect of nature in the letters of Petrarch, either when, in 1345, (three years, therefore, before the death of Laura), he attempted the ascent of Mont Ventour from Vaucluse, hoping and longing to behold from its summit a part of his native land; or, when he visited the gulf of Baiæ, or the banks of the Rhine to Cologne. His mind was occupied by the classical remembrances of Cicero and the

Roman poets, or by the emotions of his ascetic melancholy, rather than by surrounding nature. (Vid. Petrarchæ Epist. de rebus familiaribus, lib. iv. 1; v. 3 and 4: pag. 119, 156, and 161, ed. Lugdun. 1601). I find, however, an exceedingly picturesque description of a great tempest which Petrarch observed near Naples in 1343 (lib. v. 5, p. 165): but it is a solitary instance.

(83) p. 54.—Humboldt, Examen critique de l'histoire de la Géographie du nouveau Continent, T. iii. p. 227—248.

(84) p. 55.—Kosmos, Bd. i. S. 296 and 469 (English translation, vol. i. pp. 272 and 447).

(85) p. 56.—Journal of Columbus on his first voyage (Oct. 29, 1492; Nov. 25—29; Dec. 7—16; Dec. 21); also his letter to Doña Maria de Guzman, ama del Principe D. Juan, Dec. 1500, in Navarrete, Coleccion delos Viages que hiciéron por mar los Españoles, T. i. p. 43, 65, 72, 82, 92, 100, and 266.

(86) p. 56.—Navarrete, Coleccion de los Viages, T. i. p. 303—304 (Carta del Almirante a los Reyes escrita en Jamaica a 7 de Julio, 1503); Humboldt, Examen crit. T. iii. p. 231—236.

(87) p. 56.—Tasso, canto xvi. stanze 9—16.

(88) p. 57.—See Friedrich Schlegel's sammtl. Werke, Bd. ii. S. 96; and on the disturbing mythological dualism, and the mixture of antique fable with Christian contemplations, see Bd. x. S. 54. Camoens has tried, in stanzas which have not been sufficiently attended to (82—84), to justify this mythological dualism. Tethys avows, in a somewhat naïve manner, but in verses which are a noble flight of poetry, "that she herself, Saturn, Jupiter, and all the host of gods, are vain fables, born to mortals by blind delusion, and serving only to embellish the poet's song—"A Sancta Providencia que em Jupiter aqni se representa."

(89) p. 57.—Os Lusiadas de Camões, canto i. est. 19; canto vi. est. 71—82. See also the comparison in the fine description of a tempest raging in a forest, canto i. est. 35.

(90) p. 58.—The fire of St. Elmo: "o lume vivo que a maritima gente tem por santo, em tempo de tormenta" (Canto v. est. 18). One flame, the Helena of the Greek mariners, brings misfortune (Plin. ii. 37); two flames, Castor and Pollux, appearing with a rustling sound, "like the fluttering wings of birds," are good omens (Stob. Eclog. Phys. i. p. 514; Seneca, Nat. Quæst. i. 1). On the eminently graphical character of Camoens' descriptions of nature, and the peculiar manner in which their subjects are brought as it were visibly before the mind's eye, see the great Paris edition of 1818, in the Vida de Camões, by Dom Joze Maria de Souza. p. cii.

([91]) p. 58.—Compare the waterspout in Canto v. est. 19—22, with the also highly poetic and faithful description of Lucretius, vi. 423—442. On the fresh water, which, towards the close of the phænomenon, falls apparently from the upper part of the column of water, see Ogden on Waterspouts (from Observations made in 1820, during a voyage from Havannah to Norfolk), in Silliman's American Journal of Science, Vol. xxix. 1836, p. 254—260.

([92]) p. 58.—Canto iii. est. 7—21, of the text of Camoens in the editio princeps of 1572, which has been given afresh in the excellent and splendid edition of Dom Joze Maria de Souza-Botelho (Paris, 1818). In the German quotations I have usually followed the translation of Donner (1833). The principal aim of the Lusiad of Camoens is the honour and glory of his nation. Would it not be a monument, well worthy of his fame, if a hall were constructed in Lisbon, after the noble examples of the halls of Schiller and Göthe in the Grand Ducal palace of Weimar, and if the twelve grand compositions of my deceased friend Gérard, which adorn the Souza edition, were executed in large dimensions, in fresco, on well lit walls? The dream of the king Dom Manoel, in which the rivers Indus and Ganges appear to him, the Giant Adamastor hovering over the Cape of Good Hope ("Eu sou aquelle occulto e grande Cabo, et quem chamais vós outros Tormentorio"), the murder of Ignes de Castro, and the lovely Ilha de Venus, would all have the finest effect.

([93]) p. 58.—Canto x. est. 79—90. Camoens, like Vespucci, terms the part of the heavens nearest to the southern pole, poor in stars (Canto v. est. 14). He is also acquainted with the ice of the southern seas (Canto v. est. 27).

([94]) p. 59.—Canto x. est. 91—141.

([95]) p. 59.—Canto ix. est. 51—63. (Consult Ludwig Kriegk, Schriften zur allgemeinen Erdkunde, 1840, S. 338.) The whole Ilha de Venus is an allegorical fable, as is clearly indicated in Est. 89; but the beginning of the relation of Dom Manoel's dream depicts an Indian mountain and forest district (Canto iv. est. 70).

([96]) p. 60.—Fondness for the old literature of Spain, and for the enchanting region in which the Araucana of Alonso de Ercilla y Zuñiga was composed, has led me to read conscientiously through the whole of this poem of 22000 lines on two occasions, once in Peru, and again very recently in Paris, when, by the kindness of a learned traveller, M. Ternaux Compans, I received a very scarce book, printed in 1596, at Lima, and containing the nineteen cantos of the Arauco domado compuesto por el Licenciado Pedro de Oña natural de los Infantes de Engol en Chile. Of the epic poem of Ercilla, in which Voltaire sees an Iliad, and Sismondi a newspaper in rhyme, the first fifteen cantos were composed between 1555 and 1563, and were published in 1569; the later cantos were first printed in 1590, only six years before the miserable poem of Pedro de Oña, which bears the same title as one of the master works of Lope de Vega, in which the Cacique Caupolican is the principal personage. Ercilla is naïve and true-hearted; especially in those parts of his composition which he wrote in the field, mostly on bark of trees and skins of beasts for want of paper. The description of his poverty, and of the ingratitude which he experienced at the court of King Philip, is extremely touching, particularly at the close of the 37th canto:

> "Climas passè, mudè constelaciones,
> Golfos inavegables navegando,
> Estendiendo Señor, vuestra corona
> Hasta la austral frigida zona."

"The flower of my life is past; late instructed, I will renounce earthly things, weep, and no longer sing." The natural descriptions of the garden of the sorcerer, of the tempest raised by Eponamon, and of the ocean (P. i. p. 80, 135, and 173; P. ii. p. 130 and 161, in the edition of 1733), are cold and lifeless: geographical registers of words are accumulated in such manner, that, in Canto xxvii., twenty-seven proper names follow each other in immediate succession in a single stanza of eight lines. Part II. of the Araucana is not by Ercilla, but is a continuation, in twenty cantos, by Diego de Santistevan Osorio, appended to the thirty-seven cantos of Ercilla.

(97) p. 60.—In the Romancero de Romances caballeresco é historicos ordenado, por D. Augustin Duran, P. i. p. 189, and P. ii. p. 237, see the fine strophes commencing "Yba declinando el dia"—"Su curso y ligeros horas"—and on the flight of King Roderick, beginning

> "Quando las pintadas aves
> Mudas estan y la tierra
> Atenta esucha los rios."

(98) p. 60.—Fray Luis de Leon, Obras proprias y traducciones, dedicadas a Don Pedro Portocarero, 1681, p. 120: Noche serena. A deep feeling of nature also reveals itself at times in the ancient mystic poetry of the Spaniards (Fray Luis de Granada, Santa Teresa de Jesus, Malon de Chaide); but the natural pictures are usually only the external veil, symbolising ideal contemplations.

(99) p. 61.—Calderon, in the "Steadfast Prince:" on the approach of the Spanish fleet, Act i. scene 1; and on the sovereignty of the wild beasts in the forest, Act iii. scene 2.

([100]) p. 62.—The passages in the text relating to Calderon and Shakspeare, which are distinguished by marks of quotation, are taken from unpublished letters, addressed to myself, by Ludwig Tieck.

([101]) p. 65.—The works referred to were published in the following order of time:—Jean Jacques Rousseau, Nouvelle Héloise, 1759; Buffon, Epoques de la Nature, 1778, but his Histoire naturelle, 1749—1767; Bernardin de St.-Pierre, Etudes de la Nature, 1784, Paul et Virginie, 1788, Chaumière Indienne, 1791; George Forster, Reise nach der Südsee, 1777, Kleine Schriften, 1794. More than half a century before the publication of the Nouvelle Héloise, Madame de Sévigné had already manifested, in her charming Letters, a vivid sense of natural beauty, such as can rarely be traced in the age of Louis XIV. See the fine natural descriptions in the letters of April 20, May 31, August 15, September 16, and November 6, 1671, and October 23 and December 28, 1689 (Aubenas, Hist. de Madame de Sévigné, 1842, p. 201 and 427). I have referred in the text to the old German poet, Paul Flemming, who, from 1633 to 1639, accompanied Adam Olearius on his journeys to Muscovy and to Persia, because, according to the authority of my friend Varnhagen von Ense (Biographische Denkw. Bd. iv. S. 4, 75, and 129), "Flemming's compositions are characterised by a fresh and healthful vigour," and because his images drawn from external nature are tender and full of life.

([102]) p. 68.—Letter of the Admiral from Jamaica, July 7, 1503: "El mundo es poco; digo que el mundo no es tan grande como dice el vulgo" (Navarrete, Coleccion de Viages esp. T. i. p. 300).

([103]) p. 70.—See Journal and Remarks, by Charles Darwin, 1832—1836, in the Narrative of the Voyages of the Adventure and Beagle, Vol. iii. p. 479—490, where an exceedingly beautiful description of Tahiti is given.

([104]) p. 70.—On George Forster's merit as a man and a writer, see Gervinus, Gesch. der poet. National-Litteratur der Deutschen, Th. v. S. 390—392.

([105]) p. 71.—Freytag's Darstellung der arabischen Verskunst, 1830, S. 402.

([106]) p. 75.—Herod. iv. 88.

([107]) p. 75.—A portion of the works of Polygnotus and Mikon (the painting of the battle of Marathon in the Pokile at Athens) might still be seen, according to the testimony of Himerius, at the end of the fourth century (of our era), or 850 years after their execution (Letronne, Lettres sur la Peinture historique murale, 1835, p. 202 and 453).

([108]) p. 76.—Philostratorum Imagines, ed. Jacobs et Welcker, 1825, p. 79 and 485. Both the learned editors defend, against former suspicions, the authenticity of the description of the paintings in the ancient Neapolitan Pinacothek (Jacobs, p. xvii. and xlvi.; Welcker, p. lv. and xlvi.). Otfried Müller supposes that Philostratus's picture of the islands (ii. 17), as well as that of the marsh district (i. 9), of the Bosphorus, and of the fishermen (i. 12 and 13), had much resemblance in their manner of representation to the mosaic of Palestrina. Plato, in the introductory part of Critias (p. 107), mentions landscape painting as representing mountains, rivers, and forests.

([109]) p. 76.—Particularly through Agatharcus, or at least according to the rules laid down by him. Aristot. Poet. iv. 16; Vitruv. Lib. v. cap. 7, Lib. vii. in Præf. (ed. Alois Maxinius, 1836, T. i. p. 292, T. ii. p. 56); compare Letronne's work, before cited, p. 271—280.

([110]) p. 76.—On "Objects of Rhopographia," vide Welcker ad Philostr. Imag. p. 397.

([111]) p. 76.—Vitruv. Lib. vii. cap. 5 (T. ii. p. 91).

([112]) p. 76.—Hirt, Gesch. der bildenden Künste bei den Alten, 1833, S. 332; Letronne, p. 262 and 468.

([113]) p. 76.—Ludius qui primus (?) instituit amœnissimam parietum picturam (Plin. xxxv. 10). The topiaria opera of Pliny, and varietates topiorum of Vitruvius, were small landscape decorative paintings. The passage of Kalidasa is in the 6th act of Sacontala.

([114]) p. 77.—Otfried Müller, Archäologie der Kunst, 1830, S. 609. Having before spoken in the text of the paintings found in Pompeii and Herculaneum as being but little allied to nature in her freedom, I must here notice some exceptions, which may be considered strictly as landscapes in the modern sense of the word. See Pitture d' Ercolano, Vol. ii. tab. 45, Vol. iii. tab. 53; and, as backgrounds in charming historical compositions, tab. 61, 62, and 63, Vol. iv. I do not refer to the remarkable representation in the Monumenti dell' Instituto di Corrispondenza Archeologica, Vol. iii. tab. 9, because its genuine antiquity is considered doubtful by an archæologist of much acumen, Raoul Rochette.

([115]) p. 77.—Against the supposition maintained by Du Theil (Voyage en Italie, par l'Abbé Barthélemy, p. 284) of Pompeii having still existed in splendour under Adrian, and not having been completely destroyed until the end of the fifth century, see Adolph von Hoff, Geschichte der Veränderungen der Erdoberflache, Th. ii. 1824, S. 195—199.

([116]) p. 78.—See Waagen, Kunstwerke und Künstler in England und Paris, Th. iii. 1839, S. 195—201; and particularly S. 217—224, where he describes the celebrated Psalter of the Paris Bibliothèque (of the tenth century), which shews how long the "antique mode of composition" maintained itself in Constantinople. I was indebted, at the time of my public lectures in 1828, to the kind and valuable communications of this profound connoisseur of art (Professor Waagen, Director of the Gallery of Paintings of my native city), for interesting notices on the history of art after the time of the Roman empire. What I afterwards wrote on the gradual development of landscape painting, I communicated in the winter of 1835, in Dresden, to the distinguished and lamented author of the Italienischen Forschungen, Baron von Rumohr; and I received from him a great number of historical illustrations, which he gave me permission to publish entire in case the form of my work should permit.

(117) p. 78.—Waagen, in the work above referred to, Th. i. 1837, S. 59; Th. iii. 1839, S. 352—359.

(118) p. 79.—"Already Pinturicchio painted rich and well-composed landscapes in the Belvidere of the Vatican as independent decorations. He influenced Raphael, in whose paintings many *landscape peculiarities* cannot be traced to Perugino. In Pinturicchio and his friends we also already find those singular pointed forms of mountains which, in your lectures, you were inclined to derive from the Tyrolese dolomitic cones, which Leopold von Buch has rendered so celebrated, and by which travelling artists might have become impressed in the transit between Italy and Germany. I rather believe that these conical forms in the earliest Italian landscapes must be regarded either as very old conventional mountain forms, in antique bas-reliefs and mosaic works, or as unskilfully foreshortened views of Soracte and similarly isolated mountains in the Campagna of Rome" (from a letter addressed to me by Carl Friedrich von Rumohr, in October 1832). To indicate more precisely the conical and pointed mountains which are here in question, I recal the fanciful landscape which forms the background in Leonardo da Vinci's universally admired picture of Mona Lisa (the wife of Francesco del Giocondo). Among the artists of the Flemish school, who more particularly formed landscape into a separate branch, we should name further Patenier's successor, Herry de Bles, named Civetta from his animal monogram, and subsequently the brothers Matthew and Paul Bril, who, during their sojourn in Rome, produced a strong impression in favour of this particular branch of art. In Germany, Albrecht Altdorfer, Durer's scholar, practised landscape painting even somewhat earlier and more successfully than Patenier.

(119) p. 80.—Painted for the church of San Giovanni e Paolo at Venice.

(120) p. 81.—Wilhelm von Humboldt, gesammelte Werke, Bd. iv. S. 37.

Compare also, on the different gradations of the life of nature, and on the tone of mind and feeling awakened by landscape, Carus, in his interesting letters on landscape painting (Briefen über die Landschaftmalerei, 1831, S. 45).

([121]) p. 81.—We find concentrated in the seventeenth century the works of Johann Breughel, 1569—1625; Rubens, 1577—1640; Domenichino, 1581—1641; Philippe de Champaigne, 1602—1674; Nicolas Poussin, 1594—1655; Gaspar Poussin (Dughet), 1613—1675; Claude Lorraine, 1600—1682; Albert Cuyp, 1606—1672; Jan Both, 1610—1650; Salvator Rosa, 1615—1673; Everdingen, 1621—1675; Nicolaus Berghem, 1624—1683; Swanevelt, 1620—1690; Ruysdael, 1635—1681; Minderhoot Hobbema, Jan Wynants, Adriaen van de Velde, 1639—1672; Carl Dujardin, 1644—1687.

([122]) p. 81.—An old picture of Cima da Conegliano, of the school of Bellino (Dresdner Gallerie, 1835, No. 40), has some extraordinarily fanciful representations of date palms with a knob in the middle of the leafy crown.

([123]) p. 82.—Dresdner Gallerie, No. 917.

([124]) p. 83.—Franz Post, or Poost, was born at Harlem, in 1620, and died there in 1680. His brother likewise accompanied Count Maurice of Nassau as architect. Of the paintings, some representing the banks of the Amazons are to be seen in the picture gallery at Schleisheim, and others at Berlin, Hanover, and Prague. The engravings (in Barläus, Reise des Prinzen Moritz von Nassau, and in the royal collection of copperplate prints at Berlin) evidence a fine sense of natural character in the form of the coast, the shape and nature of the ground, and the aspect of vegetation, as displayed in musacæ, cactuses, palms, different species of ficus with board-like excrescences at the foot of the stem, rhizophoras, and arborescent grasses. The picturesque Brazilian series of views terminates singularly enough with a German forest of pineasters surrounding the castle of Dillenburg (Plate lv.) The remark in the text (p. 82), on the influence which the establishment of botanic gardens in Upper Italy, towards the middle of the sixteenth century, may have exercised on the knowledge of the physiognomy of tropical forms of vegetation, induces me to recal in this note that, in the thirteenth century, Albertus Magnus, who was equally active and influential in promoting natural knowledge and the study of the Aristotelian philosophy, possessed a hothouse in the convent of the Dominicans at Cologne. This celebrated man, who had already fallen under the suspicion of sorcery on account of his speaking machine, entertained the King of the Romans, Wilhelm of Holland, on the 6th of January, 1249, in a large space in the convent-garden, where he kept up an agreeable warmth, and preserved fruit trees and plants in flower throughout the winter. We find the account of this banquet exaggerated into a tale of wonder in the Chronica Joannis de Beka, written in the middle of the 14th century (Beka et Heda de Episcopis Ultrajectenis, recogn. ab Arn. Buchelio, 1643, p. 79; Jourdain, Recherches critiques sur l'Age des Traductions d'Aristote, 1819, p. 331: Buhle, Gesch. der Philosophie, Th. v. S. 296). Although some remains discovered in the excavations at Pompeii shew that the ancients made use of panes of glass, yet nothing has yet been found to indicate the use of glass or forcing houses in ancient horticulture. The conduction of heat by the caldaria in baths might have led to an arrangement of artificially warmed places for growing or forcing plants; but the shortness of the Greek and Italian winters no doubt rendered such arrangements less necessary. The Adonis gardens (κηποι Αδωνιδος), so indicative of the *meaning* of the festival of Adonis, consisted, according to Böckh, of plants in small pots, which were no doubt intended to represent the garden where Aphrodite and Adonis met. Adonis was the symbol of the quickly fading flower of youth—of all that flourishes luxuriantly and perishes rapidly; and the festivals which bore his name, the celebration of which was accompanied by the lamentations of women, were amongst those in which the ancients had reference to the decay of nature. I have spoken in the text of hothouse plants as contrasted with those which grow naturally; the ancients used the term "Adonis-gardens" proverbially, to express something which had sprung up rapidly, but gave no promise of full maturity or substantial duration. The plants, which were not many coloured flowers, but lettuce, fennel, barley, and wheat, were not forced in winter, but in summer, being made to grow by artificial means in an unusually short space of time, viz. in eight days. Creuzer (Symbolik und Mythologie, 1841, Th. ii. S. 427, 430, 479, and 481) supposes that the growth of the plants of the Adonis garden was accelerated by the application both of strong natural and artificial heat in the room in which they were placed. The garden of the Dominican convent at Cologne recals the Greenland (?) convent of St. Thomas, where the garden was kept free from snow during the winter, being constantly warmed by natural hot springs, as is told by the brothers Zeni, in the account of their travels (1388—1404), the geographical locality of which is, however, very problematical. (Compare Zurle, Viaggiatori Veneziani, T. ii. p. 63—69; and Humboldt, Examen critique de l'Hist. de la Géographie, T. ii. p. 127.) Regular hothouses seem to have been of very late introduction in our botanic gardens. Ripe pineapples were first obtained at the end of the seventeenth century (Beckmann,

Geschichte der Erfindungen, Bd. iv. S. 287); and Linnæus even asserts, in the Musa Cliffortiana florens Hartecampi, that the first banana which flowered in Europe was at Vienna, in the garden of Prince Eugene, in 1731.

([125]) p. 83.—These views of tropical vegetation, illustrative of the "physiognomy of plants," form, in the Royal Museum at Berlin (in the department of miniatures, drawings, and engravings), a treasure of art which, for its peculiarity and picturesque variety, is as yet without a parallel in any other collection. The sheets edited by the Baron von Kittlitz are entitled, "Vegetations Ansichten der Küstenländer und Inseln des stillen Oceans, aufgenommen 1827—1829 auf der Entdeckungs-reise der kais. russ. Corvette Senjawin, (Siegen, 1844). There is also great truth to nature in the drawings of Carl Bodmer, which are engraved in a masterly manner, and illustrate the great work of the travels of Prince Maximilian zu Wied in the interior of North America.

([126]) p. 88.—Humboldt, Ansichten der Natur, 2te Ausgabe, 1826, Bd. i. S. 7, 16, 21, 36, and 42. Compare also two very instructive memoirs, Friedrich von Martius, Physionomie des Pflanzenreiches in Brasilien, 1824, and M. von Olfers, allgemeine Uebersicht von Brasilien, in Feldners Reisen, 1828, Bd. i. S. 18—23.

([127]) p. 94.—Wilhelm von Humboldt, in his Briefwechsel mit Schiller, 1830, S. 470.

([128]) p. 95.—Diodor. ii. 13. He however gives to the celebrated gardens of Semiramis a circumference of only twelve stadia. The district near the pass of Bagistanos is still called the "bow or circuit of the garden"—Tauk-i-bostan (Droysen, Gesch. Alexanders des Grossen, 1833, S. 553).

([129]) p. 95.—In the Schahnameh of Firdusi it is said, "a slender cypress, sprung from Paradise, did Zerdusht plant before the gate of the temple of fire" (at Kishmeer in Khorasan). "He had written on the tall cypress tree, that Gushtasp had embraced the true faith, that the slender tree was a testimony thereof, and that thus did God extend righteousness. When many years had passed over, the tall cypress became so large that the hunter's cord could not go round its circumference. When its top was furnished with many branches, he encompassed it with a palace of pure gold......and caused it to be said abroad in the world, Where is there on the earth a cypress like that of Kishmeer? God sent it me from Paradise, and said, Bow thyself from thence to Paradise." (When the Caliph Motewekkil had the sacred cypresses of the Magians cut down, this one was said to be 1450 years old.) Compare Vuller's Fragmente über die Religion des Zoroaster, 1831, S. 71 and 114;

and Ritter, Erdkunde, Th. vi., 1. S. 242. The cypress (in Arabic arar wood, in Persian serw kohi) appears to be originally a native of the mountains of Busih, west of Herat (vide Géographie d'Edrisi, traduit par Jaubert, 1836, T. i. p. 464).

([130]) p. 95.—Achill. Tat. i. 25; Longus, Past. iv. p. 108, Schäfer. "Gesenius (Thes. Linguæ Hebr. T. ii. p. 1124) suggests, very justly, the view that the word Paradise belonged originally to the ancient Persian language, but that its use has been lost in the modern Persian. Firdusi, although his own name was taken from it, usually employs only the word behischt; the ancient Persian origin of the word is, however, expressly witnessed by Pollux, in the Onomast. ix. 3, and by Xenophon, Œcon. 4, 13, and 21; Anab. i. 2, 7, and i. 4, 10; Cyrop. i. 4, 5. In the sense of 'pleasure-garden' or 'garden,' the word was probably transferred from the Persian into the Hebrew (*pardês*, Cant. iv. 13; Nehem. ii. 8; and Eccl. ii. 5), into the Arabic (*firdaus*, plur. *faradisu*, compare Alcoran, xxiii. 11, and Luc. 23, 43), into the Syrian and Armenian (*partes*, vide Ciakciak, Dizionario Armeno, 1837, p. 1194; and Schröder, Thes. Ling. Armen. 1711, præf. p. 56). The derivation of the Persian word from the Sanscrit (*pradêsa* or *paradêsa*, circuit, or district, or foreign land), noticed by Benfey (Griech. Wurzellexikon, Bd. i. 1839, S. 138), and previously by Bohlen and Gesenius, suits perfectly well in form, but only indifferently in sense."—Buschmann.

([131]) p. 96.—Herod. vii. 31 (between Kallatebus and Sardes).

([132]) p. 96.—Ritter, Erdkunde, Th. iv. 2. S. 237, 251, and 681; Lassen, indische Alterthumskunde, Bd. i. S. 260.

([133]) p. 96.—Pausanius, i. 21, 9. Compare also Arboretum Sacrum, in Meursii Opp. ex recensione Joann. Lami, Vol. x. Florent. 1753, p. 777—844.

([134]) p. 97.—Notice historique sur les Jardins des Chinois, in the Mémoires concernant les Chinois, T. viii. p. 309.

([135]) p. 97.—Idem, p. 318—320.

([136]) p. 97.—Sir George Staunton, Account of the Embassy of the Earl of Macartney to China, Vol. ii. p. 245.

([137]) p. 97.—Fürst v. Pückler-Muskau, Andeutungen über Landschaftsgärtnerei, 1834. See also his Picturesque Descriptions of the Old and New English Parks, as well as that of the Egyptian Garden of Schubra.

([138]) p. 98.—Eloge de la Ville de Moukden, Poême composé par l'Empereur Kien-long, traduit par le P. Amiot, 1770, p. 18, 22—25, 37, 63—68, 73—87, 104, and 120.

([139]) p. 99.—Mémoires concernant les Chinois, T. ii. p. 643—650,

([140]) p. 99.—Ph. Fr. von Siebold, Kruidkundige Naamlijst van japansche en chineesche Planten, 1844, p. 4. How great a difference between the variety of vegetable forms cultivated for so many centuries past in Eastern Asia, and the comparative poverty of the list given by Columella, in Poem de Cultu Hortorum (v. 95—105, 174—176, 255—271, 295—306), and to which the celebrated garland-weavers of Athens were confined! It was not until the time of the Ptolemies, that in Egypt, and particularly in Alexandria, somewhat greater pains were taken by the more skilful gardeners to obtain variety, particularly for winter cultivation. (Compare Athen. v. p. 196.)

([141]) p. 101.—Kosmos, Bd. i. S. 50—57 (Engl. edit. Vol. i. p. 43—51).

([142]) p. 107.—Niebuhr, röm. Geschichte, Th. i. S. 69; Droysen, Gesch. der Bildung des hellenistischen Staatensystems, 1843, S. 31—34, 567—573; Fried. Cramer de studiis quæ veteres ad aliarum gentium contulerint linguas, 1844, p. 2—13.

([143]) p. 109.—In Sanscrit, rice is *vrihi*, cotton *karpâsa*, sugar *'sarkara*, and nard *nanartha;* vide Lassen, indische Alterthumskunde, Bd. i. 1843, S. 245, 250, 270, 289, and 538. On *'sarkara* and *kanda* (whence our sugar-candy), see my Prolegomena de distributione geographica plantarum, 1817, p. 211:—"Confudisse videntur veteres saccharum verum cum Tebaschiro Bambusæ, tum quia utraque in arundinibus inveniuntur, tum etiam quia vox sanscradana *scharkara*, quæ hodie (ut pers. *schakar* et hindost. *schukur*) pro saccharo nostro adhibetur, observante Boppio, ex auctoritate Amarasinhæ, proprie nil dulce (madu) significat, sed quicquid lapidosum et arenaceum est, ac vel calculum vesicæ. Verisimile igitur, vocem scharkara initio dumtaxat tebaschirum (saccar mombu) indicasse, posterius in saccharum nostrum humilioris arundinis (ikschu, kandekschu, kanda) ex similitudine aspectus translatam esse. Vox Bambusæ ex mambu derivatur; ex kanda nostratium voces candis zuckerkand. In tebaschiro agnoscitur Persarum schir, h. e. lac, sanscr. kschiram." The Sanscrit name for tabaschir (see Lassen, Bd. i. S. 271—274) is *tvakkschira*, bark milk; milk from the bark (*tvatsch*). Compare also Pott, Kurdische Studien in der Zeitschrift für die Kunde des Morgenlandes, Bd. vii. S. 163—166, and the able discussion by Carl Ritter, in his Erdkunde von Asien, Bd. vi. 2, S. 232—237.

([144]) p. 112.—Ewald, Geschichte des Volkes Israel, Bd. i. 1843, S. 332—334; Lassen, ind. Alterthumskunde, Bd. i. S. 528. Compare Rödiger, in the Zeitschrift für die Kunde des Morgenlandes, B. iii. S. 4, on Chaldeans and Kurds, which latter Strabo terms *Kyrti.*

([145]) p. 112.—Bordj, the watershed of Ormuzd, nearly where the chain of the Thian-schan (or heaven mountains), at its western termination, abuts against the Bolor (Belur-tagh), or rather intersects it, under the name of the Asferah chain, north of the highland of Pamer (Upa-Mêru, or country above Meru). Compare Burnouf, Commentaire sur le Yacna, T. i. p. 239, and Addit. p. clxxxv. with Humboldt, Asie centrale, T. i. p. 163, T. ii. pp. 16, 377, and 390.

(146) p. 112.—Chronological data for Egypt:—"Menes, 3900 B. C. at least, and probably tolerably exact;—commencement of the 4th dynasty (comprising the Pyramid builders, Chephren-Schafra, Cheops-Chufu, and Mykerinos or Memkera), 3430;—invasion of the Hyksos under the 12th dynasty, to which belongs Amenemha III. the builder of the original Labyrinth, 2200. A thousand years at least before Menes, and probably still more, must be allowed for the gradual growth of a civilisation which had reached its completion, and had in part become fixed, at least 3430 years before our era."—(Lepsius, in several letters to myself, in March 1846, after his return from his memorable expedition.) Compare also Bunsen's considerations on the commencement of Universal History, (which, strictly speaking, does not include the earliest history of mankind), in his ingenious and learned work, Ægyptens Stelle in der Weltgeschichte, 1845, 1st book, S. 11—13. The history and regular chronology of the Chinese go back to 2400, and even to 2700, before our era, much beyond Ju to Hoang-ty. There are many literary monuments of the 13th century B. C.; and in the 12th, Thscheu-li records the measurement of the length of the solstitial shadow by Tscheu-kung, in the town of Lo-yang, south of the Yellow River, which is so exact that Laplace found it quite accordant with the theory of the alteration of the obliquity of the ecliptic, which was only propounded at the close of the last century; so that there can be no suspicion of a fictitious measurement obtained by calculating back. See Edouard Biot sur la Constitution politique de la Chine au 12ème siècle avant notre ère (1845), pp. 3 and 9. The building of Tyre and of the original temple of Melkarth, the Tyrian Hercules, would reach back to 2760 years before our era, according to the account which Herodotus received from the priests (II. 44). Compare also Heeren, Ideen über Politik und Verkehr der Völker, Th. i. 2, 1824, S. 12. Simplicius, from a notice transmitted by Porphyry, estimates the antiquity of Babylonian astronomical observations which were known to Aristotle at 1903 years before Alexander the Great; and the profound and cautious chronologist Ideler considers this datum by no means improbable. Compare

his Handbuch der Chronologie, Bd. i. S. 207; the Abhandlungen der Beriner Akad. auf das J. 1814, S. 217; and Böckh, metrol. Untersuchungen über die Masse des Alterthums, 1838, S. 36. It is a question still wrapped in obscurity, whether there is historic ground in India earlier than 1200 B. C., according to the Chronicles of Kashmeer (Radjatarangini, trad. par Troyer), while Megasthenes (Indica, ed. Schwanbeck, 1846, p. 50) reckons from 60 to 64 centuries from Manu to Chandragupta, for 153 kings of the dynasty of Magadha; and the astronomer Aryabhatta places the beginning of his Chronology 3102 B. C. (Lassen, ind. Alterthumsk. Bd. i. S. 473, 505, 507, and 510). For the purpose of rendering the numbers contained in this note more significant in respect to the history of civilization, it may not be superfluous to recal, that the destruction of Troy is placed 1184—Homer 1000 or 950—and Cadmus the Milesian, the first historical writer among the Greeks, 524 years before our era. This comparison of epochs shews how unequally the desire for an exact record of events and enterprises made itself felt among the nations most highly susceptible of culture: it reminds us involuntarily of the sentence which Plato, in the Timæus, places in the mouth of the priests of Sais: "O Solon, Solon! you Greeks still remain ever children; nowhere in Hellas is there an aged man. Your souls are ever youthful; you have in them no knowledge of antiquity, no ancient faith, no wisdom grown hoar by age."

(147) p. 112.—Compare Kosmos, Bd. i. S. 92 and 160 (Engl. ed. Vol. i. p. 79 and 144).

(148) p. 112.—Wilhelm von Humboldt über eine Episode des Maha-Bharata, in his Gesammelten Werken, Bd. i. S. 73.

(149) p. 116.—Kosmos, Bd. i. S. 309 and 351 (Eng. ed. Vol. i. p. 283 and 322); Asie centrale, T. iii. p. 24 and 143.

(150) p. 117.—Plato, Phædo, pag. 109, B. (compare Herod. ii. 21). Cleomedes also depressed the surface of the earth in the middle to receive the Mediterranean (Voss, krit. Blätter, Bd. ii. 1828, S. 144 and 150).

(151) p. 117.—I first developed this idea in my Rel. hist. du voyage aux régions équinoxiales, T. iii. p. 236; and in the Examen crit. de l'hist. de la gèogr. au 15ème siècle, T. i. p. 36—38. Compare also Otfried Müller, in the Göttingischen gelehrten Anzeigen, 1838, Bd. i. S. 375. The westernmost basin, to which I apply the general name of Tyrrhenian, includes, according to Strabo, the Iberian, Ligurian, and Sardinian seas. The Syrtic basin, east of Sicily, includes the Ausonian or Siculian, the Lybian, and the Ionian seas. The southern and south-western part of the Ægean sea was

called Cretic, Saronic, and Myrtoic. The remarkable passage in Aristot. de Mundo, cap. iii. (pag. 393, Bekk.) relates merely to the sinuous form of the coasts of the Mediterranean, and its effect on the inflowing ocean.

([152]) p. 118.—Kosmos, Bd. i. S. 253 and 454 (Engl. ed. Vol. i. p. 231 and 435).

([153]) p. 119.—Humboldt, Asie centrale, T. i. p. 67. The two remarkable passages of Strabo are the following:—"Eratosthenes names three, and Polybius five points of projecting land in which Europe terminates. The peninsulas named by Eratosthenes are, first, the one which extends to the pillars of Hercules, to which Iberia belongs; next, that which terminates at the Sicilian straits, on which is Italy; and thirdly, that which extends to Malea, and contains all the nations between the Adriatic, the Euxine, and the Tanais."—(Lib. ii. p. 109.) "We begin with Europe because it is of irregular form, and is the part of the world most favourable to the ennoblement of men and of citizens. It is every where habitable, except some lands near the Tanais, which are desert on account of the cold."—(Lib. ii. pag. 126.)

([154]) p. 119.—Ukert, Geogr. der Griechen und Römer, Th. i. Abth. 2, S. 345—348, and Th. ii. Abth. 1, S. 194; Johannes v. Müller, Werke, Bd. i. S. 38; Humboldt, Examen critique, T. i. pp. 112 and 171; Otfried Müller, Minyer, S. 64; and the same in a critical notice (only too kind) of my memoir on the Mythic Geography of the Greeks (Gött. gelehrte Anzeigen, 1838, Bd. i. S. 372 and 383). I expressed myself generally thus:—"En soulevant des questions qui offriraient déjà de l'importance dans l'intérêt des études philologiques, je n'ai pu gagner sur moi de passer entièrement sous silence ce qui appartient moins à la description du monde réel qu'au cycle de la géographie mythique. Il en est de l'espace comme du tems; on ne sauroit traiter l'histoire sous un point de vue philosophique, en ensevelissant dans un oubli absolu les tems heroïques. Les mythes des peuples, mêlés à l'histoire et à la géographie, ne sont pas *en entier du domaine du monde idéal.* Si le vague est un de leurs traits distinctifs, si le symbole y couvre la réalité d'un voile plus ou moins épais, les mythes intimement liés entr'eux, n'en révèlent pas moins la souche antique des premiers aperçus de cosmographie et de physique. Les faits de l'histoire et de la géographie primitives ne sont pas seulement d'ingénieuses fictions, les opinions qu'on s'est formées sur le monde réel s'y reflètent." The great investigator of antiquity, whom I have named, whose early death on the soil of Greece, to which he devoted such profound and varied research, has been universally lamented, thought, on the contrary, that, "in the poetic idea of the earth, such as it appears in Greek poetry, the chief part is by no means to be ascribed to the results of actual experience, invested by credulity and the love of the marvellous, with a fabulous appearance, (as is supposed to have been particularly the case in the maritime legends of the Phœnician sailors); we should, on the contrary, seek the bases of the imaginary picture rather in certain ideal presuppositions and requirements of the feelings, on which a true geographical knowledge has only gradually begun to work. From this there has often resulted the interesting phenomenon, that purely subjective creations of a fancy working under the guidance of certain ideas, pass almost imperceptibly into real countries, and well-known objects of scientific geography. We may infer from these considerations, that all pictures of the imagination, either mythical or arrayed in mythical forms, belong, in their proper groundwork, to an ideal world, and have no original connexion with the actual extension of the knowledge of the earth, or of navigation beyond the pillars of Hercules." The opinion expressed by me in the French work was more accordant with the earlier views of Otfried Müller, for in the Prolegomenon zu einer wissenschaftlichen Mythologie, S. 68 and 109, he said very distinctly, that, "in mythical narratives, what is done and what is imagined, the real and the ideal, are most often closely combined with each other." Compare also, on the Atlantis and Lyktonia, Martin, Etudes sur le Timée de Platon, T. i. p. 293—326.)

(155) p. 120.—Naxos by Ernst Curtius, 1846, S. 11; Droysen, Geschichte der Bildung des hellenistischen Staatensystems, 1843, S. 4—9.

(156) p. 121.—Leopold von Buch über die geognostischen Systeme von Deutschland, S. xi.; Humboldt, Asie centrale, T. i. p. 284—286.

(157) p. 121.—Kosmos, Bd. i. S. 479 (Engl. edit. Vol. i. p. 461).

(158) p. 122.—All relating to Egyptian chronology and history (from p. 122 to p. 125), which is distinguished by marks of quotation, rests on manuscript communications received from my friend Professor Lepsius, in March 1846.

(159) p. 122.—With Otfried Müller, I place the Doric immigration into the Peloponnesus 328 years before the first Olympiad (Dorier, Abth. ii. S. 436.)

(160) p. 123.—Tac. Annal. ii. 59. In the Papyrus of Sallier (Campagnes de Sesostris), Champollion found the names of the Javani or Jouni and the Luki (Ionians and Lycians?). Compare Bunsen, Ægypten, Buch i. S. 60.

(161) p. 124.—Herod. ii. 102 and 103; Diod. Sic. i. 55 and 56. Of the memorial pillars (stelæ) or tokens of victory which Ramses Miamoun set up in the countries which he traversed, three are expressly named by Herodotus

(ii. 106)—"one in Palestinian Syria, and two in Ionia—on the passage from the Ephesian territory to Phocæa, and from that of Sardis to Smyrna." A rock inscription, in which the name of Ramses presents itself several times, has been found in Syria, near the Lycus, not far from Beirut (Berytus), as well as another ruder one in the valley of Karabel, near Nymphio, and, according to Lepsius, on the way from the Ephesian territory to Phocæa. Lepsius, in the Ann. dell' Institut archeol. Vol. x. 1838, p. 12; and in his letter from Smyrna, Dec. 1845, published in the archäologischen Zeitung, Mai 1846, No. 41, S. 271—280. Kiepert, in idem, 1843, No. 3, S. 35. The now rapidly advancing discoveries in archæology and phonetic languages will hereafter decide whether, as Heeren believes (Geschichte der Staaten des Alterthums, 1828, S. 76), the great conqueror penetrated as far as Persia and Hindostan, "as Western Asia did not then as yet contain any great empire" (the building of Assyrian Nineveh is placed only in 1230 B.C.). Strabo (lib. xvi. p. 760) speaks of a memorial pillar of Sesostris near the Strait of Deire, now called Bab-el-Mandeb. It is, however, also very probable, that in "the old kingdom," above 900 years before Ramses Miamoun, Egyptian kings may have made similar military expeditions into Asia. It was under the Pharaoh Setos II. the second successor of the great Ramses Miamoun, and belonging to the 19th dynasty, that Moses went out of Egypt, according to Lepsius about 1300 years before our era.

(162) p. 125.—According to Aristotle, Strabo, and Pliny; but not according to Herodotus. See Letronne, in the Révue des deux Mondes, 1841, T. xxvii. p. 219; and Droysen, Bildung des hellenist. Staatensystems, S. 735.

(163) p. 125.—To the important opinions of Rennell, Heeren, and Sprengel, which are favourable to the reality of the circumnavigation of Lybia, we must now add that of a profound philologist, Etienne Quatremère (Mémoires de l'Acad. des Inscriptions, T. xv. P. 2, 1845, p. 380—388). The most convincing argument for the truth of the account given by Herodotus (iv. 42) appears to me to be the observation which seems to him so incredible, viz. "that those who sailed round Lybia, in sailing from east to west, had had the sun on their *right hand*." In the Mediterranean, in sailing from east to west, the sun at noon was always seen to the left only. It would seem as if a more accurate knowledge of the possibility of such a navigation had existed in Egypt previous to the time of Neku II. (Nechos), as Herodotus makes him distinctly command the Phœnicians "to make their return to Egypt by the Pillars of Hercules." It is singular that Strabo, who (lib. ii. p. 98), discusses at such length the attempted circumnavigation of Eudoxus of Cyzicus under

Cleopatra, and mentions fragments of a ship from Gadeira found on the Ethiopian (eastern) coast, declares the accounts given of earlier circumnavigations actually accomplished to be *Bergaii fables* (lib. ii. p. 100); but he by no means denies the possibility of the circumnavigation itself (lib. i. p. 38), and affirms that from the east to the west there is but little remaining wanting to its completion (lib. i. p. 5). Strabo did not at all concur in the extraordinary isthmus-hypothesis of Hipparchus and Marinus of Tyre, according to which Eastern Africa joined on to the south-east end of Asia, making the Indian Ocean a Mediterranean Sea (Humboldt, Examen crit. de l'Hist. de la Géographie, T. i. p. 139—142, 145, 161, and 229; T. ii. p. 370—373). Strabo quotes Herodotus, but does not name Nechos, whose expedition he altogether confounds with one directed by Darius round Southern Persia and Arabia (Herod. iv. 44). Gosselin has even proposed, with too great boldness, to change the reading from Darius to Nechos. A counterpart for the horses' head of the ship of Gadeira, which Eudoxus is said to have exhibited in a market-place in Egypt, may be found in the remains of a ship of the Red Sea, brought to the coast of Crete by westerly currents, according to the account of a very trustworthy Arabian historian (Masudi, in the Morudj-al-dzeheb, Quatremère, p. 389, and Reinaud, Relation des Voyages dans l'Inde, 1845, T. i. p. xvi. and T. ii. p. 46).

(164) p. 125.—Diod. lib. i. cap. 67, 10; Herodotus, ii. 154, 178, and 182. On the probability of intercourse between Egypt and Greece before the time of Psammetichus, see the ingenious observations of Ludwig Ross, in Hellenika, Bd. i. 1846, S. v. and x. "In the times immediately preceding Psammetichus," says the last named writer, "there was in both countries a period of internal disorder, which could not but entail a diminution and partial interruption of intercourse.

(165) p. 126.—Böckh, metrologische Untersuchungen über Gewichte, Munzfüsse und Masse des Alterthums in ihrem Zusammenhang, 1838, S. 12 und 273.

(166) p. 126.—See the passages collected in Otfried Müller's Minyer, S. 115, and in his Dorier, Abth. i. S. 129; Franz, Elementa Epigraphices Græcæ, 1840, p. 13, 32, and 34.

(167) p. 127.—Lepsius, in his important memoir, über die Anordnung und Verwandtschaft des semitischen, indischen, alt-persischen, alt-ægyptischen und æthiopischen Alphabets, 1836, S. 23, 28 und 57; Gesenius, Scripturæ Phœniciæ Monumenta, 1837, p. 17.

(168) p. 128.—Strabo, lib. xvi. p. 757.

(169) p. 128.—It is easier to determine the locality of the "land of tin" (Britain and the Scilly Islands) than that of the "amber coast;" for it seems to me very improbable that the old Greek denomination κασσιτερος, which was in use even in the Homeric times, is to be derived from a stanniferous mountain in the south-west of Spain, called Mount Cassius, and which Avienus, who was well acquainted with the country, placed between Gaddir and the mouth of a small southern Iberus (Ukert, Geogr. der Griechen und Römer, Theil ii. Abth. i. S. 479). Kassiteros is the ancient Indian Sanscrit word kastîra. Zinn in German, dën in Icelandic, tin in English, and tenn in Swedish, is in the Malay and Javanese language, timah; a similarity of sound which reminds us of that of the old German word glessum (the name given to transparent amber) to the modern "glas," glass. The names of articles of commerce pass from nation to nation, and become adopted into the most different languages (see above, p. 109, and Note 143.) Through the intercourse which the Phœnicians, by means of their factories in the Persian Gulf, maintained with the east coast of India, the Sanscrit work kastîra, expressing a most useful product of further India, and still existing among the old Aramaic idioms in the Arabian word kasdir, became known to the Greeks even before Albion and the British Cassiterides had been visited (Aug. Wilh. v. Schlegel, in the indischen Bibliothek, Bd. ii, S. 393; Benfey, Indica, S. 307; Pott, etymol. Forschungen, Th. ii. S. 414; Lassen, indische Alterthumskunde, Bd. i. S. 239). A name often becomes an historical monument, and the etymological analysis of languages, which is sometimes ignorantly derided, is not without its fruit. The ancients were also acquainted with the existence of tin (one of the rarest metals on the globe) in the country of the Artabri and the Callaici, in the north-west part of the Iberian continent (Strabo, lib. iii. p. 147; Plin. xxxiv. c. 16); nearer of access, therefore, for navigators from the Mediterranean than the Cassiterides (Œstrymnides of Avienus). When I was in Galicia, in 1799, before embarking for the Canaries, mining operations were still carried on, on a very poor scale, in the granitic mountains (see my Rel. hist. T. i. p. 51 and 53). The occurrence of tin in this locality is of some geological importance, on account of the former connection of Galicia, the peninsula of Brittany, and Cornwall.

(170) p. 128.—Etienne Quatremère, Mém. de l'Acad. des Inscript. T. xv. P. ii. 1845, p. 363—370.

(171) p. 128.—The early expressed opinion (Heinzen's neues Kielisches Magazin, Th. ii. 1787, S. 339; Sprengel, Gesch. der geogr. Entdeckungen, 1792; S. 51; Voss, krit. Blätter, Bd. ii. S. 392—403) is now gaining

ground, that the amber was brought *by sea*, at first only from the *west* Cimbrian coast, and that it reached the Mediterranean chiefly by land, being brought across the intervening countries by means of inland traffic and barter. The most thorough and acute investigation of this subject is contained in Ukert's memoir über das Electrum, in der Zeitschrift für Alterthumswissenschaft, Jahr. 1838, No. 52—55. (Compare with it the same author's Geographie der Griechen und Römer, Th. ii. Abth. ii. 1832, S. 26—36, Th. iii. i. 1843, S. 86, 175, 182, 320, and 349.) The Massilians, who, according to Heeren, penetrated, after the Phœnicians, as far as the Baltic, under Pytheas, hardly went beyond the mouths of the Weser and the Elbe. The amber islands (Glessaria, also called Austrania) are placed by Pliny (iv. 16) decidedly west of the Cimbrian promontory in the German Sea; and the connection with the expedition of Germanicus sufficiently shews that an island in the Baltic is not meant. Moreover, the effects of the ebb and flood tides in the estuaries which throw up amber, where, according to the expression of Servius, "mare vicissim tum accedit, tum recedit," suits the coasts between the Helder and the Cimbrian peninsula, but does not suit the Baltic, in which Timæus places the island of Baltia (Plin. xxxvii. 2). Abalus, a day's journey from an æstuarium, cannot, therefore, be the Kurische Nehrung. On the voyage of Pytheas to the west shores of Jutland, and on the amber trade along the whole coast of Skagen, as far as the Netherlands, see also Werlauff, Vidrag til den nordiscke Ravhandels Historie (Copenh. 1835). Tacitus, not Pliny, is the first writer acquainted with the glessum of the shores of the Baltic, in the land of the Æstyans (Æstuorum gentium) and the Venedi, concerning whom the great ethnologist Schaffarik (slawische Alterthümer, Th. i. S. 151—165), is uncertain whether they were Slavonians or Germans. The more active direct connection with the Samland coast of the Baltic, and with the Æstyans by means of the overland route through Pannonia, by Carnuntum, which was opened by a Roman knight under Nero, appears to me to have belonged to the later times of the Roman Cæsars (Voigt, Gesch. Preussen's, Bd. i. S. 85.) The relations between the Prussian coasts and the Greek colonies on the Black Sea are evidenced by fine coins, struck probably before the 85th Olympiad, which have been recently found in the Netz district (Lewezow, in den Abhandl. der Berl. Akad. der Wiss. aus dem Jahr 1833, S. 181—224). No doubt the amber stranded or buried on coasts (Plin. xxxvii. cap. 2),—the electron, the *sun stone* of the very ancient mythus of the Eridanus,—came to the south, both by land and by sea, from very different districts. The "amber *dug up* at two places in Scythia was in part very dark coloured." Amber is still collected near Kaltschedansk, not far from Kamensk, on the Ural; fragments imbedded in lignite were given to us in Katharinenburg. See G. Rose, Reise nach dem Ural, Bd. i. S. 481; and Sir Roderick Murchison, in the Geology of Russia, Vol. i. p. 366. The fossil wood which often surrounds the amber had early attracted the attention of the ancients. This resin, which was at that time so highly valued, was ascribed to the black poplar (according to the Chian Scymnus, v. 396, p. 367, Letronne), or to a tree of the cedar or pine kind (according to Mithridates, in Plin. xxxvii. cap. 2 and 3). The recent excellent investigations of Prof. Goppert, at Breslau, have shewn that the conjecture of the Roman collector was the more correct. Respecting the fossil amber tree (Pinites succinifer) belonging to an earlier vegetation, compare Kosmos, Bd. i. S. 298 (Engl. edit. Vol. i. p. 273) and Berendt, organische Reste im Bernstein, Bd. i. Abth. i. 1845, S. 89.

([172]) p. 129.—Respecting the Chremetes, see Aristot. Meteor. lib. i. p. 350, Bekk.; and respecting the southern stars, of which Hanno makes mention in his ship's journal, see my Rel. Hist. t. i. p. 172; and Examen Crit. de la Geogr. t. i. p. 39, 180, and 288; t. iii. p. 135. (Gosselin Recherches sur la Géogr. Systém. des Anciens, t. i. p. 94 and 98; Ukert, Th. i. S. 61-66.)

([173]) p. 129.—Strabo, lib. xvii. p. 826. The destruction of Phœnician colonies by Nigritians (lib. ii. p. 131) appears to indicate a very southern locality; more so, perhaps, than the crocodiles and elephants mentioned by Hanno, as both these were certainly found north of the desert of Sahara, in Maurusia, and in the whole western country near the chain of Mount Atlas, as is plain from Strabo, lib. xvii. p. 827; Ælian de Nat. Anim. vii. 2; Plin. v. 1, and from many occurrences in the wars between Rome and Carthage. On this important subject, as respects the geography of animals, see Cuvier, Ossemens fossiles, 2 éd. t. i. p. 74; and Quatremère's work, already cited (Mém. de l'Acad. des Inscriptions, t. xv. p. 2, 1845), p. 391—394.)

([174]) p. 130.—Herod. iii. 106.

([175]) p. 131.—In another work (Examen Crit. t. i. p. 130—139; t. ii. p. 158 and 169; t. iii. p. 137—140) I have treated in detail this often contested subject, as well as the passages of Diodorus (v. 19 and 20), and of the Pseudo-Aristot. (Mirab. Auscult. cap. lxxxv. p. 172, Bekk.) The compilation of the Mirab. Auscult. appears to be older than the end of the first Punic war, as in cap. cv. p. 211, it describes Sardinia as under the dominion of the Carthaginians. It is also remarkable, that the wood-clothed island mentioned in this work is said to be uninhabited, not therefore peopled with

Guanches. Guanches inhabited the whole group of the Canary Islands, but not the island of Madeira, in which no inhabitants were found either by John Gonzalves and Tristan Vaz in 1519, or at an earlier period by Robert Macham and Anna Dorset (supposing their romantic story to be historically true.) Heeren applies the description of Diodorus to Madeira only, yet he thinks that in the account of Festus Avienus (v. 164), so conversant with Punic writings, he can recognise the frequent volcanic earthquakes of the Peak of Teneriffe. (Vide his Ideen über Politik und Handel, Th. II. Abth. 1, 1826, S. 106.) From the geographical connection, the description of Avienus appears to me to refer to a more northern locality, perhaps even to the Kronic sea. (Examen Crit. t. iii. p. 138.) Ammianus Marçellinus (xxii. 15), also notices the Punic sources which Juba used. Respecting the probability of the Semitic origin of the name of the Canary Islands (the dog islands of Pliny's Latin etymology!), see Credner's biblische Vorstellung vom Paradiese, in Illgen's Zeitschr. für die historische Theologie, Bd. vi. 1836, S. 166—186. All that has been written from the most ancient times to the middle ages, respecting the Canary Islands, has been recently brought together in the fullest manner by Joaquim Jose da Costa de Macedo, in a work entitled, Memoria em que se pretende provar que os Arabes não conhecerão as Canarias antes dos Portuguezes, 1844. Where history, so far as it is founded on certain and distinctly expressed testimony is silent, there remain only different degrees of probability; but an absolute denial of every fact in the world's history of which the evidence is not perfectly distinct, appears to me no happy application of philologic and historic criticism. The many indications which have come down to us from antiquity, and careful considerations of the geographical relations of proximity to ancient undoubted settlements on the African coast, lead me to believe that the Canary group was known to the Phœnicians, Carthaginians, Greeks and Romans, and perhaps even to the Etruscans.

([176]) p. 131.—Compare the calculations in my Rel. Hist. t. i. p. 140 and 287. The Peak of Teneriffe is distant 2° 49′ of arc from the nearest point of the African coast. Assuming a mean refraction of 0·08, the summit of the Peak may therefore be seen from a height of 202 toises, and thus from the Montañas Negras, not far from Cape Bojador. In this calculation the elevation of the Peak above the level of the sea has been taken at 1904 toises. It has been recently determined trigonometrically by Captain Vidal at 1940 toises, and barometrically by Messrs. Coupvent and Dumoulin (D'Urville, Voyage au Pole Sud, Hist. t. i. 1842, p. 31 and 32) at 1900 toises. But

Lancerote, with a volcano, la Corona, of 300 toises elevation (Leop. v. Buch, Canarische Inseln, S. 104), and Fortaventura, are much nearer to the main-land than Teneriffe: the distance of the first-named island being 1° 15′, and that of the second 1° 2′.

(177) p. 132.—Ross only mentions this assertion as a report. (Hellenika, Bd. i. S. 11.) May the supposed observation have rested on a mere illusion? If we take the elevation of Etna above the sea at 1704 toises (lat. 37° 45′, long. from Paris 12° 41′), and that of the place of observation, on the Taygetos (the Elias Mountain), at 1236 toises (lat. 36° 57′, long. from Paris 20° 1′), and the distance between the two at 352 geographical miles, we have for the point above Etna, receiving light from it, and being visible on Taygetos (or for the cloud perpendicularly above the luminous column of smoke, and reflecting its light), an elevation of 7612 toises, or 4½ times greater than that of Etna. But if, as my friend Professor Encke has remarked, we might assume the reflecting surface to be that of a cloud placed nearly intermediately between Etna and Taygetos, then its height above the sea would only require to be 286 toises.

(178) p. 133.—Strabo, lib. xvi. p. 767, Casaub. According to Polybius, both the Euxine and the Adriatic could be seen from the Aimon mountains; Strabo was already aware of the inadmissibility of such a supposition (lib. vii. p. 313.) Compare Seymnus, p. 93.

(179) p. 133.—On the synonymes of Ophir, see my Examen Crit. de l'Hist. de la Geographie, t. ii. p. 42. Ptolemy, in lib. vi. cap. 7, p. 156, speaks of a Sapphara, metropolis of Arabia; and in lib. vii. cap. 1, p. 168, of Supara, in the Gulf of Camboya (Barigazenus sinus, according to Hesychius), "a country rich in gold"! Supara signifies in Indian, fair shore (Schönufer.) (Lassen, Diss. de Tapobrane, p. 18, indische Altersthumkunde, Bd. i. S. 107; Keil, Professor in Dorpat, über die Hiram-Salomonische Schiffahrt nach Ophir und Tarsis, S. 40—45.)

(180) p. 133.—Whether ships of Tarshish mean ocean ships, or whether, as Michaelis contends, they have their name from the Phœnician Tarsus, in Cilicia? see Keil, S. 7, 15—22, and 71—84.

(181) p. 133.—Gesenius, Thesaurus Linguæ Hebr. t. i. p. 141; and the same in the Encycl. of Ersch and Gruber, Sect. III. Th. iv. S. 401; Lassen, ind. Alterthumskunde, Bd. i. S. 538; Reinaud, Relation des Voyages faits par les Arabes dans l'Inde et en Chine, t. i. 1845, p. xxviii. The learned Quatremère, who, in a very recently published treatise (Mém. de l'Acad. des Inscriptions, t. xv. pt. 2, 1845, p. 349—402), again considers, with Heeren,

Ophir to be the east coast of Africa, explains the thukkiim (thukkiyyim) to mean not peacocks, but parrots, or Guinea-fowls (p. 375.) Respecting Socotora, compare Bohlen, das alte Indien, Th. ii. S. 139, with Benfey, Indien, S. 30—32. Sofala is described as a country rich in gold by Edrisi (in Amédée Jaubert's translation, t. i. p. 67), and subsequently by the Portuguese, after Gama's voyage of discovery (Barros, Dec. I. liv. x. cap. i; P. ii. p. 375; Külb, Geschichte der Entdeckungs reisen, Th. i. 1841, S. 236.) I have called attention elsewhere to the circumstance that Edrisi, in the middle of the 12th century, speaks of the employment of quicksilver in the goldwashings made by the negroes in this country, as a long known practice. Remembering the great frequency of the interchange of r and l, we find the name of the east African Sofala perfectly equivalent to that of Sophara, which is used in the Septuagint, with several other forms, for the Ophir of Solomon's and Hiram's fleet. Ptolemy also, as has been noticed above (Note 179), speaks of a Sapphara, in Arabia (Ritter, Asien, Bd. viii. 1846, S. 252), and a Supara in India. The significant Sanscrit names of the mother country had been repeated, or, as it were, reflected on neighbouring or opposite coasts: we find similar relations in the present day in the Spanish and English Americas. The range of the trade to Ophir might thus, according to my view, be extended over a wide space, just as a Phœnician voyage to Tartessus might include touching at Cyrene and Carthage, Gadeira and Cerne; and one to the Cassiterides might embrace the Artabrian, British, and East Cimbrian coasts. It is, however, remarkable, that we do not find incense, spices, and silk and cotton cloth, named among the wares from Ophir, together with ivory, apes, and peacocks. The latter are exclusively Indian, although, from their gradual extension to the westward, they were often called by the Greeks "Median and Persian birds:" the Samians even supposed them to have been originally belonging to Samos, on account of the peacocks kept by the priests in the sanctuary of Hera. From a passage in Eustathius (Comm. in Iliad. t. iv. p. 225, ed. Lips. 1827), on the sacredness of peacocks in Libya, it has been unduly inferred that the *ταως* also belonged to Africa.

(182) p. 999.—See Columbus on Ophir, and el Monte Sopora, "which Solomon's fleet could only reach in three years," in Navarrete, viages y descubrimientos que hicieron los Españoles, t. i. p. 103. The great discoverer says elsewhere, still in the hope of reaching Ophir, "the excellence and power of the gold of Ophir are indescribable; he who possesses it does what he wills in this world; nay, it even avails him to draw souls from purgatory to paradise" ("llega à que echa las animas al paraiso.")—Carta del Almirante

escrita en la Jamaica, 1503; Navarrete, t. i. p. 309. Compare my Examen Critique, t. i. p. 70, and 109; t. ii. p. 38—44, and on the proper duration of the Tarshish voyage, Keil, S. 106.

(183) p. 133.—Ctesiæ Cnidii Operum Reliquiæ, ed. Felix Baehr, 1824, cap. iv. and xii. p. 248, 271, and 300. But the accounts collected by the physician at the Persian court from native sources, and therefore not altogether to be rejected, relate to districts in the north of India, and from these the gold of the Daradas must have come to Abhira, the mouth of the Indus, and the coast of Malabar, by many circuitous routes. (Compare my Asie Centrale, t. i. p. 157, and Lassen, ind. Alterthumskunde, Bd. i. S. 5.) Is it not probable that the wonderful story repeated by Ctesias, of an Indian spring, at the bottom of which malleable iron was found when the fluid gold had run off, was based on a misunderstood account of a foundry? The molten iron was taken for gold from its colour; and when the yellow colour had disappeared in cooling, the black mass of iron was found underneath.

(184) p. 134.—Aristot. Mirab. Auscult. cap. 86 and 111, p. 175 and 225, Bekk.

(185) p. 135.—Die Etrusker, by Otfried Müller, Abth. ii. S. 350; Niebuhr, Römische Geschichte, Th. ii. S. 380.

(186) p. 135.—A story was formerly repeated in Germany after Father Angelo Cortenovis, that the tomb of the hero of Clusium, Lars Porsena, described by Varro, ornamented with a bronze hat and bronze pendent chains, was an apparatus for atmospherical electricity, or for conducting lightning; (as were, according to Michaelis, the metal points on Solomon's temple;) but the tale obtained currency at a time when men were much inclined to attribute to ancient nations the remains of a supernaturally revealed primitive knowledge which was soon after obscured. The most important ancient notice of the relations between lightning and conducting metals (a fact not difficult of discovery), still appears to me to be that of Ctesias (Indica, cap. 4, p. 169, ed. Lion; p. 248, ed. Baehr). He had possessed two iron swords, presents from the king Artaxerxes Mnemon, and from his mother Parysatis, which, when planted in the earth, averted clouds, hail, and strokes of lightning. He had himself seen the operation, for the king had twice made the experiment before his eyes." The exact attention paid by the Etruscans to the meteorological processes of the atmosphere in all that deviated from the ordinary course of phenomena, makes it certainly to be lamented that nothing has come down to us from their Fulgural books. The epochs of the appearance of great comets, of the fall of meteoric stones, and of showers of falling stars, would

no doubt have been found recorded in them, as in the more ancient Chinese annals, of which Edouard Biot has made use. Creuzer (Symbolik und Mythologie der alten Völker, Th. iii. 1842, S. 659) has attempted to show, that the natural features of Etruria may have influenced the peculiar turn of mind of its inhabitants. A "calling forth" of the lightning, which is ascribed to Prometheus, reminds us of the pretended "drawing down" of lightning by the Fulguratores. This operation consisted in a mere conjuration, and may well have been of no more efficacy than the skinned ass's head, which, in the Etruscan rites, was considered a preservative from danger in thunder storms.

(187) p. 135.—Otfr. Müller, Etrusker, Abth. ii. S. 162 to 178. In the very complicated Etruscan augural theory, a distinction was made between the "soft reminding lightnings sent by Jupiter from his own perfect power, and the violent electrical explosions or chastening thunderbolts which he might only send constitutionally after consultation with the other twelve gods." (Seneca, Nat. Quæst. ii. p. 41.)

(188) p. 135.—Joh. Lydus de Ostentis, ed. Hase, p. 18 in præfat.

(189) p. 136.—Strabo, lib. iii. p. 139, Casaub. Compare Wilhelm von Humboldt, über die Urbewohner Hispaniens, 1821, S. 123 and 131—136. M. de Saulcy has been recently engaged, with success, in deciphering the Iberian alphabet; the ingenious discoverer of cuneiform writing, Grotefend, the Phrygian; and Sir Charles Fellowes, the Lycian alphabet. (Compare Ross, Hellenika, Bd. i. S. 16.)

(190) p. 137.—Herod. iv. 42 (Schweighäuser ad Herod. T. v. p. 204). Compare Humboldt, Asie Centrale, T. i. p. 54 and 577.

(191) p. 138.—On the most probable etymology of Kaspapyrus of Hecatäus (Fragm. ed. Klausen, No. 179, v. 94), and Kaspatyrus of Herodotus (iii. 102, and iv. 44), see my Asie centrale, T. i. p. 101—104.

(192) p. 138.—Psemetek and Achmes. See above, Kosmos, Bd. ii. S. 159 (Engl. ed. Vol. ii. p. 125).

(193) p. 138.—Droysen, Geschichte der Bildung des hellenistischen Staatensystems, 1843, S. 23.

(194) p. 138.—See above, Kosmos, Bd. ii. S. 10 (Engl. ed. Vol. ii. p. 10).

(195) p. 139.—Völker, Mythische Geographie der Griechen und Römer, Th. i. 1832, S. 1—10; Klausen, über die Wanderungen der Io und des Herakles, in Niebuhr und Brandis rheinischen Museen für Philologie, Geschichte und griech. Philosophie, Jahrg. iii. 1829, S. 293—323.

(196) p. 139.—In the mythus of Abaris (Herod. iv. 36), the man does not travel through the air on an arrow, but carries the arrow "which Pythagoras

gave him (Iambl. de Vita Pythag. xxix. p. 194, Kiessling), in order that it might be useful to him in all difficulties during long wanderings." Creuzer, Symbolik, Th. ii. 1841, S. 660—664. On the repeatedly disappearing and reappearing Arimaspian bard, Aristeas of Proconnesus, vide Herod. iv. 13—15.

(197) p. 139.—Strabo, lib. i. p. 38, Casaub.

(198) p. 140.—Probably the valley of the Don or of the Kuban; compare my Asie Centrale, T. ii. p. 164. Pherecydes says expressly (fragm. 37 ex Schol. Apollon. ii. 1214), that the Caucasus burned, and therefore Typhon fled to Italy; from which Klausen, in the work above referred to (S. 298), explains the ideal relation of the "fire kindler" (πυρχαευς), Prometheus, to the burning mountain. Although the geological constitution of the Caucasus, which has been very recently well examined by Abich, and its connection with the volcanic chain of the Thian-schan, in the interior of Asia (which connection has, I think, been shown by me in my Asie Centrale, T. ii. p. 55—59), render it by no means improbable that very early traditions may have preserved reminiscences of great volcanic eruptions; yet it is rather to be assumed, that the Greeks may have been led to the hypothesis of the "burning" by etymological circumstances. On the Sanscrit etymologies of Graucasus (Glansberg?) (or shining mountain), see Bohlen's and Burnouf's statements, in my Asie Centrale, T. i. p. 109.

(199) p. 140.—Otfried Müller, Minyer, S. 247, 254, and 274. Homer was not acquainted with the Phasis, or with Colchis, or with the pillars of Hercules; but Hesiod names the Phasis. The mythical narrations concerning the return of the Argonauts by the Phasis into the Eastern Ocean, and the "double" Triton lake, formed either by the pretended bifurcation of the Ister, or by volcanic earthquakes (Asie Centrale, T. i. p. 179, T. iii. p. 135—137; Otfr. Müller, Minyer, S. 357), are particularly important towards a knowledge of the earliest views entertained regarding the form of the continents. The geographical fancies of Peisandros, Timagetus, and Apollonius of Rhodes, were propagated until late in the middle ages, operating sometimes as bewildering and deterring obstacles, and sometimes as stimulating incitements to actual discoveries. This reaction of antiquity upon later times, when men were almost more led by opinions than by actual observations, has not been hitherto sufficiently regarded in the history of geography. The object of the notes to Cosmos is not merely to present bibliographical sources from the literature of different nations, for the elucidation or illustration of statements contained in the text, but I have also desired to deposit in these notes, which permit greater freedom, such abundant materials for reflection as I

have been able to gather from my own experience, and from long-continued literary studies.

(200) p. 141.—Hecatæi fragm. ed. Klausen, p. 39, 92, 98, and 119. See also my investigations on the history of the geography of the Caspian Sea, from Herodotus down to the Arabian El-Istachri, Edrisi, and Ibn-el-Vardi, on the sea of Aral, and on the bifurcation of the Oxus and the Araxes, in my Asie Centrale, T. ii. p. 162—297.

(201) p. 141.—Cramer de Studiis quæ veteres ad aliarum gentium contulerint linguas, 1844, p. 8 and 17. The ancient Colchians appear to have been identical with the tribe of Lazi (Lazi, gentes Colchorum, Plin. vi. 4; the Λαζοι of Byzantine writers); see Vater (Professor in Kasan), der Argonautenzug aus den Quellen dargestellt, 1845, Heft i. S. 24; Heft ii. S. 45, 57, and 103. In the Caucasus, the names Alani (Alanethi for the land of the Alani), Ossi, an ass, may still be heard. According to the investigations commenced with philosophic and linguistic acumen in the valleys of the Caucasus by George Rose, the language spoken by the Lazi would appear to contain remains of the ancient Colchian idiom. The Iberian and Grusic group of languages includes Lazian, Georgian, Suanian, and Mingrelian, all belonging to the family of the Indo-Germanic languages. The language of the Ossates is nearer to the Gothic than to the Lithuanian.

(202) p. 141.—On the relationship of the Scythians (Scolotes or Sacæ), Alani, Goths, Massa-Getæ, and the Yueti of the Chinese historians, see Klaproth, in the commentary to the Voyage du Comte Potocki, T. i. p. 129, as well as my Asie Centrale, T. i. p. 400; T. ii. p. 252. Procopius himself says very distinctly (De Bello Gothico, iv. 5 ed. Bonn, 1833, Vol. ii. p. 476,) that the Goths were formerly called Scythians. The identity of the Getæ and the Goths has been shewn by Jacob Grimm in his recently-published work, über Jornandes, 1846, S. 21. Niebuhr believed (see his Untersuchungen über die Geten und Sarmaten, in his kleinen histor. und philologischen Schriften, 1te Sammlung, 1828, S. 362, 364, and 395,) that the Scythians of Herodotus belong to the family of the Mongolian tribes; but this opinion has the less probability, since these tribes, partly under the yoke of the Chinese, and partly under that of the Hakas or Kirghis (Χερχις of Menander) still lived far in the east of Asia round Lake Baikal in the beginning of the 13th century. Herodotus distinguishes, moreover, the bald-headed Argippæans (iv. 23) from the Scythians; and if the first-named are said to be "flat-nosed," they have at the same time "a long chin," which, according to my experience, is by no means a physiognomic characteristic of the Calmucks

or other Mongolian races, but rather characterises the blonde (Germanising?) Ousun and Tingling, to whom the Chinese historians attribute "long horse faces."

(203) p. 141.—On the dwelling-place of the Arimaspes and the gold trade of north-western Asia in the time of Herodotus, see my Asie Centrale, T. i. p. 389—407.

(204) p. 141.—"Les Hyperboréens sont un mythe météorologique. Le vent des montagnes (B'Oreas) sort des Monts Rhipéens. Au delà de ces monts, doit regner un air calme, un climat heureux, comme sur les sommets alpins dans la partie qui dépasse les nuages. Ce sont là les premiers aperçus d'une physique qui explique la distribution de la chaleur et la difference des climats par des causes locales, par la direction des vents qui dominent, par la proximité du soleil, par l'action d'un principe humide ou salin. La conséquence de ces idées systematiques était une certaine indépendence qu'on supposait entre les climats et la latitude des lieux: aussi le mythe des Hyperboréens lié par son origine au culte dorien et primitivement boréal d'Apollon, a pu se déplacer du nord vers l'ouest, en suivant Hercule dans ses courses aux sources de l'Ister, à l'île d'Erythia et aux Jardins des Hespérides. Les Rhipes, ou Monts Rhipéens, sont aussi un nom significatif météorologique: ce sont les montagnes de "l'impulsion," ou du souffle glacé (ῥιπη) celles d'où se dechaînent les tempêtes boréales" (Asie Centr. T. i. p. 392 and 403).

(205) p. 142.—In Hindostanee, as Wilford has already remarked, there are two words which might easily be confounded; one of which, tschiûntâ, a large black kind of ant (whence the diminutive tschiunti, tschinti, the small common ant); the other tschitâ, a spotted kind of panther, the little hunting leopard (cheetah; the Felis jubata, Schreb). The latter word (tschitâ) is the Sanscrit word tschitra, signifying variegated or spotted, as is shewn by the Bengalee name for the animal (tschitâbâgh and tschitibâgh, from bâgh, Sanscrit wyâghra, tiger.)—Buschmann. A passage has been recently discovered in the Mahabharata (ii. 1860) in which there is question of the ant-gold. "Wilso invenit (Journ of the Asiat. Sec. vii. 1843, p. 143,) mentionem fieri etiam in Indicis litteris bestiarum aurum effodientium, quas, quum terram effodiant, eodem nomine (pipilica) atque formicas Indi nuncupant." Compare Schwanbeck, in Megasth; Indicis, 1846, p. 73. I have been struck by seeing that in the basaltic districts of the Mexican highlands the ants carry to their heaps shining grains of hyalite, which I could collect out of the ant-hills.

(206) p. 145.—Strabo, lib iii. p. 172 (Bökh, Pind. Fragm. v. 155). The voyage

of Colæus of Samos is placed in Ol. 31, according to Otfr. Müller (Prolegomena zu einer wissenschaftlichen Mythologie); and in Ol. 35, 1, or the year 640, according to Letronne's investigation (Essai sur les idées cosmographiques qui se rattachent au nom d'Atlas, p. 9). The epoch is, however, dependent on the foundation of Cyrene, which Otfr. Müller places between Ol. 35 and 37 (Minyer, S. 344, Prolegomena, S. 63); for in the time of Colæus (Herod. iv. 152), the way from Thera to Lybia was still unknown. Zumpt places the foundation of Carthage in 878, and that of Gades in 1100 B.C.

(207) p. 146.—According to the manner of the ancients (Strabo, lib. ii. p. 126, I reckon (as indeed physical and geological views require) the whole Euxine, together with the Mæotis, as forming part of the common basin of the great "Interior Sea."

(208) p. 146.—Herod. iv. 152.

(209) p. 146.—Herod. . 163, where even the discovery of Tartessus is attributed to the Phocæans; but according to Ukert (Geogr. der Griechen und Römer, Th. I. i. S. 40), the commercial enterprise of the Phocæans was seventy years later than Colæus of Samos.

(210) p. 146.—According to a fragment of Phavorinus, the words (ωκεανος, and therefore ωγην also) are not Greek, but are borrowed from the barbarians (Spohn de Nicephor. Blemm. duobus opusculis, 1818, p. 23). My brother thought that they were connected with the Sanscrit roots ogha and ogh (see my Examen critique de l'hist. de la Géogr. T. i. p. 33 and 182).

(211) p. 147.—Aristot. de Cœlo, ii. 14 (p. 298, b, Bekk.); Meteor. ii. 5 (p. 362, Bekk.) Compare my Examen critique, T. i. p. 125—130. Seneca ventures to say (Nat. Quæst. in præfat. 11), contemnet curiosus spectator domicilii (terræ) angustias. Quantum enim est quod ab ultimis littoribus Hispaniæ usque ad Indos jacet? Paucissimorum dierum spatium, si navem suus ventus implevit (Examen critique, T. i. p. 158).

(212) p. 147.—Strabo, lib. i. p. 65 and 118, Casaub. (Examen critique, T. i. p. 152.)

(213) p. 147.—In the Diaphragma (the dividing line of the Earth) of Dicearchus, the elevation passes through the Taurus, the chains of Demavend and Hindoo-koosh, the Kuen-lün of Northern Thibet, and the perpetually snow-clad cloud mountains of the Chinese provinces, Sse-tschuan and Kuang-si. See my orographic researches on these lines of elevation in my Asie Centrale, T. i. p. 104—114, 118—164; T. ii. p. 413 and 438.

(214) p. 148.—Strabo, lib. iii. p. 173 (Examen. crit. T. iii. p. 98).

([215]) p. 150.—Droysen, Gesch. Alexanders des Grossen, S. 544; the same in his Gesch. der Bildung des hellenistischen Staatensystems, S. 23—34, 588—592, 748—755.

([216]) p. 150.—Aristot. Polit. VII. vii. p. 1327, Bekker. (Compare also III. xvi., and the remarkable passage of Eratosthenes in Strabo, lib. i. p. 66 and 97, Casaub.)

([217]) p. 151.—Stahr, Aristotelia, Th. ii. S. 114.

([218]) p. 151.—Ste. Croix, Examen critique des historiens d'Alexandre, p. 731. (Schlegel, Ind. Bibliothek, Bd. i. S. 150.)

([219]) p. 153.—Compare Schwanbeck "de fide Megasthenis et pretio," in his edition of that writer, p. 59—77. Megasthenes often visited Palibothra, the court of the King of Magadha. He was fully initiated in the system of Indian chronology, and relates "how, in the past, the All had three times come to freedom; how three ages of the world had run their course, and in his own time the fourth had begun." (Lassen, indische Alterthumskunde, Bd. i. S. 510.) The Hesiodic doctrine of four ages of the world, connected with four great elementary destructions, which together occupy a period of 18028 years, existed also among the Mexicans. (Humboldt, Vues de Cordilleres et Monumens des peuples indigènes de l'Amerique, T. ii. p. 119—129.) In modern times a remarkable proof of the accuracy of Megasthenes has been afforded by the study of the Rigveda and the Mahabharata. Consult what Megasthenes says respecting "the land of the long-living happy persons" in the extreme north of India,—the land of Uttara-kuru (probably north of Kashmeer, towards Belurtagh), which, according to his Grecian views, he connects with the supposed "thousand years of life of the Hyperboreans." (Lassen, in the Zeitschrift für die kunde des Morgenlandes, Bd. ii. S. 62.) We may notice, in connection with this, a tradition mentioned in Ctesias, of a sacred place in the Northern Desert. (Ind. cap. viii. ed. Baehr, p. 249 and 285.) Ctesias has been long too little esteemed: the martichoras mentioned by Aristotle (Hist. de Animal. II. iii. § 10; T. i. p. 51, Schneider), the griffin, half eagle half lion, the kartazonon spoken of by Ælian, and a one-horned wild ass, are indeed referred to by him as real animals; but this was not an invention of his own, but arose, as Heeren and Cuvier have remarked, from his taking pictured forms of symbolical animals, seen on Persian monuments, for the representation of strange beasts still living in India. The acute Guignaut has, however, noticed that there is much difficulty in identifying the martichoras with Persepolitan symbols. (Creuzer, Religions de l'Antiquité; notes et éclaircissements, p. 720.)

([220]) p. 154.—I have illustrated these intricate orographical relations in my Asie Centrale, T. ii. p. 429—434.

([221]) p. 154.—Lassen, in the Zeitschrift für die Kunde des Morgenl. Bd. i. S. 230.

([222]) p. 155.—The district between Bamian and Ghori. See Carl Zimmermann's excellent orographical tabular view of Afghanistan, 1842. (Compare Strabo, lib. xv. p. 725; Diod. Sicul. xvii. 82; Menn, Meletem. hist. 1839, p. 25 and 31; Ritter über Alexanders Feldzug am Indischen Kaukasus, in the Abhandl. der Berl. Akad. of the year 1829, S. 150; Droysen, Bildung des hellenist. Staatensystems, S. 614.) I write Paropanisus, like all the good codices of Ptolemy, and not Paropamisus. I have given the reasons in my Asie Centrale, T. i. p. 114—118. (See also Lassen zur Gesch. der Griechischen und Indoskythischen Könige, S. 128.)

([223]) p. 155.—Strabo, lib. xv. p. 717, Casaub.

([224]) p. 155.—Tala, as the name of the palm Borassus flabelliformis (very characteristically termed by Amarasinha "a king of the grasses"); Arrian, Ind. vii. 3.

([225]) p. 155.—The word tabaschir is referred to the Sanscrit tvak-kschira (bark milk); see above, Note 143. In 1817, in the historical addenda to my work De Distributione Geographica Plantarum, secundum Cœli, Temperiem et Altitudinem Montium, p. 215, I called attention to the fact, that the companions of Alexander became acquainted with the true sugar of the sugar cane of the Indians, as well as with the tabaschir of the bamboo. (Strabo, lib. xv. p. 693; Peripl. maris Erythr. p. 9.) Moses of Chorene, who lived in the middle of the 5th century, was the first who described circumstantially the preparation of sugar from the juice of the Saccharum officinarum, in the province of Khorasan. (Geogr. ed. Whiston, 1736, p. 364.)

([226]) p. 155.—Strabo, lib. xv. p. 694.

([227]) p. 155.—Ritter, Erdkunde von Asien, Bd. IV. i. S. 437; Bd. VI. i. S. 698; Lassen, ind. Alterthumskunde, Bd. I. S. 317—323. The passage in Aristotle's Hist. de Animal. v. 17 (T. i. p. 209, ed. Schneider), respecting the web of a great horned spider, relates to the island of Cos.

([228]) p. 155.—So λαχχος χρωματινος, in the Peripl. maris Erythr. p. 5 (Lassen, S. 316.)

([229]) p. 155.—Plin. Hist. Nat. xvi. 32. (On the introduction of rare plants from Asia into Egypt by the Lagidæ; see Pliny, xii. 14 and 17.)

([230]) p. 156.—Humboldt, De Distrib. Geogr. Plantarum, p. 178.

([231]) p. 156.—Since the year 1847, I have often corresponded with Lassen on the remarkable passage in Pliny, xii. 6:—"Major alia (arbor) pomo et suavitate præcellentior, quo sapientes Indorum vivunt. Folium alas avium imitatur, longitudine trium cubitorum, latitudine duum. Fructum cortice mittit, admirabilem succi dulcedine ut uno quaternos satiet. Arbori nomen *palæ* pomo arienæ." The following is the result of the examination of my learned friend;—"Amarasinha places the banana (musa) at the head of all nutritive plants. Among the many Sanscrit names which he mentions, are, varanabuscha, bhanuphala (sun fruit), and moko, whence the Arabic mauza. Phala (pala) is fruit in general; and it is therefore only by a misunderstanding that it has been taken for the name of the plant. In Sanscrit varana without buscha is not the name of the banana, although the abbreviation may have belonged to the popular language. Varana would be in Greek *ουαρενα*, which is certainly not very far removed from ariena." (Compare Lassen, ind. Alterthumskunde, Bd. i. S. 262; my Essai politique sur la Nouv. Espagne, T. ii. 1827, p. 382; Relation hist. T. i. p. 491.) The chemical connection of the nourishing amylum with saccharin was divined alike by Prosper Alpinus and Abd-Allatif, since they sought to explain the origin of the banana by the insertion of the sugar cane, or the sweet date fruit, into the root of the colocasia. (Abd-Allatif, Relation de l'Egypte, traduit par Silvestre de Sacy, p. 28 and 105.)

(232) p. 156.—Respecting this epoch, consult Wilhelm von Humboldt in his work, über die Kawi-Sprache und die Verschiedenheit des menschlichen Sprachbaues, Bd. i. S. ccl. and ccliv.; Droysen, Gesch. Alexanders des Gr. S. 547; and Hellenist. Staatensystem, S. 24.

(233) p. 157.—Dante, Inf. iv. 130.

(234) p. 157.—Compare Cuvier's assertions in the Biographie universelle, T. ii. 1811, p. 458 (and unfortunately again repeated in the edition of 1843, T. ii. p. 219), with Stahr's Aristotelia, Th. i. S. 15 and 108.

(235) p. 157.—Cuvier, when engaged on the Life of Aristotle, believed that the philosopher had accompanied Alexander to Egypt, "whence," he says, "the Stagyrite must have brought back to Athens (Ol. 112, 2) all the materials for the Historia Animalium." Subsequently, in 1830, Cuvier abandoned this opinion; for after more examination he remarked, "that the descriptions of Egyptian animals were not taken from the life, but from notices by Herodotus." (See also Cuvier, Histoire des sciences naturelles, publiée par Magdeleine de Saint Agy, T. i. 1841, p. 136.)

(236) p. 157.—Among these internal indications may be enumerated,—the statement of the perfect insulation of the Caspian; the notice of the great comet which appeared when Nicomachus was Archon, Ol. 109, 4 (according

to Corsini), and which is not to be confounded with that which Herr von Boguslawski has named the comet of Aristotle (seen when Asteus was Archon, Ol. 101, 4; Aristot. Meteor. lib. i. cap. 6, 10; vol. i. p. 395, Ideler; and supposed to be identical with the comets of 1695 and 1843?); and also the mention of the destruction of the temple at Ephesus, as well as of a lunar rainbow, seen on two occasions in the course of fifty years. (Compare Schneider ad Aristot. Hist. de Animalibus, Vol. i. p. xl. xlii. ciii. and cxx.; Ideler ad Aristot. Meteor. Vol. I. p. x.; and Humboldt, Asie Cent. T. ii. p. 168.) We know that the "History of Animals," was written later than the "Meteorologica," since the last-named work alludes to the former as soon to follow. (Meteor. i. 1, 3; and iv. 12, 13.)

(237) p. 158.—The five animals named in the text, and especially the hippelaphus (horse-stag with a long mane), the hippardion, the Bactrian camel, and the buffalo, are adduced by Cuvier as proofs of the Historía Animalium having been written by Aristotle at a later period. (Hist. des Sciences Nat. T. i. p. 154.) Cuvier, in the fourth volume of the Recherches sur les Ossemens fossiles, 1823, p. 40-43 and p. 502, distinguishes between two Asiatic stags with manes, which he calls Cervus hippelaphus and Cervus aristotelis. At first he regarded the Cervus hippelaphus, of which he had seen a living example, and of which Diard had sent him skins and antlers from Sumatra, as Aristotle's hippelaphus from Arachosia, (Hist. de Animal., ii. 2, § 3, and 4, T. i. p. 43-44, Schneider): subsequently he judged that a stag's head sent to him from Bengal by Duvaucel, and the drawing of the entire large animal, agreed still better with the Stagirite's description of the hippelaphus; and this stag, which is indigenous in the mountains of Sylhet in Bengal, in Nepaul, and in the country east of the Indus, then received the name of Cervus aristotelis. If, in the same chapter in which Aristotle treats generally of animals with manes, he names together with the horse-stag (Equicervus), the Indian Guepard or hunting tiger (Felis jubata). Schneider (T. iii. p. 66) considers the reading παρδιον to be preferable to that of το ιππαρδιον. The latter reading, as Pallas also thinks, (Spicileg. Zool. fasc. i. p. 4), would be best interpreted to mean the giraffe. If Aristotle had himself seen the Guepard, and not merely heard it described, how can we suppose that he would have failed to notice non-retractile claws in a feline animal? It is equally surprising how Aristotle, who is always so accurate, if, as August Wilhelm von Schlegel maintains, he had a menagerie near his residence at Athens, and had himself dissected an elephant which had been taken at Arbela, could have failed to describe a small opening near the temples

of the elephant, which, at certain seasons particularly, secretes a strong smelling fluid, often alluded to by the Indian poets. (Schlegel's Indische Bibliothek, Bd. i. S. 163-166.) I notice this apparently trifling circumstance thus particularly, because this small aperture was made known by accounts given by Megasthenes, to whom, nevertheless, no one would be led to attribute anatomical knowledge. (Strabo, lib. xv. p. 704 and 705, Casaub.) I do not find in the different zoological works of Aristotle which have come down to us anything which necessarily implies his having had the opportunity of observing living elephants, or of his having dissected a dead one. Although it is most probable that the Historia Animalium was completed before Alexander's campaigns in Asia Minor, yet it is undoubtedly *possible* that the work may, as Stahr supposes (Aristotelia, Th. ii. S. 98), have continued to receive additions until the end of the Author's life, Ol. 114, 3, three years after the death of Alexander; but direct evidence of such being the case is wanting. The correspondence of Aristotle which we possess is not genuine, (Stahr, Th. i. S. 194—208, Th. ii. S. 169-234), and Schneider says very confidently, (Hist. de Animal. T. i. p. 40), "hoc enim tanquam certissimum sumere mihi licebit scriptas comitum Alexandri notitias post mortem demum regis fuisse vulgatas."

(238) p. 158.—I have shewn elsewhere that although the decomposition of sulphuret of mercury by distillation is described in Dioscorides, (Mat. Med. v. 110, p. 667, Saracen); yet the first description of the distillation of a fluid, (the distillation of fresh water from sea water), is to be found in the Commentary of Alexander of Aphrodisias to Aristotle de Meteorol.; see my Examen critique de l'histoire de la Géographie, T. ii. p. 308-316, and Joannis (Philoponi) Grammatici in libro de Generat. et Alexandri Aphrod. in Meteorol. Comm. Venet. 1527, p. 97, b. Alexander of Aphrodisias in Caria, the learned commentator of the Meteorologica of Aristotle, lived under the reigns of Septimius Severus and Caracalla; and although in his writings chemical apparatuses are called χυικα οϱγανα, yet a passage in Plutarch (de Iside et Osir, c. 33), proves that the word Chemie, applied by the Greeks to the Egyptian art, is not to be derived from χεω, (Hoefer, Histoire de la Chimie, T. i. p. 91, 195 and 219, T. ii. p. 109).

(239) p. 158.—Compare Sainte-Croix, Examen des historiens d'Alexandre, 1810, p. 207, and Cuvier, Histoire des Sciences naturelles, T. i. p. 137, with Schneider ad Aristot. de Historiâ Animalium, T. i. p. 42-46, and Stahr, Aristotelia, Th. i. S. 116-118. If the transmission of specimens from Egypt and the interior of Asia appears according to these authorities to be very

improbable, yet the latest writings of our great anatomist Johannes Müller shew with what wonderful acuteness and delicacy Aristotle dissected the fishes of the Greek seas. See the learned treatise of Johannes Müller on the adherence of the egg to the uterus in one of the two species of the genus Mustelus living in the Mediterranean, which in its fœtal state possesses a placenta of the vitelline vesicle which is connected with the uterine placenta of the mother; and his researches on the γαλεος λειος of Aristotle in the Abhandl. der Berliner Akad. aus d. j. 1840, S. 192-197. (Compare Aristot. Hist. Anim. vi. 10, and de Gener. Anim. iii. 3.) The fineness of Aristotle's own anatomical examinations is testified by the distinction and detailed analysis of the species of cuttle-fish, the description of the teeth of snails, and the organs of other Gasteropodes. Compare Hist. Anim. iv. 1 and 4, with Lebert in Müller's Archiv der Physiologie, 1846, S. 463 and 467. I have myself in 1797 called the attention of modern naturalists to the form of snails' teeth. See my Versuche über die gereizte Muskel und Nervenfaser, Bd. i. S. 261.

(240) p. 159.—Valer. Maxim. vii. 2; "ut cum rege aut rarissime aut quam jucundissime loqueretur."

(241) p. 160.—Aristot. Polit. i. 8, and Eth. ad Eudemum, vii. 14.

(242) p. 160.—Strabo, lib. xv. p. 690 and 695. Herod. iii. 101.

(243) p. 160.—Thus says Theodectes of Phaselis; see Kosmos, Bd. i. S. 380 and 491, (Engl. trans. Vol. i. p. 352 and note 437). Northern countries were placed to the West, and southern countries to the East. Consult Völcker über Homerische Geographie und Weltkunde, S. 43 and 87. The indefiniteness, even at that period, of the word Indies, as connected with geographical position, with the complexion of the inhabitants, and with precious natural productions, contributed to the extension of these meteorological hypotheses, for it was given at once to Western Arabia, to the countries between Ceylon and the mouth of the Indus, to Troglodytic Ethiopia, and to the African myrrh and cinnamon lands south of Cape Aromata, (Humboldt, Examen crit. T. ii. p. 35).

(244) p. 161.—Lassen ind. Alterthumskunde, Bd. i. S. 369, 372-375, 379 and 389; Ritter, Asien, Bd. iv. 1, S. 446.

(245) p. 161.—The geographical distribution of mankind is not more determinable in entire continents by degrees of latitude than that of plants and animals. The axiom propounded by Ptolemy, (Geogr. lib. cap. 9), that north of the parallel of Agisymba neither elephants, rhinoceroses, nor negroes are to be met with, is entirely unfounded. (Examen critique, T. i. p. 39.) The doctrine of the universal influence of soil and climate on the intellectual capacities and dispositions, and on the civilisation of mankind, was peculiar to

the Alexandrian school of Ammonius Sakkas, and especially to Longinus. See Proclus, Comment. in Tim. p. 50.

([246]) p. 161.—See Georg. Curtius, die Sprachvergleichung in ihrem Verhältniss zur classichen Philologie, 1845, S. 5-7, and the same author's Bildung der Tempora und Modi, 1846, S. 3-9. (Compare also Pott's Article entitled Indogermanischer Sprachstamm in the Allgem. Encyklopädie of Ersch and Gruber, Sect. ii. Th. xviii. S. 1-112.) Investigations on language in general, as touching upon the fundamental relations of thought, are, however, to be found in Aristotle, where he develops the connection of categories with grammatical relations. See the luminous statement of this comparison in Adolf Treudelenburg's histor. Beitragen zur Philosophie, 1846, Th. i. S. 23—32.

([247]) p. 162.—The schools of the Orchenes and Vorsipenes (Strabo, lib. xvi. p. 739). In this passage, in conjunction with the Chaldean astronomers, four Chaldean mathematicians are cited by name. This circumstance is of the greater historical importance, because Ptolemy always designates the observers of the heavenly bodies by the collective name of Χαλδαιοι, as if the Babylonish observations were only made "collegiately" (Ideler, Handbuch der Chronologie, Bd. i. 1825, S. 198).

([248]) p. 162.—Ideler, Handbuch der Chronologie, Bd. i. S. 202, 206, and 218. When doubts are raised respecting the fact of Callisthenes having sent astronomical observations from Babylon to Greece, on the ground of "no trace of these observations of a Chaldean priestly caste being found in the writings of Aristotle," (Delambre, Hist. de l'Astron. anc. T. i. p. 308), it is forgotten that Aristotle, where he speaks (De Cœlo, lib. ii. cap. 12), of an occultation of Mars by the moon observed by himself, expressly adds, that "similar observations had been made for many years on the other planets by the Egyptians and the Babylonians, many of which have come to our knowledge." On the probable use of astronomical tables by the Chaldeans, see Chasles, in the Comptes rendus de l'Academie des Sciences, T. xxiii. 1846, p. 852—854.

([249]) p. 163.—Seneca, Nat. Quæst. vii. 17.

([250]) p. 163.—Compare Strabo, lib. xvi. p. 739, with lib. iii. p. 174.

([251]) p. 163.—These investigations belong to the year 1824 (see Guignaut, Religions de l'Antiquité, ouvrage traduit de l'Allemand de F. Creuzer, T. i. P. 2, p. 928). See farther, Letronne, in the Journal des Savans, 1839, p. 338 and 492 ; as well as the Analyse critique des Représentations zodiacales en Egypte, 1846, p. 15 and 34. (Compare with these Ideler über den Ursprung des Thierkreises, in den Abhandlungen der Akademie des Wissenschaften zu Berlin aus dem Jahr 1838, S. 21.)

(252) p. 163.—The magnificent Cedrus deodvara (Kosmos, Bd. i. S. 43; Engl. trans. Vol. i. p. 363, note 4), which is most abundant at an elevation of from eight to eleven thousand feet on the upper Hydaspes (Behut), which flows through the lake of the Alpine valley of Kashmeer, supplied the materials for the fleet of Nearchus (Burnes' Travels, Vol. i. p. 59). The trunk of this cedar has often a circumference of forty feet, according to Dr. Hoffmeister, of whom science has unhappily been deprived, by his death on a field of battle, when accompanying Prince Waldemar of Prussia.

(253) p. 163.—Lassen, in his Pentapotamia indica, p. 25, 29, 57—62, and 77; and also in his indischen Alterthumskunde, Bd. i. S. 91. Between the Sarasvati, to the north-west of Delhi, and the rocky Drischadvati, there is situated, according to Menu's code of laws, Brahmavarta, a priestly district of Brahma, established by the gods themselves; on the other hand, in the more extensive sense of the word, Aryavarta, the land of the worthy, signifies, in the ancient Indian geography, the whole country east of the Indus, between the Himalaya and the Vindhya chain; to the south of which the ancient non-Arian aboriginal population commences. Madhya-Desa, the central land referred to in Kosmos, Bd. i. S. 15 (English trans. Vol. i. p. 14), was only a portion of Aryavarta. Compare my Asie centrale, T. i. p. 204, and Lassen, ind. Alterthumsk. Bd. i. S. 5, 10, and 93. The ancient Indian free states, the countries of the kingless, (condemned by the orthodox eastern poets), were situated between the Hydraotes and the Hyphasis, *i. e.* the present Ravi and the Beas.

(254) p. 164.—Megasthenes, Indica, ed. Schwanbeck, 1846, p. 17.

(255) p. 167.—See above, Kosmos, Bd. ii. S. 155 (English trans. Vol. ii. p. 121).

(256) p. 167.—Compare my geographical researches, Asie centrale, T. i. p. 145, and 151—157; T. ii. p. 179.

(257) p. 168.—Plin. vi. 26 (?).

(258) p. 168.—Droysen, Gesch. des hellenistischen Staatensystems, S. 749.

(259) p. 169.—Compare Lassen, indische Alterthumskunde, Bd. i. S. 107, 153, and 158.

(260) p. 169.—"Mutilated from Tâmbapanni. This Pali form sounds in Sanscrit Tâmraparni. The Greek Taprobane gives half the Sanscrit (Tâmbra, Tapro), and half the Pali" (Lassen, indische Alterthumskunde, S. 201; com-

pare Lassen, Diss. de Taprobane insula, p. 19). The Laccadives (lakke for lakscha, and dive for dwîpa, oṇe hundred thousand islands), as well as the Maldives (Malayadiba, *i. e.* islands of Malabar), were known to Alexandrian navigators.

([261]) p. 170.—Hippalus is supposed to have lived no earlier than the reign of the Emperor Claudius; but this is improbable, even though under the first Lagidæ great part of the Indian products were only purchased in Arabian markets. The south-west monsoon was itself called Hippalus, and a portion of the Erythrean or Indian Ocean is also called the Sea of Hippalus. Letronne, in the Journal des Savans, 1818, p. 405; Reinaud, Relation des Voyages dans l'Inde, T. i. p. xxx.

([262]) p. 171.—See the researches of Letronne, on the construction of the canal between the Nile and the Red Sea from Neku to the Caliph Omar, or an interval of more than 1300 years, in the Revue des deux Mondes, T. xxvii. 1841, p. 215—235. Compare also Letronne, de la Civilisation égyptienne depuis Psammitichus jusqu'à la conquête d'Alexandre, 1845, p. 16—19.

([263]) p. 171.—Meteorological speculations on the distant causes of the swelling of the Nile gave occasion to some of these journies; Philadelphus, as Strabo expresses it (lib. xvii. p. 789 and 790), "continually seeking new diversions and interests out of curiosity and bodily weakness."

([264]) p. 171.—Two hunting inscriptions, one of which "principally records the elephant hunts of Ptolemy Philadelphus," were discovered and copied by Lepsius from the colossi of Abusimbel (Ipsambul). (Compare, on this subject, Strabo, lib. xvi. p. 769 and 770; Ælian, De Nat. Anim. iii. 34, and xvii. 3; Athenæus, v. p. 196.) Although, according to the "Periplus maris Erythræi," Indian ivory was an article of export from Barygaza, yet, according to the notices of Cosmas, ivory was also exported from Ethiopia to the western peninsula of India. Since ancient times, elephants have withdrawn more to the south in eastern Africa also. According to the testimony of Polybius (v. 84), when African and Indian elephants encountered each other on fields of battle, the sight, the smell, and the cries of the larger and stronger Indian elephants drove the African ones to flight. The latter were never employed as war elephants in such large numbers as were used in Asiatic expeditions, where Chandragupta had assembled 9000, the powerful king of the Prasii 6000, and Akbar as many (Lassen, ind. Alterthumskunde, Bd. i. S. 305—307).

([265]) p. 171.—Athen. xiv. p. 654; compare Parthey, das alexandrinische Museum, eine Preisschrift, S. 55, and 171.

([266]) p. 172.—The library at Bruchium was the more ancient; it was destroyed in the burning of the fleet under Julius Cæsar. The library at Rhakotis made part of the "Serapeum," where it was combined with the museum. By the liberality of Antoninus, the collection of books at Pergamos was incorporated with the library of Rhakotis.

([267]) p. 173.—Vacherot, Histoire critique de l'Ecole d'Alexandrie, 1846, T. i. p. v. and 103. We find much evidence in antiquity, that the institute of Alexandria, like all academical corporations, together with much good arising from the concurrence of many workers, and from the power of obtaining material aids, had also some disadvantageous narrowing and restraining influence. Hadrian made his tutor, Vestinus, High Priest of Alexandria, and at the same time Head of the Museum (or President of the Academy) (Letronne, Recherches pour servir à l'Histoire de l'Egypte pendant la domination des Grecs et des Romains, 1823, p. 251).

([268]) p. 173.—Fries, Geschichte der Philosophie, Bd. ii. S. 5.; and the same author's Lehrbuch der Naturlehre, Th. i. S. 42. Compare also Considerations on the Influence which Plato exercised on the Foundation of the Experimental Sciences by the application of Mathematics, in Brandis, Geschichte der griechisch-römischen Philosophie, Th. ii. Abth. i. S. 276.

([269]) p. 174.—On the physical and geognostical opinions of Eratosthenes, see Strabo, lib. i. p. 49—56, lib. ii. p. 108.

([270]) p. 174.—Strabo, lib. xi. p. 519; Agathem. in Hudson, Geogr. Græc. Min. Vol. ii. p. 4. On the correctness of the grand orographic views of Eratosthenes, see my Asie centrale, T. i. p. 104—150, 198, 208—227, 413—415, T. ii. p. 367, and 414—435; and Examen critique de l'Hist. de la Géogr. T. i. p. 152—154. I have purposely called Eratosthenes' measurement of a degree the first *Hellenic* one, as a very ancient Chaldean determination of the magnitude of a degree in camels' paces is not improbable. See Chasles, Recherches sur l'Astronomie indienne et chaldeenne, in the Comptes rendus de l'Acad. des Sciences, T. xxiii. 1846, p. 851.

([271]) p. 175.—The latter appellation appears to me the more correct, as Strabo, lib. xvi. p. 739, cites "Seleucus of Seleucia, among several very honourable men, as a Chaldean well acquainted with the heavenly bodies." Probably Seleucia on the Tigris, a flourishing commercial city, is here meant. It is indeed singular, that the same Strabo speaks of a Seleucus as an exact observer of the ebb and flood, calling him also a Babylonian (lib. i. p. 6), and subsequently (lib. iii. p. 174), perhaps from carelessness, an Erythrean. (Compare Stobäus, Ecl. phys. p. 440.)

(272) p. 175.—Ideler, Handbuch der Chronologie, Bd. i. S. 212 and 329.

(273) p. 176.—Delambre, Histoire de l'Astronomie ancienne, T. i. p. 290.

(274) p. 176.—Bökh has examined in his Philolaos, S. 118, whether the Pythagoreans were early acquainted, through Egyptian sources, with the precession, under the name of the motion of the heaven of the fixed stars. Letronne (Observations sur les Representations zodiacales qui nous restent de l'Antiquité, 1824, p. 62) and Ideler (Handbuch der Chronol. Bd. i. S. 192) vindicate Hipparchus's exclusive claim to this discovery.

(275) p. 177.—Ideler on Eudoxus, S. 23.

(276) p. 177.—The planet discovered by Le Verrier.

(277) p. 177.—Compare Kosmos, Bd. ii. S. 141, 146, 149 and 170 (Engl. trans. Vol. ii. p. 106, 111, 114, and 145).

(278) p. 179.—Wilhelm von Humboldt über die Kawi-Sprache, Bd. i. S. xxxvii.

(279) p. 180.—The superficial extent of the Roman Empire under Augustus (according to the boundaries assumed by Heeren, in his Geschichte der Staaten des Alterthums, S. 403—470) has been calculated by Professor Berghaus, the author of the excellent Physical Atlas, at rather more than 100000 (German) geographical square miles. This is about a quarter greater than the extent of 1600000 square miles assigned by Gibbon, in his History of the Decline and Fall of the Roman Empire, Vol. i. Chap. i. p. 39, but which he indeed says must be taken as a very uncertain estimate.

(280) p. 181.—Veget. de Re Mil. iii. 6.

(281) p. 181.—Act. ii. v. 371, in the celebrated prophecy which, from the time of Columbus' son, was interpreted to relate to the discovery of America.

(282) p. 182.—Cuvier, Hist. des Sciences naturelles, T. i. p. 312—328.

(283) p. 182.—Liber Ptholemei de Opticis sive Aspectibus; the rare manuscript of the Royal Library at Paris (No. 7310), was examined by me on the occasion of discovering a remarkable passage on the refraction of rays in Sextus Empiricus (adversus Astrologos, lib. v. p. 351, Fabr.) The extracts which I made from the Parisian manuscript in 1811 (therefore before Delambre and Venturi) are given in the introduction to my Recueil d'Observations astronomiques, T. i. p. lxv.—lxx. The Greek original has not come down to us; we have only a Latin translation of two Arabic manuscripts of Ptolemy's Optics. The Latin translator gives his name as Amiracus Eugenius, Siculus. Compare Venturi, Comment. sopra la Storia e le Teorie dell' Ottica, Bologna, 1814, p. 227; Delambre, Hist. de l'Astronomie ancienne, 1817, T. i. p. 51, and T. ii. p. 410—432.

(284) p. 182.—Letronne shews, from the fanatical murder of the daughter of Theon of Alexandria, that the much contested period of Diophantus cannot fall later than the year 389 (Sur l'Origine grecque des Zodiaques prétendus égyptiens, 1837, p. 26).

(285) p. 184.—This beneficial influence of the extension of a language was finely noticed in Pliny's praise of Italy: "omnium terrarum alumna eadem et parens, numine Deûm electa, quæ sparsa congregaret imperia, ritusque molliret, et tot populorum discordes ferasque linguas sermonis commerceo contraheret, colloquia, et *humanitatem homini daret,* breviterque una cunctarum gentium in toto orbe patria fieret" (Plin. Hist. nat. iii. 5).

(286) p. 186.—Klaproth, Tableaux historique de l'Asie, 1826, p. 65—67.

(287) p. 186.—To this fair-haired, blue-eyed, Indo-germanic, Gothic, or Arian race of eastern Asia, belong the Usün, Tingling, Hutis, and great Yueti. The last are called by the Chinese writers a Thibetian Nomade race, who, 300 years before our era, migrated between the upper course of the Hoang-ho and the snowy mountains of Nanschau. I here recal this descent, as the Seres are also described as "rutilis comis et cæruleis oculis" (compare Ukert, Geogr. der Griech. und Römer, Th. ii i. Abth. ii. 1845, S. 275). We owe to the researches of Abel Remusat and Klaproth, which are among the brilliant historical discoveries of our age, the knowledge of these fair-haired races, which, in the most eastern part of Asia, gave the first impulse to what has been called "the great migration of nations."

(288) p. 187.—Letronne, in the Observations crit. et archéol. sur les Réprésentations zodiacales de l'Antiquité, 1824, p. 99, as well as in his later work, Sur l'Origine grecque des Zodiaques prétendus égyptiens, 1837, p. 27.

(289) p. 187.—The sound investigator, Colebrooke, places Warahamihira in the fifth, Brahmagupta at the end of the sixth century, and Aryabhatta rather undecidedly between 200 and 400 of our era. (Compare Holtzmann über den griechischen Ursprung des indischen Thierkreises, 1841, S. 23.)

(290) p. 187.—On the reasons on which the assertion in the text of the exceedingly late commencement of Strabo's work rests, see Groskurd's German translation, 1831, Th. i. S. xvii.

(291) p. 188.—Strabo, lib. i. p. 14; lib. ii. p. 118; lib. xvi. p. 781; lib. xvii. p. 798 and 815.

(292) p. 188.—Compare the two passages of Strabo, lib. i. p. 65, and lib. ii. p. 118 (Humboldt, Examen critique de l'Hist. de la Géographie, T. i. p. 152—154). In the important new edition of Strabo published by Gustav Kramer, 1844, Th. i. p. 100, "the parallel of Athens is read instead of the

parallel of Thinæ, as if Thinæ had first been named in the Pseudo-Arrian, in the Periplus Maris Rubri." Dodwell places the writing of the Periplus under Marcus Aurelius and Lucius Verus, but according to Letronne it was written under Septimius Severus and Caracalla. Although in five passages in Strabo all our manuscripts read Thinæ, yet lib. ii. p. 79, 86, 87, and above all 82, in which Eratosthenes himself is named, are decisive in favour of the parallel of Athens and Rhodes. Athens and Rhodes were thus confounded, as old geographers made the peninsula of Attica extend too far towards the south. It would also appear surprising, supposing the usual reading Θινων κυκλος to be the correct one, that a particular parallel, the Diaphragm of Dicearchus, should be called after a place so little known as Sinæ (Tsin). However, Cosmas Indicopleustes connects his Tzinitza (Thinæ) with the chain of mountains which divides Persia and the Romanic lands and the whole habitable world into two parts, adding the remarkable observation, that this is according to the "belief of the Indian philosophers and Brahmins." Compare Cosmas, in Montfaucon, Collect. nova Patrum, T. ii. p. 137; and my Asie centrale, T. i. p. xxiii. 120—129, and 194—203, T. ii. p. 413. The Pseudo-Arrian, Agathemeros, according to the learned investigations of Professor Franz, and Cosmas, decidedly ascribe to the metropolis of the Sinæ a very northern latitude, nearly in the parallel of Rhodes and Athens; whereas Ptolemy, misled by the accounts of mariners, speaks solely of a Thinæ three degrees south of the equator (Geogr. i. 17). I suspect that Thinæ merely meant, generally, a Chinese emporium, a harbour in the land of Tsin; and that therefore one Thinæ (Tziuitza) may have been intended north of the equator, and another south of the equator.

(293) p. 188.—Strabo, lib. i. p. 49—60, lib. ii. p. 95 and 97, lib. vi. p. 277; lib. xvii. p. 830. On the elevation of islands, and of the continent, see particularly lib. i. p. 51, 54, and 59. The old Eleat Xenophanes was led, by the numerous fossil marine productions found at a distance from the sea, to conclude that "the present dry ground had been raised from the bottom of the sea" (Origen, Philosophumena, cap. 4). Appuleius, in the time of Antoninus, collected fossils from the Gætulian (Mauritanian) mountains, and ascribed them to the flood of Deucalion, considering it to have been universal. Professor Franz, by means of very careful investigation, has refuted Beckmann's and Cuvier's belief, that Appuleius possessed a collection of specimens of natural history (Beckmann's Gesch. der Erfindungen, Bd. ii. S. 370; and Cuvier's Hist. des Sciences naturelles).

(294) p. 189.—Strabo, lib. xvii. p. 810.

(295) p. 190.—Carl Ritter, Asien, Th. v. S. 560.

(296) p. 190.—See a collection of the most striking instances of Greek and Roman errors, in respect to the directions of different chains of mountains, in the introduction to my Asie centrale, T. i. p. xxxvii.—xl. Most satisfactory investigations, respecting the uncertainty of the numerical bases of Ptolemy's positions, are to be found in a treatise of Ukert, in the Rheinischen Museum für Philologie, Jahrg. vi. 1838, S. 314—324.

(297) p. 191.—For examples of Zend and Sanscrit words which have been preserved to us in Ptolemy's Geography, see Lassen, Diss. de Taprobane insula, p. 6, 9, and 17; Burnouf's Comment. sur le Yaçna, T. i. p. xciii.—cxx. and clxxxi.—clxxxv.; and my Examen crit. de l'Hist. de la Géogr. T. i. p. 45—49. In few cases Ptolemy gives both the Sanscrit names and their significations, as for the island of Java "barley island," *Ιαβαδιου, ο σημαινει κριθης νησος*, Ptol. vii. 2 (Wilhelm v. Humboldt über die Kawi-Sprache, Bd. i. S. 60—63). The two-stalked barley (Hordeum distichon) is, according to Buschmann, still termed in the principal Indian languages (Hindustanee, Bengalee and Nepaulese, Mahratta, Cingalese, and the language of Guzerat), as well as in Persian and Malay, yava, djav, or djau, and in the language of Orissa, yaa. (Compare the Indian versions of the Bible in the passage John vi. 9 and 13; and Ainslie, Materia Medica of Hindostan, Madras, 1813, p. 217.)

(298) p. 191.—See my Examen crit. de l'Hist. de la Géographie, T. ii. p. 147—188.

(299) p. 192.—Strabo, lib. xi. p. 506.

(300) p. 192.—Menander de Legationibus Barbarorum ad Romanos, et Romanorum ad Gentes, e rec. Bekkeri et Niebuhr, 1829, p. 300, 619, 623, and 628.

(301) p. 192.—Plutarch de Facie in Orbe Lunæ, p. 921, 19 (compare my Examen crit. T. i. p. 145—191). I have met, among highly-informed Persians, with a repetition of the hypothesis of Agesianax, according to which, the marks on the lunar surface, in which Plutarch (p. 935, 4) thought he saw "a peculiar kind of shining mountains" (? volcanoes), were merely the reflected images of terrestrial lands, seas, and isthmuses. My Persian friends said, "what they shew us through telescopes on the surface of the moon are only the reflected images of our own countries."

(302) p. 192.—Ptolem. lib. iv. cap. 9; lib. vii. cap. 3 and 5. Compare Letronne, in the Journal des Savans, 1831, p. 476—480, and 545—555; Humboldt, Examen crit. T. i. p. 144, 161, and 329; T. ii. p. 370—373.

(303) p. 193.—Delambre, Hist. de l'Astronomie ancienne, T. i. p. liv.; T. ii. p. 551. Theon never makes any mention of Ptolemy's Optics, although he lived fully two centuries after him.

(304) p. 193.—In reading ancient works on physics, it is often difficult to decide whether a particular result followed from a phenomenon purposely called forth, or accidentally observed. When Aristotle (De Cœlo, iv. 4) treats of the weight of the atmosphere, which, however, Ideler appears to deny his having done (Meteorologia Veterum Græcorum et Romanorum, p. 23), he says distinctly that a "bladder when blown out is heavier than an empty bladder." The experiment, if actually tried, must have been made with condensed air.

(305) p. 193.—Aristot. de Anim. ii. 7; Biese, die Philosophie des Aristot. Bd. ii. S. 147.

(306) p. 194.—Joannis (Philoponi) Grammatici in Libr. de Generat. and Alexandri Aphrodis. in Meteorol. Comment. (Venet. 1527,) p. 97, b. Compare my Examen critique, T. ii. p. 306—312.

(307) p. 194.—The Numidian Metellus had 142 elephants killed in the circus. In the games given by Pompey, 600 lions and 406 panthers were shewn. Augustus sacrificed 3500 wild beasts in the festivities which he gave to the people; and a tender husband laments that he could not celebrate the day of his wife's death by a sanguinary gladiatorial fight at Verona, "because contrary winds detained in port the panthers which had been bought in Africa"! (Plin. Epist. vi. 34.)

(308) p. 195.—Compare Note 293. Yet Appuleius, as Cuvier recals (Hist. des Sciences naturelles, T. i. p. 287), was the first to describe accurately the bony hook in the second and third stomach of the Aplysiæ.

(309) p. 198.—"Est enim animorum ingeniorumque naturale quoddam quasi pabulum consideratio contemplatioque naturæ. Erigimur, elatiores fieri videmur, humana despicimus, cogitantesque supera atque cœlestia hæc nostra, ut exigua et minima, contemnimus" (Cic. Acad. ii. 41).

(310) p. 198.—Plin. xxxvii. 13 (ed. Sillig. T. v. 1836, p. 320). All earlier editions terminated with the words "Hispaniam quacunque ambitu mari." The conclusion of the work was discovered in 1831 in a Bamberg Codex, by Herr Ludwig v. Jan, Professor at Schweinfurt.

(311) p. 199.—Claudian in secundum consulatum Stilichonis, v. 150—155.

(312) p. 200.—Kosmos, Bd. i. S. 385 and 492, Bd. ii. S. 25 (Eng. trans. Vol. i. p. 356, and note 443, Vol. ii. p. 25). Compare also Wilhelm von Humboldt über die Kawi-Sprache, Bd. i. S. xxxviii.

([313]) p. 204.—If, as has often been said, Charles Martell's victory at Tours protected middle Europe against the Mussulman invasion, it cannot be maintained with equal justice that it was the retreat of the Mogols after the battle of Liegnitz, which prevented Buddhism from penetrating to the banks of the Elbe and the Rhine. The battle which was fought in the plain of Wahlstatt, near Liegnitz, and in which Duke Henry the Pious fell heroically, was fought on the 9th of April, 1241, four years after the Asiatic hordes under Batu, the grandson of Ghengis Khan, had subjected the Kaptschak and Russia. But the first introduction of Buddhism among the Mogols took place in the year 1247, when, at Leang-tscheu, in the Chinese province of Schensi, the sick Mongolian Prince Godan sent for the Sakya Pandita, a Thibetan arch-priest, to cure and convert him (Klaproth, in a manuscript fragment "über die Verbreitung des Buddhismus im östlichen und nordlichen Asien"). It should also be remarked, that the Moguls have never occupied themselves with the conversion of conquered nations.

([314]) p. 204.—Kosmos, Bd. i. S. 308 and 471 (English trans. Vol. i. p. 283, and note 342).

([315]) p. 206.—Hence the contrast between the tyrannical measures of Motewekkel, the tenth Caliph of the house of the Abassides, against Jews and Christians (Joseph von Hammer über die Länderverwaltung unter dem Chalifate, 1835, S. 27, 85, and 117), and the mild tolerance of wiser rulers in Spain (Conde, Hist. de la Dominacion de los Arabes en España, T. i. 1820, p. 67). It should also be remembered, that Omar, after the taking of Jerusalem, permitted every rite of Christian worship, and concluded an agreement with the Patriarch very favourable to the Christians (Fundgruben des Orients, Bd. v. S. 68).

([316]) p. 206.—"There is a tradition of a branch of the Hebrews having migrated to southern Arabia, under the name of Jokthan (Oachthan), before the time of Abraham, and of having founded there flourishing kingdoms" (Ewald, Geschichte des Volkes Israel, Bd. i. S. 337 and 450).

([317]) p. 206.—The tree which furnishes the "incense of Hadramaut," celebrated from the earliest times, has not yet been discovered and determined by any botanist, not even by the laborious and far-searching Ehrenberg; it is entirely wanting in the island of Socotora. An article resembling this incense is found in India, and particularly in Bundelcund; and is an object of considerable export from the port of Bombay, to China. This Indian kind of incense is obtained, according to Colebrooke (Asiatic Researches, Vol. ix. p. 377), from a plant made known by Roxburgh, Boswellia thurifera or serrata, of Kunth's family of Burseraceæ. As from the very ancient commercial connections between the coasts of southern Arabia and western India (Gildemeister, Scriptorum Arabum Loci de rebus Indicis, p. 35) it might be doubted whether the λιβανος of Theophrastus, (the Thus of the Romans), belonged originally to the Arabian peninsula, Lassen's remark (indische Alterthumskunde, Bd. i. S. 286), that incense is called "yawana, Javanese, *i. e.* Arabian, in Amara-Koscha itself," apparently implying that it is brought to India from Arabia, becomes very important. It is called in Amara-Koscha, "turuschka', pindaka', sihlô, (three names signifying incense), yâwanô" (Amarakocha, publ. par A. Loiseleur Deslongchamps, P. i. 1839, p. 156). Dioscorides distinguishes Arabian from Indian incense. Carl Ritter, in his excellent monograph on the kinds of incense (Asien, Bd. viii. Abth. i. S. 356—372), remarks very justly, that, from the similarity of climate, this species of plant (Boswellia thurifera) may well extend over a region reaching from India, through the south of Persia, to Arabia. The American incense (Olibanum americanum of our Pharmacopœia) is obtained from Icica gujanensis, Aubl. and Icica tacamahaca, which Bonpland and myself found growing abundantly in the vast grassy plains (Llanos) of Calaboso in South America. Icica, like Boswellia, belongs to the family of Burseraceæ. The red pine (Pinus abies, Linn.) produces the common incense of our churches. The plant which bears myrrh, and which Bruce thought he had seen (Ainslie, Materia Medica of Hindostan, Madras, 1813, p. 29) has been discovered near el-Gisan in Arabia, by Ehrenberg, and has been described, from the specimens collected by him, under the name of Balsamodendron myrrha, by Nees von Esenbeck. Balsamodendron kotaf of Kunth, an Amyris of Forskăl, was long erroneously supposed to be the true myrrh tree.

(318) p. 207.—Wellsted, Travels in Arabia, 1838, Vol. i. p. 272—289.

(319) p. 207.—Jomard, Etudes géogr. et hist. sur l'Arabie, 1839, p. 14 and 32.

(320) p. 207.—Kosmos, Bd. ii. S. 167 (English trans. Vol. ii. p. 133.)

(321) p. 208.—Isaiah, lx. 6.

(322) p. 209.—Ewald, Gesch. des Volkes Israel, Bd. i. S. 300 and 450; Bunsen, Ægypten, Buch iii. S. 10 and 32. The traditions of Medes and Persians in northern Africa indicate very ancient migrations to the westward. They have been connected with the variously related myth of Hercules, and the Phœnician Melkarth. (Compare Sallust, Bellum Jugurth. cap. 18, drawn from Punic writings, by Hiempsal; and Pliny, v. 8.) Strabo even calls the

Maurusians (inhabitants of Mauritania) "Indians who had come with Hercules."

([323]) p. 209.—Diod. Sic. lib. ii. cap. 2 and 3.

([324]) p. 209.—Ctesiæ Cnidii Operum reliquiæ, ed. Baehr., Fragmenta Assyriaca, p. 421; and Carl Müller, in Dindorf's edition of Herodotus, Par. 1844, p. 13—15.

([325]) p. 210.—Gibbon, Hist. of the Decline and Fall of the Roman Empire, Vol. ix. chap. l. p. 200, Leips. 1829.

([326]) p. 210.—Humboldt, Asie centr. T. ii. p. 128.

([327]) p. 211.—Jourdain, Recherches critiques sur l'Age des Traductions d'Aristote, 1819, p. 81 and 87.

([328]) p. 214.—Respecting the knowledge which the Arabians derived from the Hindoos, in the study of the materia medica, see Wilson's important investigations, in the Oriental Magazine of Calcutta, 1823, Feb. and March; and those of Royle, in his Essay on the Antiquity of Hindoo Medicine, 1837, p. 56—59, 64—66, 73, and 92. Compare an account of Arabic pharmaceutic writings, translated from Hindostanee, in Ainslie (Madras edition), p. 289.

([329]) p. 215.—Gibbon, Vol. ix. chap. li. p. 392; Heeren, Gesch. des Studiums der classischen Litteratur, Bd. i. 1797, S. 44 and 72; Sacy, Abd-Allatif, p. 240; Parthey, das alexandrinische Museum, 1838, S. 106.

([330]) p. 216.—Heinrich Ritter, Gesch. der christlichen Philosophie, Th. iii. 1844, S. 669—676.

([331]) p. 217.—The learned Orientalist, Reinaud, in three late writings, which shew how much may still be derived from Arabic and Persian, as well as Chinese sources; Fragments arabes et persans inédits relatifs à l'Inde antérieurement au 11ème Siècle de l'Ere chrétienne, 1845, p. xx.—xxxiii.; Relation des Voyages faits par les Arabes et les Persans dans l'Inde et à la Chine dans le 9ème Siecle de notre Ere, 1845, T. i. p. xlvi.; Mémoire géog. et hist. sur l'Inde d'après les Ecrivains arabes, persans et chinois, antérieurement au milieu du onzième Siècle de l'Ere chrétienne, 1846, p. 6. The second of these memoirs is based on the far less complete treatise of the Abbé Renaudot, entitled "Anciennes Relations des Indes, et de la Chine, de deux Voyageurs mahométans," 1718. The Arabic manuscript contains only one notice of a voyage, viz. that of the merchant Soleiman, who embarked on the Persian Gulf in the year 851; to which is added, what Abu-Zeyd-Hassan, of Syraf in Farsistan, who had never travelled to India or China, could learn from other well-informed merchants.

([332]) p. 217.—Reinaud et Favé du Feu grégeois, 1845, p. 200.

([333]) p. 217.—Ukert, über Marinus Tyrius und Ptolemäus, die Geographen, in the Rheinischen Museum für Philologie, 1839, S. 329—332; Gildemeister de rebus Indicis, Pars 1, 1838, p. 120; Asie centrale, T. ii. p. 191.

([334]) p. 217.—The "Oriental Geography of Ebn-Haukal," which Sir William Ouseley published in London in 1800, is that of Abu-Ishak el-Istachri, and, as Frähn has shewn (Ibn Fozlan, p. ix. xxii. and 256—263), is half a century older than Ebn-Haukal. The maps which accompany the "Book of Climates" of the year 920, and of which there is a fine manuscript copy in the library of Gotha, have been very useful to me in what I have written on the Caspian Sea and the Sea of Aral (Asie centrale, T. ii. p. 192—196). We now possess an edition of Istachri, and a German translation; (Liber Climatum, ad similitudinem codicis Gothani delineandum, cur. J. H. Moeller, Goth. 1839; Das Buch der Länder, translated from the Arabic by A. D. Mordtmann, Hamb. 1845).

([335]) p. 217.—Compare Joaquim Jose da Costa de Macedo, Memoria em que se pretende provar que os Arabes não conhecerão as Canarias antes dos Portuguezes, Lisboa, 1844, p. 86—99, 205—227, with Humboldt, Examen crit. de l'Hist. de la Géographie, T. ii. p. 137—141.

([336]) p. 218.—Leopold von Ledebur, über die in den baltischen Ländern gefundenen Zeugnisse eines Handels-Verkehrs mit dem Orient zur Zeit der arabischen Weltherrschaft, 1840, S. 8 and 75.

([337]) p. 218.—The determinations of longitude which Abul-Hassan Ali of Morocco, an astronomer of the 13th century, has incorporated with his work on the astronomical instruments of the Arabs, are all computed from the first meridian of Arin. M. Sédillot fils first directed the attention of geographers to this meridian; it has also been an object of careful research to myself, because Columbus, being as always guided by Cardinal d'Ailly's Imago Mundi, in his phantasies respecting the difference of form which he supposes between the eastern and western hemisphere, speaks of an Isla de Arin: centro de el hemispherio del qual habla Toloméo y què́s debaxo la linea equinoxial entro el Sino Arabico y aquel de Persia. (Compare J. J. Sédillot, Traité des Instrumens astronomiques des Arabes, publ. par L. Am. Sédillot, T. i. 1834, p. 312—318, T. ii. 1835, preface, with Humboldt's Examen crit. de l'Hist. de la Géogr. T. iii. p. 64, and Asie centrale, T. iii. p. 593—596, where will be found the data which I derived from the Mappa Mundi of Alliacus of 1410, in the "Alphonsine Tables," 1483, and in Madrignano's Itinerarium Portugallensium, 1508. It is singular that Edrisi appears to know nothing of Khobbet Arin (Cancadora, more properly Kank-der). Sédillot fils (in the Mémoire sur les Systèmes géographiques des Grecs et des Arabes, 1842, p. 20—25) places the meridian of Arin in the group of the Azores; whereas the learned commentator of Abulfeda, Reinaud (Mémoire sur l'Inde antérieurement au 11eme Siècle de l'Ere chrétienne, d'après les Ecrivains arabes et persans, p. 20—24), assumes "Arin to have been a name originating by confusion with Azyn, Ozein, and Odjein, an old seat of cultivation: according to Burnouf, Udjijayani in Malwa Οζηνη of Ptolemy; and that this Ozene is in the meridian of Lanka, and that in later times Arin was believed to be au island on the coast of Zanguebar, perhaps Εσσυνον of Ptolemy." Compare also Am. Sédillot, Mém. sur les Instr. astron. des Arabes, 1841, p. 75.

([338]) p. 218.—The Caliph Al-Mamun caused many valuable Greek manuscripts to be purchased in Constantinople, Armenia, Syria, and Egypt, and to be translated direct from Greek into Arabic, the earlier Arabic versions having long been founded on Syrian translations (Jourdain, Recherches crit. sur l'Age et sur l'Origine des Traductions latines d'Aristote, 1819, p. 85, 88, and 226). Al-Mamun's exertions have rescued much which, without the Arabians, would have been lost to us. A similar service has been rendered by Armenian translations, as Neumann of Munich has first shewn. Unhappily a notice by the historian Geuzi of Bagdad, preserved to us by the celebrated geographer Leo Africanus, in a memoir entitled "De Viris inter Arabes illustribus," gives reason to believe, that at Bagdad itself many Greek originals, supposed to be useless, were burnt; but no doubt this passage does not relate to important manuscripts already translated. It is capable of more interpretations than one, as has been shewn by Bernhardy (Grundriss der griechen Litteratur, Th. i. S. 489), in opposition to Heeren's Geschichte der classischen Litteratur, Bd. i. S. 135. The Arabic translations of Aristotle have often been made useful in executing Latin ones (*e. g.* the eight books of Physics, and the History of Animals); but the larger and better part of the Latin translations have been made direct from the Greek (Jourdain, Rech. crit. sur l'Age des Traductions d'Aristote, p. 230—236). We may recognise an allusion to the same twofold source in the memorable letter which the Emperor Frederick II. of Hohenstaufen sent with translations of Aristotle to his universities, and especially to that of Bologna in 1232. This letter contains the expression of noble sentiments, and shews that it was not only the love of natural history which taught Frederick II. to appreciate the philosophical value of the "Compilationes varias quæ ab Aristotele aliisque philosophis sub Græcis Arabicisque Vocabulis Antiquitus editæ sunt." He writes: "We have from our earliest youth desired a closer acquaintance with science, although the cares of government have withdrawn us therefrom. As far as we could, we delighted in spending our time in the careful reading of excellent works, to the end that the mind might be enlightened and strengthened by exercises, without which the life of man is wanting both in rule and in freedom (ut animæ clarius vigeat instrumentum in acquisitione scientiæ, sine qua mortalium vita non regitur liberaliter). Libros ipsos tamquam præmium amici Cæsaris gratulantur accipite, et ipsos antiquis philosophorum operibus, qui vocis vestræ ministerio reviviscunt, aggregantes in auditorio vestro." . . . . (Compare Jourdain, p. 169—178, and Friedrich von Raumer's excellent Geschichte der Hohenstaufen, Bd. iii. 1841, S. 413.) The Arabs formed a uniting link between ancient and modern science: without their love of translation, succeeding ages would have lost great part of that which the Greeks had either formed themselves, or derived from other nations. It is in this point of view that the subjects which have been touched upon, though seemingly purely linguistic, have a general cosmical interest.

([339]) p. 218.—Michael Scot's translation of Aristotle's Historia Animalium, and a similar work by Avicenna (Manuscript No. 6493 in the Paris Library), are spoken of by Jourdain, Traductions d'Aristote, p. 135—138, and by Schneider, Adnot. ad Aristotelis de Animalibus Hist. lib. ix. cap. 15.

([340]) p. 218.—On Ibn-Baithar, see Sprengel, Gesch. der Arzneykunde, Th. ii. 1823, S. 468; and Royle on the Antiquity of Hindoo Medicine, p. 28. We possess, since 1840, a German translation of Ibn-Baithar, under the title Grosse Zusammenstellung über die Kräfte der bekannten einfachen Heil- und Nahrungs-mittel, translated from the Arabic by J. v. Sontheimer, 2 vols.

([341]) p. 219.—Royle, p. 35—65. Susruta, son of Visvamitra, is considered by Wilson to have been a cotemporary of Rama. We have a Sanscrit edition of his works (The Sus'ruta, or System of Medicine taught by Dhanwantari, and composed by his disciple Sus'ruta, ed. by Sri Madhusūdana Gupta, Vol. i. ii. Calcutta, 1835, 1836), and a Latin translation (Sus'rutas Ayurvédas, id est Medicinæ Systema a venerabili D'havantare demonstratum, a Susruta discipulo compositum, nunc pr. ex Sanskrita in Latinum sermonem vertit Franc. Hessler, 2 vols. Erlangæ, 1844, 1847.

([342]) p. 219.—Avicenna says, "Deiudar (Deodar), of the genus 'abhel (juniperus); also an Indian pine which yields a peculiar milk, syr deiudar (fluid turpentine)."

([343]) p. 219.—Spanish Jews from Cordova carried the lessons of Avicenna to Montpellier, and contributed in a principal degree to the establishment of its celebrated medical school, belonging to the 12th century, which was modelled according to Arabian patterns (Cuvier, Hist. des Sciences naturelles, T. i. p. 387).

([344]) p. 219.—Respecting the gardens of the palace of Rissafah, which was built by Abdurrahman Ibn-Moawijeh, see History of the Mohammedan Dynasties in Spain, extracted from Ahmed Ibn Mohammed Al-Makkari, by Pascual de Gayangos, Vol. i. 1840, p. 209—211. En su Huerta plantò el Rey Abdurrahman una palma que era entonces (756) unica, y de ella procediéron todas las que hay en España. La vista del arbol acrecentaba mas que templaba su melancolia" (Antonio Conde, Hist. de la Dominacion de los Arabes en España, T. i. p. 169).

([345]) p. 220.—The preparation of nitric acid and aqua regia by Djaber (whose proper name was Abu-Mussah-Dschafar) is more than 500 years anterior to Albertus Magnus and Raymond Lully, and almost 700 years anterior to the Erfurt Monk, Basilius Valentinus. Nevertheless, the discovery of these decomposing (dissolving) acids, which constitutes an epoch in chemical knowledge, was long ascribed to the three last named Europeans.

([346]) p. 220.—Respecting the rules given by Razes for the vinous fermentation of amylum and sugar, and for the distillation of alcohol, see Höfer, Hist. de la Chimie, T. i. p. 325. Although Alexander of Aphrodisias (Joannis Philoponi Grammatici in Libr. de Generatione et Interitu Comm. Venet. 1527, p. 97), properly speaking, only describes circumstantially distillation from sea-water, yet he also indicates that wine may also be distilled. This is the more remarkable, because Aristotle had put forward (Meteorol. ii. 3, p. 358, Bekker) the erroneous opinion, that in natural evaporation fresh water only rose from wine, as from the salt water of the sea.

([347]) p. 220.—The chemistry of the Indians, comprising alchemistic arts, is called rasâyana (rasa, juice or fluid, also quicksilver; and âyana, march or proceeding), and forms, according to Wilson, the seventh division of the Ayur-Veda, the "science of life, or of the prolongation of life" (Royle, Hindoo Medicine, p. 39—48). The Indians have been acquainted from the earliest times (Royle, p. 131) with the application of mordants in calico or cotton printing, an Egyptian art which we find most clearly described in Pliny, lib. xxxv. cap. 11, No. 150. The name "chemistry" indicates literally "Egyptian art," the art of the black land; for Plutarch (de Iside et Osir, cap. 33) knew that the Egyptians called their country Χημια, from the black earth. The inscription on the Rosetta stone has Chmi. I find the word chemie, as applied to the decomposing art, first in the deerees of Diocletian against "the old writings of the Egyptians which treat of the 'chemie' of gold and silver (*περι χημιος αργυρου και χρυσου*)." Compare my Examen crit. de l'Hist. de la Géographie et de l'Astronomie nautique, T. ii. p. 314.

(348) p. 221.—Reinaud et Favé du Feu grégeois, des Feux de Guerre, et des Origines de la Poudre à Canon, in their Histoire de l'Artillerie, T. i. 1845, p. 89—97, 201, and 211; Piobert, Traité d'Artillerie, 1836, p. 25; Beckmann, Technologie, S. 342.

(349) p. 221.—Laplace, Précis de l'Hist. de l'Astronomie, 1821, p. 60; and Sédillot, Mémoire sur les Instrumens astr. des Arabes, 1841, p. 44. Also Thomas Young (Lectures on Natural Philosophy and the Mechanical Arts, 1807, Vol. i. p. 191) does not doubt that Ebn-Junis, at the end of the tenth century, applied the pendulum to the measurement of time, but he ascribes the first combination of the pendulum with wheel-work to Sanctorius, in 1612 (44 years before Huyghens). Respecting the very skilfully made timepiece which was among the presents which Haroun Al-Raschid, or rather the Caliph Abdallah, sent, two centuries before, from Persia to Charlemagne at Aix-la-Chapelle, Eginhard says distinctly, that it was moved by water (horologium ex aurichalco arte mechanica mirifice compositum, in quo duodecim horarum cursus ad Clepsidram vertebatur); Einhardi Annales, in Pertz's Monumenta Germaniæ Historica Scriptorum, T. i. 1826, p. 195. Compare H. Mutius, de Germanorum Origine, Gestis, &c. Chronic. lib. viii. p. 57, in Pistorii Germanicorum Scriptorum, T. ii. Francof. 1584; Bouquet, Recueil des Historiens des Gaules, T. v. p. 333 and 354. The hours were marked by the sound of the fall of small balls, as well as by the coming forth of small horsemen from as many opening doors. The manner in which the water acted in such timepieces may indeed have been very different among the Chaldeans, who "weighed time" (determined it by the weight of fluids), and among the Greeks and the Indians in Clepsydras; for the hydraulic clockwork of Ctesibius, under Ptolemy Euergetes II. which gave the civil hours throughout the year at Alexandria, according to Ideler was never known under the common denomination of *κλεψυδρα* (Ideler's Handbuch der Chronologie, 1825, Bd. i. S. 231). According to Vitruvius's description (lib. ix. cap. 4), it was a real astronomical clock, a "horologium ex aqua," a very complicated "machina hydraulica," working by means of toothed wheels (versatilis tympani denticuli æquales alius alium impellentes). It is thus not improbable, that the Arabians, acquainted with the accounts of improved mechanical constructions under the Roman Empire, succeeded in making an

hydraulic clock with wheel-work (tympana quæ nonnulli rotas appellant, Græci autem περιτροχα, Vitruvius, x. 4). Leibnitz (Annales Imperii Occidentis Brunsvicensis, ed. Pertz, T. i. 1843, p. 247) expresses his admiration of the construction of the clock of Haroun Al-Raschid (Abd-Allatif, trad. par Silvestre de Sacy, p. 578). A much more remarkable piece of skilful work was that which the Sultan sent from Egypt, in 1232, to the Emperor Frederic II. It was a large tent, in which the sun and moon were made to move by mechanism, so as to rise and set, and to shew the hours of the day and of the night at correct intervals of time. In the Annales Godefridi Monachi S. Pantaleonis apud Coloniam Agrippinam, it is described as "tentorium, in quo imagines solis et lunæ artificialiter motæ cursum suum certis et debitus spaciis peragrant, et horas diei et noctis infallibiliter indicant (Freheri Rerum Germanicarum Scriptores, T. i. Argentor. 1717, p. 398). The monk Godefridus, or whoever else may have treated of those years in the chronicle which was, perhaps, written by many different authors for the convent of St. Pantaleon at Cologne (see Böhmer, Fontes Rerum Germanicarum, Bd. ii. 1845, S. xxxiv.—xxxvii.), lived in the time of the great Emperor Frederic II. himself. The emperor had this curious work, the value of which was estimated at 20000 marks, preserved at Venusium, with other treasures (Fried. von Raumer, Gesch. der Hohenstaufen, Bd. iii. S. 430). That the whole tent was given a movement like that of the vault of heaven, as has often been asserted, appears to me very improbable. The Chronica Monasterii Hirsaugiensis, edited by Trithemius, contains scarcely any thing more than a mere repetition of the passage in the Annales Godefridi, without giving any information about the mechanical construction (Joh. Trithemii Opera Historica, P. ii. Francof. 1601, p. 180). Reinaud says that the movement was effected "par des ressorts cachés" (Extraits des Historiens Arabes relatifs aux Guerres des Croisades, 1829, p. 435).

(350) p. 223.—On the Indian tables which Alphazari and Alkoresmi translated into Arabic, see Chasles, Recherehes sur l'Astronomie indienne, in the Comptes rendus des Séances de l'Acad. des Sciences, T. xxiii. 1846, p. 846—850. The substitution of the sine for the arc, which is usually ascribed to Albategnius, in the beginning of the tenth century, also belongs originally to the Indians: tables of sines are to be found in the Surya-Siddhanta.

(351) p. 223.—Reinaud, Fragments Arabes relatifs à l'Inde, p. xii.—xvii. 96—126, and especially 135—160. Albiruni's proper name was Abul-Ryhan. He was a native of Byrun in the valley of the Indus, was a friend of Avicenna, and lived with him at the Arabian academy which had been formed

in Charezm. His sojourn in India, and the writing of his history of India (Tarîkhi-Hind), the most remarkable fragments of which have been made known by Reinaud, belong to the years 1030—1032.

(352) p. 224.—Sédillot, Materiaux pour servir à l'Histoire comparée des Sciences mathematiques chez les Grecs et les Orientaux, T. i. p. 50—89; the same, in the Comptes rendus de l'Acad. des Sciences, T. ii. 1836, p. 202, T. xvii. 1843, p. 163—173, T. xx. 1845, p. 1308. M. Biot maintains, in opposition to this opinion, that the fine discovery of Tycho Brahe by no means belongs to Abul-Wefa; that the latter was acquainted, not with the "variation," but only with the second part of the "evection" (Journal des Savans, 1843, p. 513—532, 609—626, 719—737; 1845, p. 146—166; and Comptes rendus, T. xx. 1845, p. 1319—1323).

(353) p. 224.—Laplace, Expos. du Système du Monde, Note 5, p. 407.

(354) p. 225.—On the observatory of Meragha, see Delambre, Histoire de l'Astronomie du Moyen Age, p. 198—203; and Am. Sédillot, Mem. sur les Instrumens arabes, 1841, p. 201—205, where the gnomon is described with a circular opening. On the peculiarities of the star catalogue of Ulugh Beig, see J. J. Sédillot, Traité des Instruments astronomiques des Arabes, 1834, p. 4.

(355) p. 225.—Colebrooke, Algebra, with Arithmetic and Mensuration from the Sanscrit of Brahmegupta and Bhascara, Lond. 1817. Chasles, Aperçu historique sur l'Origine et le Développement des Méthodes en Géométric, 1837, p. 416—502; Nesselmann, Versuch einer kritischen Geschichte der Algebra, Th. i. S. 30—61, 273—276, 302—306.

(356) p. 225.—Algebra of Mohammed Ben-Musa, edited and translated by F. Rosen, 1831, p. 8, 72, and 196—199. The mathematical knowledge of India was extended to China about the year 720; but this was at a period when many Arabians were already settled in Canton and other Chinese cities. Reinaud, Relation des Voyages faits par les Arabes dans l'Inde et à la Chine, T. i. p. 109; T. ii. p. 36.

(357) p. 226.—Chasles, Histoire de l'Algèbre, in the Comptes rendus, T. xiii. 1841, p. 497—524, 601—626; compare also Libri, in the same, p. 559—563.

(358) p. 226.—Chasles, Aperçu historique des Méthodes en Géométrie, 1837, p. 464—472. The same, in the Comptes rendus de l'Acad. des Sciences, T. viii. 1839, p. 78; T. ix. 1839, p. 449; T. xvi. 1843, p. 156—173, and 218—246; T. xvii. 1843, p. 143—154.

(359) p. 227.—Humboldt, über die bei verschiedenen Völkern üblichen

Systeme von Zahlzeichen und über den Ursprung des Stellenwerthes in den indischen Zahlen, in Crelle's Journal für die reine und angewandte Mathematik, Bd. iv. (1829), S. 205—231; compare also my Examen crit. de l'Hist. de la Géographie, T. iv. p. 275. The simple relation of the different methods which nations, to whom the Indian arithmetic by position was unknown, employed for expressing the multiplier of the fundamental group, contains, I believe, the explanation of the gradual rise or origin of the Indian system. If we express the number 3568, either perpendicularly or horizontally, by means of "indicators," which correspond to the different divisions of the Abacus, (thus, $\overset{3}{M}\ \overset{5}{C}\ \overset{6}{X}\ \overset{8}{I}$), we shall easily perceive that the group-signs (M C X I) could be left out. But our Indian numbers are no other than these indicators; they are the multipliers of the different groups. We are also reminded of this designation (solely by means of indicators) by the ancient Indian Suanpan (the reckoning machine which the Moguls introduced into Russia), which has successive rows or wires representing the thousands, hundreds, tens, and units. These rows would present, in the numerical example just cited, 3, 5, 6, and 8 balls. In the Suanpan, no group-sign is visible: the group-signs are the positions themselves; and these positions (rows or wires) are occupied by units (3, 5, 6, and 8) as multipliers or indicators. In both ways, whether by the written or by the palpable arithmetic, we arrive at position-value, and at the simple use of nine numbers. If a row is empty, the place will be unfilled in writing. If a group (a member of the progression) is wanting, the vacuity is graphically filled by the symbol of vacuity (sûnya, sifron, tzüphra). In the "Method of Eutocius," I find, in the group of the myriads, the first trace of the exponential system of the Greeks so important for the East: M^{α}, M^{β}, M^{γ}, designate 10000, 20000, 30000. That which is here applied only to the myriads extends among the Chinese and Japanese, who derived their instruction from the Chinese 200 years before the Christian Era, to all the multipliers of the groups. In the Gobar, the Arabian "dust writing," (discovered by my deceased friend and teacher, Silvestre de Sacy, in a manuscript in the library of the old Abbey of St. Germain des Près,) the group-signs are points—therefore, noughts or ciphers; for in India, Thibet, and Persia, noughts and points are identical. In the Gobar, 3· is 30; 4·· is 400; and 6∴ is 6000. The Indian numbers, and the knowledge of the value of position, must be more modern than the separation of the Indians and the Arians; for the Zend nation only used the far less convenient Pehlvi numbers. The opinion that the Indian notation has undergone successive improvements appears to me to derive particular support from the Tamul system, which expresses units by nine characters, and all other values by group-signs for 10, 100, and 1000, having multipliers added to the left. I draw the same inference from the singular αριθμοι ενδικοι in a scholium of the monk Neophytos, discovered by Prof. Brandis in the library of Paris, and kindly communicated to me for publication. The nine characters of Neophytos are, with the exception of the 4, quite similar to the present Persian; but these nine units are raised to 10, 100, 1000 times their value by writing one, two, or three ciphers (o) above them; as $\overset{o}{2}$ for twenty, $\overset{o}{2}\ 4$ for twenty-four, $\overset{oo}{5}$ for five hundred, and $\overset{oo}{3}\ 6$ for three hundred and six. If we suppose points to be used instead of ciphers, we have the Arabic dust writing, Gobar. As my brother Wilhelm von Humboldt has often remarked of the Sanscrit, that it is very inappropriately designated by the terms "Indian" and "ancient Indian" language, since there are in the Indian peninsula several very ancient languages not at all derived from the Sanscrit,—so the expression Indian, or ancient Indian, system of notation is also vague, both in respect to the form of the characters and also to the spirit of the method, which latter sometimes consists in simple juxta-position, sometimes in the use of Coefficients and Indicators, and sometimes in proper "position-value." Even the existence of the cipher, or character for 0, is not a necessary condition of the simple position-value in Indian notation, as the scholium of Neophytos shews. The Indians who speak the Tamul language have numerical characters which appear to differ from their alphabetic characters. The 2 and the 8 have a faint resemblance to the 2 and the 5 of the Devanagari figures, (Rob. Anderson, Rudiments of Tamul Grammar, 1821, p. 135); and yet an accurate comparison shews that the Tamul numerical characters are derived from the Tamul alphabetical writing. Still more different from the Devanagari figures are, according to Carey, the Cingalese. In the latter, and in the Tamul, we find neither position-value nor zero sign, but symbols for tens, hundreds, and thousands. The Cingalese work, like the Romans, by juxta-position; the Tamuls by coefficients. Ptolemy, in his Almagest and in his Geography, uses the present zero sign to represent the descending or negative scale in degrees and minutes. The zero sign is, consequently, of more ancient use in the West than the epoch of the invasion of the Arabs. (See my work above cited, and the memoir printed in Crelle's Mathematical Journal, S. 215, 219, 223, and 227.)

[360] p. 228.—Wilhelm von Humboldt, über die Kawi-Sprache, Bd. i. S. cclxii. Compare also the excellent description of the Arabians, in Herder's Ideen zur Gesch. der Menscheit, Book xix. 4 and 5.

([361]) p. 231.—Compare Humboldt, Examen crit. de l'Hist. de la Géographie, T. i. p. viii. and xix.

([362]) p. 233.—Parts of America were seen, but not landed on, 14 years before Leif Eireksson, in the voyage which Bjarne Herjulfson undertook from Greenland to the southward in 986. He first saw the land at the island of Nantucket, a degree south of Boston; then in Nova Scotia; and, lastly, in Newfoundland, which was subsequently called "Litla Helluland," but never "Vinland." The gulf which divides Newfoundland from the mouth of the great river St. Lawrence was called by the northmen settled in Iceland and Greenland, Markland Gulf. See Caroli Christiani Rafn, Antiquitates Americanæ, 1845, p. 4, 421, 423, and 463.

([363]) p. 233.—Gunnbjörn was wrecked, in 876 or 877, on the rocks subsequently called by his name, which were lately rediscovered by Captain Graah. It was Gunnbjörn who first saw the east coast of Greenland, but without landing upon it. (Rafn, Antiquit. Amer. p. 11, 93, and 304.)

([364]) p. 234.—Kosmos, Bd. ii. S. 163 (Engl. trans. Vol. ii. p. 129).

([365]) p. 234.—These mean annual temperatures of the east coast of America, between the parallels of 42° 25′ and 41° 15′, correspond in Europe to the latitudes of Berlin and Paris, places situated 8° or 10° more to the north. Moreover, on this coast the decrease of mean annual temperature from lower to higher latitudes is so rapid that, in the interval of latitude between Boston and Philadelphia, which is 2° 41′, an increase of a degree of latitude corresponds to a decrease in the mean annual temperature of almost 2° of the Centigrade thermometer; whereas, in the European system of isothermal lines, the same difference of latitude, according to my researches, barely corresponds to a decrease of half a degree of temperature, (Asie centrale, T. iii. p. 227).

([366]) p. 234.—See Carmen Færöicum in quo Vinlandiæ mentio fit, (Rafn, Antiquit. Amer. p. 320 and 332).

([367]) p. 235.—The Runic stone was placed on the highest point of the Island of Kingiktorsoak "on the Saturday before the day of victory," *i. e.* before the 21st of April, a great Heathen festival of the ancient Scandinavians, which, at their reception of Christianity, was converted into a Christian festival. Rafn, Antiquit. Amer. p. 347—355. On the doubts which Brynjulfsen, Mohnike, and Klaproth have expressed respecting the Runic numbers, see my Examen crit. T. ii. p. 97—101; yet, from other indications, Brynjulfsen and Graah regard the important monument on the Women's Islands (as well as the Runic inscriptions found at Igalikko and Egegeit, lat. 60° 51′

and 60° 0′, and the ruins of buildings at Upernavick, lat. 72° 50′, as belonging decidedly to the 11th and 12th centuries.

(368) p. 235.—Rafn, Antiquit. Amer. p. 20, 274, and 415—418 (Wilhelmi über Island, Hvitramannaland, Greenland, and Vinland, S. 117—121). According to a very ancient Saga, the most northern part of the east coast of Greenland was also visited in 1194, under the name of Svalbard, at a part which corresponds to Scoresby's land, near the point where my friend, then Captain Sabine, made his pendulum observations, and where I possess a very dreary cape, in 73° 16′ (Rafn, Antiquit. Amer. p. 303, and Aperçu de l'ancienne Géographie des Régions arctiques de l'Amérique, 1847, p. 6.)

(369) p. 235.—Wilhelmi, work above quoted, S. 226; Rafn, Antiquit. Amer. p. 264 and 453. The settlements on the west coast of Greenland, which, until the middle of the 14th century, were in a very flourishing condition, underwent a gradual decay, from the ruinous operation of commercial monopoly, from the attacks of Esquimaux (Skrälinger), the black death which, according to Hecker, desolated the North during the years 1347 to 1351, and the invasion of a hostile fleet from some unknown quarter. At the present time, credit is no longer given to the meteorological myth of a sudden alteration of climate, and of the formation of an icy barrier, which had for its immediate consequence the entire separation of the colonies established in Greenland from their mother country. As these colonies were only on the more temperate district of the west coast of Greenland, it cannot be true that a bishop of Skalholt, in 1540, saw, on the east coast of Greenland, beyond the icy barrier, "shepherds feeding their flocks." The accumulation of masses of ice on the east coast opposite to Iceland depends on the configuration of the land, the neighbourhood of a chain of mountains having glaciers and running parallel to the line of coast, and on the direction of marine currents. This state of things did not take its origin from the close of the 14th or the beginning of the 15th centuries. As Sir John Barrow has very justly shewn, it has been subject to many accidental alterations, particularly in the years 1815—1817. (See Barrow, Voyages of Discovery within the Arctic Regions, 1846, p. 2—6). Pope Nicholas V. named a bishop for Greenland as late as 1448.

(370) p. 236.—The principal sources of information are the historic narrations of Eric the Red, Thorfinn Karlsefne, and Snorre Thorbrandsson, probably committed to writing as early as the 12th century in Greenland itself, and partly by descendants of settlers born in Vinland (Rafn, Antiquit. Amer. p. vii. xiv. and xvi.) The care with which genealogical tables were kept was

so great, that that of Thorfinn Karlsefne, whose son Snorre Thorbrandsson was born in America, has been brought down from 1007 to 1811.

([371]) p. 237.—Hvitramannaland, the land of the white men. Compare the original sources of information, in Rafn, Antiquit. Amer. p. 203—206, 211, 446—451; and Wilhelmi über Island, Hvitramannaland, &c. S. 75—81.

([372]) p. 238.—Letronne, Recherches géogr. et crit. sur le Livre de "Mensura Orbis Terræ," composed en Irlande, par Dicuil, 1814, p. 129—146. Compare my Examen crit. de l'Hist. de la Géogr. T. ii. p. 87—91.

([373]) p. 238.—I have appended to the ninth book of my travels (Relation historique, T. iii. 1825, p. 159) a collection of the stories which have been told from the time of Raleigh, of natives of Virginia speaking pure Celtic; of the Gaelic salutation, hao, hui, iach, having been heard there; of Owen Chapelain, in 1669, saving himself from the hands of the Tuscaroras, who were about to scalp him, "by addressing them in his native Gaelic." These Tuscaroras of North Carolina are now, however, distinctly recognised by linguistic investigations, as an Iroquois tribe. See Albert Gallatin on Indian Tribes, in the Archæologica Americana, Vol. ii. 1836, p. 23 and 57. A considerable collection of Tuscarora words is given by Catlin, one of the most excellent observers of manners who at any time sojourned amongst the aborigines of America. He, however, is often inclined to regard the rather fair and often blue-eyed nation of the Tuscaroras as a mixed race, descended from ancient Welsh and from the original inhabitants of the American continent. See his Letters and Notes on the Manners, Customs, and Condition of the North American Indians, 1841, Vol. i. p. 207; Vol. ii. p. 259 and 262—265. Another collection of Tuscarora words is to be found in my brother's manuscript notes respecting language, in the Royal Library at Berlin. "Comme la structure des idioms americains parait singulierement bizarre aux differens peuples qui parlent les langues modernes de l'Europe occidentale, et se laissent facilement tromper par de fortuites analogies de quelques sons, les théologiens ont cru généralement y voir de l'hébreu, les colons espagnols du basque, les colons anglais ou français du gallois, de l'irlandais ou du bas-breton. . . . . . J'ai rencontré un jour, sur les côtes du Perou, un officier de la marine espagnol et un baleinier anglais, dont l'un prétendait avoir entendu parler basque à Tahiti, et l'autre gale-irlandais aux Iles sandwich" (Humboldt, Voyage aux Régions équinoxiales, Relat. hist. T. iii. 1825, p. 160). Although, however, no connection of language has yet been proved, I by no means wish to deny that the Basques and the nations of Celtic origin inhabiting Ireland and Wales, who were early engaged in fisheries on the most remote coasts, were the constant rivals of the Scandinavians in the northern parts of the Atlantic, and even that the Irish preceded the Scandinavians in the Färoe Islands and in Iceland. It is much to be desired that in our days, when a healthy spirit of criticism, severe but not contemptuous, prevails, the old investigations of Powel and Richard Hakluyt (Voyages and Navigations, Vol. iii. p. 4) might be resumed in England, and also in Ireland itself. Are there grounds for the statement that fifteen years before Columbus's discovery, the wanderings of Madoc were celebrated in the poems of the Welsh bard Meredith? I do not participate in the rejecting spirit which has but too often thrown popular traditions into obscurity; I incline far more to the firm persuasion that, by greater diligence and perseverance, many of the historical problems which relate to the charts of the early part of the middle ages,—to the striking agreement in religious traditions, manner of dividing time, and works of art in America and Eastern Asia;—to the migrations of the Mexican nations,—to the ancient centres of dawning civilization in Aztlan, Quivira, and Upper Louisiana, as well as in the elevated table lands of Cundinamarca and Peru,—will one day be cleared up by discoveries of facts which have been hitherto entirely unknown to us. See my Examen crit. de l'Hist. de la Géogr. du Nouveau Continent, T. ii. p. 142—149.

(374) p. 240.—Whereas this circumstance of the absence of ice in February 1477 has been adduced as a proof that Columbus's Island of Thule could not be Iceland, Finn Magnusen found, in ancient historical sources, that up to March 1477 the northern part of Iceland had no snow, and that in February of the same year the southern coast was free from ice (Examen crit. T. i. p. 105; T. v. p. 213). It is very remarkable, that Columbus, in the same "Tratado de las cinco zonas habitables," mentions a more southern island, Frislanda; a name which plays a great part in the travels of the brothers Zeni (1388—1404) which are mostly regarded as fabulous, but which is wanting in the maps of Andrea Bianco (1436), and in that of Fra Mauro (1457—1470). (Compare Examen crit. T. ii. p. 114—126.) Columbus cannot have been acquainted with the travels of the Fratelli Zeni, as they even remained unknown to the Venetian family until the year 1558, in which Marcolini first published them, 52 years after the death of the great admiral. Whence was the admiral's acquaintance with the name Frislanda?

(375) p. 241.—See the proofs, which I have collected from trustworthy documents, for Columbus in the Examen crit. T. iv. p. 233, 250, and 261, and for Vespucci, T. v. p. 182—185. Columbus was so full of the idea of Cuba being part of the continent of Asia, and even the south part of Cathay

(the province of Mango), that on the 12th of June, 1494, he caused the whole crews of his squadrons (about 80 sailors) to swear that they were convinced he might go from Cuba to Spain by land ("que esta tierra de Cuba fuese la tierra firme al comienzo de las Indias y fin á quien en estas partes quisiere venir de España por tierra"); and that "if any who now swore it should at any future day assert the contrary, they would incur the punishment of perjury, in receiving one hundred stripes, and having the tongue torn out." (See Informacion del Escribano publico Fernando Perez de Luna, in Navarrete, Viages y Descubrimientos de los Españoles, T. ii. p. 143—149.) When Columbus was approaching the island of Cuba on his first expedition, he thought himself opposite the Chinese commercial cities of Zaitun and Quinsay ("y es cierto, dice el Almirante, questa es la tierra firme y que estoy, dice él, ante Zayto y Guinsay"). He designs to deliver the letters of the Catholic monarchs to the Great Mogul Khan (Gran Can) in Cathay; and having thus discharged the mission entrusted to him, to return immediately to Spain (but by sea). Subsequently he sends on shore a baptised Jew, Luis de Torres, because he understands Hebrew, Chaldee, and some Arabic, which are languages in use in Asiatic trading cities. (See Columbus's Journal of his Voyage, 1492, in Navarrete, Viages y Descubrim, T. i. p. 37, 44, and 46.) As late as 1533, the Astronomer Schoner maintains the whole of the so-called New World to be a part of Asia (superioris Indiæ), and the city of Mexico (Temistitan) conquered by Cortes to be no other than the Chinese commercial city of Quinsay, so immoderately extolled by Marco Polo. (See Joannis Schoneri Carlostadii Opusculum geographicum, Norimb. 1533, Pars ii. cap. 1—20.)

(376) p. 241.—Da Asia de João de Barros e de Diogo de Couto, Dec. i. liv. iii. cap. 11 (Parte i. Lisboa, 1778, p. 250).

(377) p, 244.—Jourdain, Rech. crit. sur les Traductions d'Aristote, p. 230, 234, and 421—423; Letronne, des Opinions cosmographiques des Pères de l'Eglise, rapprochées des Doctrines philosophiques de la Grèce, in the Revue des deux Mondes, 1834, T. i. p. 632.

(378) p. 244.—Friedrich von Raumer über die Philosophie des dreizehnten Jahrhunderts, in his Hist. Taschenbuch, 1840, S. 468. On the inclination towards Platonism in the middle ages, and on the contests of the schools, see Heinrich Ritter, Gesch. der christl. Philosophie, Th. ii. S. 159; Th. iii. S. 131—160, and 381—417.

(379) p. 245.—Cousin, Cours de l'Hist. de la Philosophie, T. i. 1829, p. 360 and 389—436; Fragmens de Philosophie cartésienne, p. 8—12 and

403. Compare also the recent ingenious work of Christian Bartholmèss, entitled Jordano Bruno, 1847, T. i. p. 308; T. ii. p. 409—416.

(380) p. 246.—Jourdain sur les Trad. d'Aristote, p. 236; and Michae Sachs, die religiöse Poesie der Juden in Spanien, 1845, S. 180—200.

(381) p. 247.—The greater share of merit in regard to the history of animals belongs to the Emperor Frederic II. Important independent observations on the internal structure of birds are due to him. (See Schneider, in Reliqua Librorum Frederici II. Imperatoris de Arte venandi cum Avibus, T. i. 1788, in the Preface.) Cuvier also calls this prince the "first independent and original zoologist of the scholastic Middle Ages." For Albert Magnus's correct view of the distribution of heat over the surface of the globe, under different latitudes and at different seasons, see his Liber cosmographicus de Natura Locorum, Argent. 1515, fol. 14 B and 23 A (Examen crit. T. i. p. 54—58). In his own observations, however, Albertus Magnus unhappily often shews the uncritical spirit of his age. He thinks he knows "that rye changes on a good soil into wheat; that from a beech wood which has been cut down, by means of the decayed matter a birch wood will spring up; and that from oak branches stuck into the earth vines arise." (Compare also Ernst Meyer über die Botanik des 13ten Jahrhunderts, in the Linnæa, Bd. x. 1836, S. 719.)

(382) p. 248.—So many passages of the Opus majus shew the respect which Roger Bacon paid to Grecian antiquity, that, as Jourdain has already remarked (p. 429), we can only interpret the wish expressed by him in a letter to Pope Clement VI. "to burn the works of Aristotle, in order to stop the propagation of error among the schools," as referring to the bad Latin translations from the Arabic.

(383) p. 248.—"Scientia experimentalis a vulgo studentium penitus ignorata; duo tamen sunt modi cognoscendi, scilicet per argumentum et experientiam (the ideal path, and the path of experiment). Sine experientia nihil sufficienter sciri potest. Argumentum concludit, sed non certificat, neque removet dubitationem; ut quiescat animus in intuitu veritatis, nisi eam inveniat via experientiæ" (Opus Majus, Pars vi. cap. 1). I have collected all the passages relating to Roger Bacon's physical knowledge, and to his proposals for invention and discovery, in the Examen crit. de l'Hist. de la Géogr. T. ii. p. 295—299. Compare also Whewell, Philosophy of the Inductive Sciences, Vol. ii. p. 323—337.

(384) p. 248.—See Kosmos, Bd. ii. S. 228 (Engl. edit. Vol. ii. p. 193). I find Ptolemy's Optics quoted in the Opus Majus (ed. Jebb, Lond. 1733), p.

79, 288, and 404. It has been justly denied (Wilde, Geschichte der Optik, Th. i. S. 92—96), that knowledge derived from Alhazen, of the magnifying power of segments of spheres, actually led Bacon to construct spectacles; that invention appears either to have been known as early as 1299, or to belong to the Florentine Salvino degli Armati, who was buried, in 1317, in the Church of Santa Maria Maggiore at Florence. If Roger Bacon, who completed his Opus Majus in 1267, speaks of instruments by means of which small letters appear large, "utiles senibus habentibus oculos debiles," his words, and the practically erroneous considerations which he subjoins, shew that he cannot himself have executed the plan which floated before his mind as possible.

(385) p. 250.—See my Examen crit. T. i. p. 61, 64—70, 96—108; T. ii. p. 349. "Il existe aussi de Pierre d'Ailly, que Don Fernando Colon nomme toujours Pedro de Helico, cinq mémoires de Concordantia Astronomiæ cum Theologia. Ils rapellent quelques essais très modernes de Géologie hébraissante publiés 400 ans après le cardinal."

(386) p. 250.—Compare Columbus's letter (Navarrete, Viages y Descubrimientos, T. i. p. 244) with the Imago Mundi of Cardinal d'Ailly, cap. 8, and Roger Bacon's Opus Majus, p. 183.

(387) p. 252.—Heeren, Gesch. der classischen Litteratur, Bd. i. S. 284—290.

(388) p. 252.—Klaproth, Mémoires relatifs à l'Asie, T. iii. p. 113.

(389) p. 252.—The Florentine edition of Homer of 1488; but the first printed Greek book was the grammar of Constantine Lascaris, in 1476.

(390) p. 252.—Villemain, Mélanges historiques et littéraires, T. ii. p. 135.

(391) p. 252.—The result of the investigations of the librarian Ludwig Wachler, at Breslau (see his Geschichte der Litteratur, 1833, Th. i. S. 12—23). Printing without moveable types does not go back, even in China, beyond the beginning of the tenth century of our era. The first four books of Confucius were printed, according to Klaproth, in the province of Szutschuen, between 890 and 925; and the description of the technical manipulation of the Chinese printing press might have been read in western countries as early as 1310, in Raschid-eddin's Persian history of the rulers of Cathay. According to the most recent results of the important researches of Stanislas Julien, however, an ironsmith in China itself would seem to have used moveable types, made of burnt clay, between the years 1041 and 1048 A.D. or almost 400 years before Guttenberg. This is the invention of the Pi-sching, which, however, remained without application.

([392]) p. 253.—See the proofs, in my Examen crit. T. ii. p. 316—320. Josafat Barbaro (1436) and Ghislin von Busbeck (1555) still found, between Tana (Asof), Caffa, and the Erdil (the Volga), Alani and Gothic tribes speaking German (Ramusio, delle Navigationi et Viaggi, Vol. ii. p. 92 b and 98 a). Roger Bacon always terms Rubruquis only frater Willielmus, quem dominus Rex Franciæ misit ad Tartaros.

([393]) p. 254.—The great and fine work of Marco Polo (Il Milione di Messer Marco Polo), as we possess it in the correct edition of Count Baldelli, is incorrectly called "Travels"; it is for the most part a descriptive, one might say a statistical, work; in which it is difficult to distinguish what the traveller saw himself, what he learned from others, and what he derived from topographic descriptions, in which the Chinese literature is so rich, and which might be accessible to him through his Persian interpreters. The striking resemblance between the narrative of the travels of Hiuan-thsang, the Buddhistic pilgrim of the seventh century, and that which Marco Polo found in 1277 (respecting the Pamir-Highland), early drew my whole attention. Jacquet, who an early decease withdrew from the investigation of Asiatic languages, and who, like Klaproth and myself, was long occupied with the great Venetian traveller, wrote to me, a short time before his death, "Je suis frappé comme vous de la forme de rédaction littéraire du Milione. Le fond appartient sans doute à l'observation directe et personnelle du voyageur, mais il a probablement employé des documents qui lui ont été communiqués soit officiellement, soit en particulier. Bien des choses paraissent avoir été empruntées à des livres Chinois et Mongols, bien que ces influences sur la composition du Milione soient difficiles à reconnaître dans les traductions successives sur lesquelles Polo aura fondé ses extraits." Whilst our modern travellers are only too well pleased to occupy their readers with their own persons, Marco Polo takes no less pains to blend his own observations with the official data communicated to him; of which, as governor of the city of Yangui, he might have many. (See my Asie centrale, T. ii. p. 395.) The compiling method of the illustrious traveller also helps to explain the possibility of his dictating his book while confined in the prison at Genoa, in 1295, to his fellow-prisoner and friend Messer Rustigielo of Pisa, as if the documents had been lying before him. (Compare Marsden, Travels of Marco Polo, p. xxxiii.)

([394]) p. 254.—Purchas, Pilgrims, Part iii. ch. 28 and 56 (p. 23 and 24).

([395]) p. 254.—Navarrete, Coleccion de los Viages y Descubrimientos que hiciéron por mar los Españoles, T. i. p. 261; Washington Irving, History of the Life and Voyages of Christopher Columbus, 1828, Vol. iv. p. 297.

(396) p. 255.—Examen crit. de l'Hist. de la Géog. T. i. p. 63 and 215; T. ii. p. 350. Marsden, Travels of Marco Polo, p. lvii. lxx. and lxxv. The first German Nuremberg version of 1477, (das puch des edeln Ritters uñ landtfarers Marcho Polo) appeared in print in Columbus's lifetime; the first Latin translation in 1490, and the first Italian and Portuguese translations in 1496 and 1502.

(397) p. 256.—Barros, Dec. i. liv. iii. cap. 4, p. 190, says expressly that "Bartholomeu Diaz, e os de sua companhia per causa dos perigos, e tormentas, que *em o dobrar* delle passáram, lhe puzeram nome Tormentoso." The merit of first doubling the Cape does not therefore belong, as usually stated, to Vasco de Gama. Diaz was at the Cape in May 1487, almost therefore at the same time that Pedro de Covilham and Alonso de Payva of Barcelona arrived from their expedition. In December of the same year (1487), Diaz brought himself to Portugal the news of his important discovery.

(398) p. 256.—The planisphere of Sanuto, who calls himself "Marinus Sanuto dictus Torxellus de Veneciis," belongs to the work, Secreta fidelium Crucis. "Marinus prêcha adroitement une croisade dans l'intérêt du commerce, voulant détruire la prospérité de l'Egypte, et diriger toutes les marchandises de l'Inde par Bagdad, Bassora et Tauris (Tebriz), à Kaffa, Tana (Azow), et aux côtes asiatiques de la Méditerranée. Contemporain et compatriote de Polo, dont il n'a pas connu le Milione, Sanuto s'élève à de grandes vues de politique commerciale. C'est le Raynal du moyen-âge, moins l'incrédulité d'un abbé philosophe du 18me siècle."—(Examen crit. T. i. p. 331, 333—348.) The Cape of Good Hope is called Capo di Diab on the map of Fra Mauro, which was compiled between 1457 and 1459: see the learned memoir of Cardinal Zurla, entitled, Il Mappamondo di Fra Mauro Camaldolese, 1806, § 54.

(399) p. 257.—Avron or avr (aur) is a less-used term for North, employed instead of the more ordinary "schemâl"; the Arabic Zohron or Zohr, from which Klaproth erroneously endeavours to derive the Spanish sur and Portuguese sul (which is, without doubt, like our Süd, a true Germanic word), does not properly belong to the particular denomination of the quarter indicated; it signifies only the time of high noon; South is dschenûb. Respecting the early knowledge of the Chinese of the south pointing of the magnetic needle, see Klaproth's important investigations in his Lettre à M. A. de Humboldt, sur l'Invention de la Boussole, 1834, p. 41, 45, 50, 66, 79 and 90; and the Memoir of Azuni of Nice, which appeared in 1805, entitled, Dissertation sur l'Origine de la Boussole, p. 35 and 65—68. Navarrete, in his Discurso

historico sobre los progresos del Arte de Navegar en España, 1802, p. 28, recals a remarkable passage in the Spanish Leyes de las Partidas (II. tit. ix. ley 18) of the middle of the 13th century:—"The needle which guides the mariner in the dark night, and shows him how to direct his course both in good and in bad weather, is the intermediary (medianera) between the loadstone (la piedra) and the North star" ..... See the passage in Las siete Partidas del sabio Rey Don Alonso el IX. (according to the usual manner of counting the Xth.) Madrid, 1829, T. i. p. 473.

(400) p. 258.—Jordano Bruno, par Christian Bartholmèss, 1847, T. ii. p. 181—187.

(401) p. 258.—Tenian los mariantes instrumento, carta, compas y aguja." —Salazar, Discurso sobre los progresos de la Hydrografia en España, 1809, p. 7.

(402) p. 258.—Kosmos, Bd. ii. S. 203 (Engl. ed. Vol. ii. p. 169.)

(403) p. 258.—Respecting Cusa (Nicolaus of Cuss, properly of Cues on the Moselle), see above, Kosmos, Bd. ii. S. 140 (Engl. ed. Vol. ii. p. 106); and Clemens' treatise, über Giordano Bruno and Nicolaus de Cusa, S. 97, where there is given an important fragment, written by Cusa's own hand, and discovered only three years ago, respecting a threefold movement of the earth. (Compare also Chasles, Aperçus aur l'origine des methodes en Géométrie, 1807, p. 529.)

(404) p. 259.—Navarrete, Dissertacion histórica sobre la parte que tuviéron los Españoles en las guerras de Ultramar ó de las Cruzadas, 1816, p. 100; and Examen crit. T. i. p. 574—277. An important improvement in observation by means of the plumb-line has been attributed to Georg von Peuerbach, the teacher of Regiomontanus. But the use of the plumb-line had long been known to the Arabs, as we learn by Abul-Hassan-Ali's compendious description of astronomical instruments, written in the 13th century: Sedillot, Traité des instrumens astronomiques des Arabes, 1835, p. 379; 1841, p. 205.

(405) p. 259.—In all the writings on the art of navigation which I have examined, I find the erroneous opinion that the Log, for the measurement of the distance passed over, has only been in use since the end of the 16th or the beginning of the 17th century. In the Encyclopædia Britannica (7th edition, 1842), Vol. xiii. p. 416, it is still said: "The author of the device for measuring the ship's way is not known, and no mention of it occurs till the year 1607, in an East India voyage, published by Purchas." This year is also named as the extreme limit in all earlier and later dictionaries.—

—(Gehler, Bd. vi. 1831, S. 450.) It is only Navarrete, in the Dissertacion sobre los progresos del Arte de Navegar, 1802, who places the use of the log-line in English ships in the year 1577.—(Duflot de Mofras, Notice biographique sur Mendoza et Navarrete, 1845, p. 64.) Subsequently he affirms in another place (Coleccion de los Viages de los Españoles, T. iv. 1837, p. 97), that "in Magellan's time the ship's speed was only estimated by the eye (á ojo), until in the 16th century the corredera (the log) was devised." The measurement of the distance sailed over by means of heaving the log, although this means must in itself be termed imperfect, has become of such great importance towards a knowledge of the velocity and direction of oceanie currents, that I have been led to make it an object of careful research. I give here the principal results which are contained in the 6th and still unpublished volume of my Examen critique de l'histoire de la Géographie et des progrès de l'Astronomie nautique. The Romans, in the time of the republic, had in their ships apparatus for measuring the distance passed over, consisting of wheels four feet high provided with paddles placed outside the ship, just as in our steamboats, and as in the apparatus for propelling vessels which Blasco de Garay had proposed in 1543 at Barcelona to the Emperor Charles V. —(Arago, Annuaire du Bureau des Longitudes, 1829, p. 152.) The ancient Roman way-measurer (ratio a majoribus tradita, qua in via rheda sedentes vel mari navigantes scire possumus quot millia numero itineris fecerimus) is described in detail by Vitruvius (lib. x. cap. 14), the credit of whose Augustan age has indeed been recently much shaken by C. Schultz and Osann. By means of three toothed wheels acting on each other, and by the falling of small round stones from a wheel-case (loculamentum) having only a single hole, the number of revolutions of the outside wheels which dipped in the sea, and the number of miles passed through in the day's course, were given. Whether these hodometers were much used in the Mediterranean, "as they might afford both use and pleasure," Vitruvius does not say. In the biography of the Emperor Pertinax by Julius Capitolinus, mention is made of the purchase of the effects left by the Emperor Commodus, among which was a travelling carriage provided with a similar hodometric apparatus.—(Cap. 8 in Hist. Augustæ Script. ed. Lugd. Bat. 1671, T. i. 554.) The wheels gave at once "the measure of the distance passed over and the duration of the journey" in hours. A much more perfect hodometer used both on the water and on land has been described by Hero of Alexandria, the pupil of Ctesibius, in his Greek still inedited manuscript on the Dioptra.—(See Venturi, Comment. sopra la Storia dell' Ottica, Bologna, 1814, T. i. p. 134—139.) We find nothing on the subject we are considering, in the literature of the middle ages, until we come to the period of several "books of Nautical Instruction," written or printed in quick succession by Antonio Pigafetta (Trattato di Navigazione, probably before 1530); Francisco Falero (1535, a brother of the astronomer Ruy Falero, who was to have accompanied Magellan on his voyage round the world, and left behind him a Regimiento para observar la longitud en la mar); Pedro de Medina of Seville (Arte de Navegar, 1545); Martin Cortes of Bujalaroz (Breve Compendio de la esfera y de la arte de navegar, 1551); and Andres Garcia de Cespedes (Regimiento de Navigacion y Hidrografia, 1606). From almost all these works, some of which have become extremely rare, as well as from the Suma de Geografia which Martin Fernandez de Enciso had published in 1519, we recognise most distinctly that navigators were taught to estimate the "distance sailed over" in Spanish and Portuguese ships, not by any distinct measurement, but only by estimation or appreciation by the eye, according to certain established principles. Medina says (libro iii. cap. 11 and 12), "to know the course of the ship as to the length of distance passed over, the pilot must set down in his register how much distance she has made according to hours (*i. e.* guiding himself by the hourglass, "ampolleta,") and for this he must know that the most a ship advances in an hour is four miles, and with feebler breezes three, or only two." Cespedes (Regimiento, p. 99 and 156) calls this mode of proceeding "echar punto por fantasia." This fantasia, as Enciso justly remarks, depends, if great errors are to be avoided, on the pilot's knowledge of the qualities of his ship: on the whole, however, every one who has been long at sea will have remarked with surprise, when the waves are not very high, how nearly the mere estimation of the ship's velocity accords with the subsequent result obtained by the log. Some Spanish pilots call the old, and it must be admitted hazardous, method of mere estimation (cuenta de estima), sarcastically, and certainly very incorrectly, "la corredera de los Holandeses, corredera de los perezosos." In Columbus's ship's journal, frequent reference is made to the contest with Alonso Pinzon as to the distance passed over since their departure from Palos. The hour or sandglasses, ampolletas, which they made use of, ran out in half an hour, so that the interval of a day and night was reckoned at 48 ampolletas. In this important journal of Columbus, it is said (for example on the 22d of January, 1493): "Andaba 8 millas por hora hasta pasadas 5 ampolletas, y 3 antes que comenzase la guardia, que eran 8 ampolletas."—(Navarrete, T. i. p. 143.) The Log (la corredera) is never mentioned. Are we to assume that Columbus was acquainted with and employed it, but that, being a means already in very general use, he did not think it necessary to name it? in the same way that Marco Polo does not mention tea, or the great wall of China. Such an assumption appears to me very improbable, even if there were no other reason, because I find in the proposals made by the pilot Don Jayme Ferrer, 1495, for the exact examination of the position of the Papal line of demarcation, that, when it is question of the determination of the distance sailed over, the appeal is made only to the accordant sentence (juicio) of 20 very experienced mariners (que apunten en su carta de 6 en 6 horas el camino que la nao fará segun su juicio.) If the log had been in use, no doubt Ferrer would have prescribed how often it should be hove. I find the first application of the log in a passage of Pigafetta's Journal of Magellan's voyage of circumnavigation, which long lay buried among the manuscripts in the Ambrosian Library at Milan. It is said in it, that, in the month of January 1521, when Magellan had already arrived in the Pacific, "Secondo la misura che facevamo del viaggio colla catena a poppa, noi percorrevamo da 60 in 70 leghe al giorno."—(Amoretti, Primo Viaggio intorno al Globo terracqueo, ossia Navigazione fatta dal Cavaliere Antonio Pigafetta sulla squadra del Cap. Magaglianes, 1800, p. 46.) What can this arrangement of a chain at the hinder part of the ship (catena a poppa), "which we used throughout the entire voyage to measure the way," have been other than an apparatus similar to our log? The "running out" log-line divided into knots, the log-ship, and the half-minute or log-glasses are not mentioned; but this silence need not surprise us in speaking of a long-known matter. In the part of the Trattato di Navigazione of the Cavaliere Pigafetta given by Amoretti in extracts, amounting indeed only to 10 pages, the "catena della poppa" is not again mentioned.

([406]) p. 259.—Barros, Dec. I. liv. iv. p. 320.

([407]) p. 261.—Examen crit. T. i. p. 3—6 and 290.

([408]) p. 262.—Compare Opus Epistolarum Petri Martyris Anglerii Mediolanensis, 1670, ep. cxxx. and clii. "Præ lætitia prosilisse te, vixque à lachrymis præ gaudio temperasse quando literas adspexisti meas, quibus de antipodium orbe, latenti hactenus, te certiorem feci, mi suavissime Pomponi, insinuasti. Ex tuis ipse literis colligo, quid senseris. Sensisti autem, tantique rem fecisti, quanti virum summa doctrina insignitum decuit. Quis namque cibus sublimibus præstari potest ingeniis isto suavior? quod condimentum gratius? à me facio conjecturam. Beari sentio spiritus meos, quando accitos alloquor prudentes aliquos ex his qui ab ea redeunt provincia (Hispaniola insula.") The expression, "Christophorus quidam Colonus," reminds us, I will not say of the too often and unjustly quoted "nescio quis Plutarchus" of Aulus Gellius (Noct. Atticæ, xi. 16), but of the "quodam Cornelio scribente," in the answer of the king Theodoric to the prince of the Æstyans, who was to be informed respecting the true origin of amber from the Germ. cap. 45, of Tacitus.

(409) p. 202.—Opus Epistol. No. ccccxxxvii. and dlxii. An extraordinary person, Hieronymus Cardanus, a fantastic enthusiast and at the same time an acute mathematician, also calls attention in his "physical problems" to how much of the knowledge of the earth consisted in facts to the observation of which one man has led. Cardani Opera, ed. Lugdun. 1663, T. ii. Probl. p. 630 and 659; "at nunc quibus te laudibus afferam Christophore Columbi, non familiæ tantum, non Genuensis urbis, non Italiæ Provinciæ, non Europæ partis orbis solum, sed humani generis decus." In comparing the "problems" of Cardanus with those of the later Aristotelian school, amidst the confusion and the feebleness of the physical explanations which prevail almost equally in both collections, I remark in Cardanus a circumstance which appears to me characteristic of the sudden enlargement of geography at that epoch; namely, that the greater part of his problems relate to comparative meteorology. I allude to the considerations on the warm insular climate of England in contrast with the winter at Milan;—on the dependence of hail on electric explosions;—on the cause and direction of oceanic currents;—on the maxima of atmospheric heat and cold not arriving until after the summer and winter solstices;—on the elevation of the region of snow under the tropics;—on the temperature dependent on the radiation of heat from the sun and from all the heavenly bodies;—on the greater intensity of light in the southern hemisphere, &c.—"Cold is merely absence of heat. Light and heat differ only in name, and are in themselves inseparable." Cardani Opp. T. i. de vita propria, p. 40; T. ii. Probl. 621, 630—632, 653 and 713; T. iii. de subtilitate, p. 417.

(410) p. 263.—See my Examen crit. T. ii. p. 210—249. According to the manuscript, Historia general de las Indias, lib. i. cap. 12, "la carta de marear que Maestro Paula Fisico (Toscanelli) envió á Colon" was in the hands of Bartholomé de las Casas when he wrote his work. Columbus's ship's journal, of which we possess an extract (Navarrete, T. i. p. 13), does not quite agree with the relation which I find in a manuscript written by Las Casas, which was kindly communicated to me by M. Ternaux-Compans. The ship's journal says, "Iba hablando el Almirante (martes 25 de Setiembre, 1492) con Martin Alonso Pinzon, capitan de la otra carabela Pinta, sobra una carta que le habia enviado tres dias hacia á la carabela, donde segun

parece *tenia pintadas el Almirante* ciertas islas por aquella mar . . . ." In the manuscript of Las Casas (lib. i. cap. 12), on the other hand, I find, "La carta de marear que embió (Toscanelli al Almirante) yo que esta historia scrivo la tango en mi poder. Creo que todo su viage sobre esta carta fundó;" (lib. i. cap. 38) "asi fué que el martes 25 de Setiembre llegase Martin Alonso Pinzon con su caravela Pinta á hablar con Christobal Colon sobre una carta de marear que Christobal Colon le avia embiado . . . . *Esta carta es la que le embió Paulo Fisico el Florentin, la qual yo tengo en mi poder* con otras cosas del Almirante y escrituras de su misma mano que traxéron á mi poder. En ella le pintó muchas islas . . . . ." Are we to assume that the Admiral had drawn upon the map of Toscanelli the islands which he expected to find, or does "tenia pintadas" merely mean "the Admiral had a map on which were painted . . . . ."?

(411) p. 264.—Navarrete, Documentos, No. 69, in T. iii. der Viages y Descubr. p. 565—571; Examen crit. T. i. p. 234—249 and 252, T. iii. p. 158—165 and 224. Respecting the contested spot of the first landing in the West Indies, see T. iii. p. 186—222. The map of the world of Juan de la Cosa, which has acquired so much celebrity, and which was discovered by Walckenaer and myself in the year 1832, during the cholera epidemic, and which was drawn six years before the death of Columbus, has thrown new light on these contested questions.

(412) p. 265.—Respecting Columbus's graphical and often poetical descriptions of nature, see above, Kosmos, Bd. ii. S. 55—57 (Engl. edit. Vol. ii. p. 54—56).

(413) p. 266.—See the results of my investigations, in the Relation hist. du Voyage aux Régions équinoxiales du nouveau Continent, T. ii. p. 702; and in the Examen crit. de l'Hist de la Géographie, T. i. p. 309.

(414) p. 266.—Biddle, Memoir of Sebastian Cabot, 1831, p. 52—61; Examen crit. T. iv. p. 231.

(415) p. 266.—In a part of Columbus's Journal (Nov. 1, 1492) which has received but little attention, it is said, "I have (in Cuba) opposite and near to me Zayto y Guinsay (Zaitun and Quinsay, Marco Polo, ii. 77) del Gran Can." Navarrete, Viages y Descubrim. de los Españoles, T. i. p. 46; and above, note 375. The curve towards the south, which Columbus on his second voyage remarked in the most western part of the coast of Cuba, had an important influence, as I have elsewhere observed, on the discovery of South America, and on that of the Delta of the Orinoco and Cape Paria; see Examen crit. T. iv. p. 246—250. Anghiera (Epist. clxviii. ed. Amst. 1670,

p. 96) writes, "Putat (Colonus) regiones has (Pariæ) esse Cubæ contiguas et adhærentes: ita quod utræque sint Indiæ Gangetidis continens ipsum . . . ."

([416]) p. 267.—See the important manuscript of Andres Bernaldez, Cura de la Villa de los Palacios (Historia de los Reyes Catholicos, cap. 123). This history comprises the years 1488 to 1513. Bernaldez had received Columbus, in 1496, on his return from his second voyage, into his house. By the particular kindness of M. Ternaux-Compans, to whom the History of the Conquista owes many important elucidations, I was enabled to make a free use, in Dec. 1838, at Paris, of this manuscript, which was in the possession of my distinguished friend the historiographer, Don Juan Bautista Muñoz (Compare Fern. Colon, Vida del Almirante, cap. 56).

([417]) p. 267.—Examen crit. T. iii. p. 244—248.

([418]) p. 268.—Cape Horn was discovered in February 1526, by Francisco de Hoces, in the expedition of the Commendador Garcia de Loaysa, which, following that of Magellan, was destined for the Moluccas. Whilst Loaysa sailed through the Straits of Magellan, Hoces, with his Caravel, the San Lesmes, was separated from the flotilla, and driven as far as 55° S. latitude. "Dijeron los del buque que les parecia que era alli acabamiento de tierra" (Navarrete, Viages de los Espanoles, T. v. p. 28 and 404—488). Fleurieu maintains that Hoces only saw the Cabo del buen Successo, west of Staten-Island. Such a strange uncertainty respecting the form of the land prevailed anew towards the end of the 16th century, that the author of the Araucana (Canto i. oct. 9) could believe that the Magellanic straits had closed by an earthquake, and by the raising of the bottom of the sea; and, on the other hand, Acosta (Historia natural y moral de las Indias, lib. iii. cap. 10) took the Terra del Fuego for the beginning of a great south polar land. (Compare also Kosmos, Bd. ii. S. 62 and 124; Engl. edit. Vol. ii. p. 60 and Note 96.)

([419]) p. 268.—The question, whether the isthmus-hypothesis, according to which Cape Prasum, on the east of Africa, joined on to an east Asiatic isthmus from Thinæ, is to be traced back to Marinus of Tyre, or to Hipparchus, or to the Babylonian Seleucus, or rather to Aristotle de Cœlo (ii. 14), has been treated by me in detail in another work (Examen crit. T. i. p. 144, 161, and 329; T. ii. p. 370—372).

([420]) p. 269.—Paolo Toscanelli was so much distinguished as an astronomer, that Behaim's teacher, Regiomontanus, dedicated to him, in 1463, his work "De Quadratura Circuli," directed against the Cardinal Nicolaus de Cusa. He constructed the great gnomon in the Church of Santa Maria Novella at Florence, and died in 1482, at the age of 85, without having lived long enough to enjoy the tidings of the discovery of the Cape of Good Hope by Diaz, and that of the tropical part of the new continent by Columbus.

([421]) p. 270.—As the old continent, from the western extremity of the Iberian peninsula to the coast of China, comprehends almost 130° of longitude, there remain about 230° as the space which Columbus should have had to traverse to reach Cathay (China); but less if he only proposed to reach Zipangi (Japan). This difference of 230° which I have taken, is between the Portuguese Cape St. Vincent (11° 20′ W. of Paris), and the far projecting part of the Chinese coast near the then so celebrated port of Quinsay, so often named by Columbus and Toscanelli (lat. 30° 28′, long. 117° 47′ E. of Paris). (Synonymes for Quinsay in the province of Tschekiang are Kanfu, Hangtscheufu, Kingszu.) The general commerce in the east of Asia was shared, in the 13th century, between Quinsay and Zaitun (Pinghai or Tseuthung) opposite to the island of Formosa (then Tungfan) in 25° 5′ N. lat. (see Klaproth, Tableau hist. de l'Asie, p. 227). The distance of Cape St. Vincent from Zipangi (Niphon) is 22° of longitude less than from Quinsay, or about 209° instead of 230° 53′. It is a striking circumstance that, through accidental compensations, the oldest statements, those of Eratosthenes, and Strabo (lib. i. p. 64), come within 10° of the above mentioned result of 129° for the difference of longitude of the *οικουμενη*. Strabo, in the very place where he alludes to the possible existence of two great habitable continents in the northern hemisphere, says that our *οικουμενη* in the parallel of Thinæ (Athens, see Kosmos, Bd. ii. S. 223; Engl. edit. Vol. ii. p. 188) takes more than one-third of the earth's circumference. Marinus of Tyre, being misled by the length of the time occupied in the navigation from Myos Hormos to India, by the erroneously assumed direction of the greater axis of the Caspian from east to west, and by the over estimation of the length of the route by land to the country of the Seres, gave to the old continent a breadth of 225° instead of 129°, thus advancing the Chinese coast to the Sandwich Islands. Columbus naturally preferred this result to that of Ptolemy, according to which Quinsay should have been found in the meridian of the eastern part of the archipelago of the Carolinas. Ptolemy, in the Almagest (ii. 1), places the coast of the Sinæ at 180°; and in his Geography (lib. i. cap. 12), at 177¼°. As Columbus estimated the navigation from Iberia to the Sines at 120°, and Toscanelli even at only 52°, they might both, estimating the length of the Mediterranean at about 40°, have naturally called the apparently so hazardous enterprise only a "brevissimo camino." Martin Behaim, also, on his "world apple" (the celebrated globe which he finished in 1492, and which is still kept in the

Behaim house at Nuremberg), places the coast of China (or the throne of the king of Mango, Cambalu, and Cathay) only 100° west of the Azores, *i. e.* as Behaim lived four years at Fayal, and probably counted the distance from that point, 119° 40′ west of Cape St. Vincent." Columbus was probably acquainted with Behaim at Lisbon, where they both lived from 1480 to 1484 (see my Examen crit. de l'Hist. de la Géographie, T. ii. p. 357—369). The many wholly erroneous numbers which are to be found in all the writings on the discovery of America, and the then supposed extent of Eastern Asia, have induced me to compare more closely the opinions of the middle ages with those of classical antiquity.

([422]) p. 270.—The eastern part of the Pacific was first navigated by white men in a boat, when Alonso Martin de Don Benito, (who had seen the sea horizon with Vasco Nunez de Balboa on the 25th September, 1513, from the little Sierra de Quarequa), descended a few days afterwards to the Golfo de San Miguel, before Balboa went through the ceremony of taking possession of the ocean! Seven months previously Balboa had announced to his court that the South Sea, of which he had heard from the natives, was very easy to navigate: "mar muy mansa y que nunca anda brava como la mar de nuestra banda" (de las Antillos). The name Oceano Pacifico, however, was, as Pigafetta tells us, first given by Magellan to the Mar del Sur (Balboa's name). In August 1519 (before Magellan's expedition), the Spanish government, which was not wanting in watchfulness and activity, had given secret orders, in November 1514, to Pedrarius Davila, Governor of the province of Castilla del Oro (the northwesternmost of South America), and to the great navigator Juan Diaz de Solis;—to the first to have four caravels built in the Golfo de San Miguel "to make discoveries in the newly discovered South Sea"; and to the second, to seek for an opening ("abertura de la tierra") from the eastern coast of America, with the view of arriving at the back ("a' aspeldas") of the new country, *i. e.* of the sea-surrounded western portion of Castilla del Oro. The expedition of Solis (October 1515 to August 1516) led him far to the south, and to the discovery of the Rio de la Plata, which was long called the Rio de Solis. (Compare, respecting the little known first discovery of the Pacific, Petrus Martyr, Epist. dxl. p. 296, with the documents of 1513—1515 in Navarrete, T. iii. p. 134 and 357; also my Examen crit. T. i. p. 320 and 350.)

([433]) p. 270.—Respecting the geographical position of the Desventuradas (San Pablo, lat. 16¼° S. long., 135¾° west of Paris; Isla de Tiburones, lat. 10¾° S., long. 145° W.), see my Examen crit. T. i. p. 286; and Navarrete,

T. iv. p. lix. 52, 218, and 267. The great epoch of geographical discoveries gave occasion to many such illustrious heraldic bearings as that mentioned in the text; (the terrestrial globe, with the inscription "Primus circumdedisti me," to Sebastian de Elcano and his descendants). The arms which, as early as May 1493, were given to Columbus, "para sublimarlo" with posterity, contain the first map of America—a range of islands in front of a gulf (Oviedo, Hist. general de las Indias, ed. de 1547, lib. ii. cap. 7, fol. 10 a; Navarrete, T. ii. p. 37; Examen crit. T. iv. p. 236). The Emperor Charles V. gave to Diego de Ordaz, who boasted of having ascended the volcano of Orizaba, the drawing of that conical mountain; and to the historian Oviedo, who resided uninterruptedly for 34 years (from 1513 to 1547) in tropical America, the four stars of the southern cross, as armorial bearings (Oviedo, lib. ii. cap. 11, fol. 16 b).

(424) p. 271.—See my Essai politique sur le Royaume de la Nouvelle Espagne, T. ii. 1827, p. 259; and Prescott, History of the Conquest of Mexico (New York, 1843), Vol. iii. p. 271 and 336.

(425) p. 273.—Gaetano discovered one of the Sandwich Islands in 1542. Respecting the voyage of Don Jorge de Menezes (1526), and that of Alvaro de Saavedra (1528), to the Ilhas de Papuas, see Barros da Asia, Dec. iv. Liv. i. cap. 16, and Navarrete, T. v. p. 125. The "Hydrography" of Joh. Rotz (1542), which is preserved in the British Museum, and has been examined by the learned Dalrymple, contains outlines of New Holland; as does also the collection of maps of Jean Valard of Dieppe (1552), for the first knowledge of which we are indebted to M. Coquebert Monbret.

(426) p. 273.—After the death of Mendaña, the command of the expedition, which did not terminate until 1596, was undertaken in the South Sea by his wife, Doña Isabela Baretos, a woman of distinguished personal courage, and great mental endowments (Essai polit. sur la Nouv. Espagne, T. iv. p. 111.) Quiros practised distillation of fresh from salt water on a considerable scale in his ship, and his example was followed in several instances (Navarrete, T. i. p. liii.) The entire operation, as I have elsewhere proved, on the testimony of Alexander of Aphrodisias, was known as early as the third century of our era, although not then practised in ships.

(427) p. 273.—See the excellent work of Professor Meinicke at Prenzlau, entitled, "Das Festland australien, eine geogr. Monographie," 1837, Th. i. S. 2—10.

(428) p. 276.—This king died in the time of the Mexican king Axyacatl, who reigned from 1464 to 1477. The learned native Mexican historian,

Fernando de Alva Ixtlilxochitl, whose manuscript chronicle of the Chichimeques, which I saw, in 1803, in the palace of the Viceroy of Mexico, and which Mr. Prescott has made such happy use of in his work (Conquest of Mexico, Vol. i. p. 61, 173, and 206; Vol. iii. p. 112), was a descendant of the poet king Nezahualcoyotl. The Aztec name of the historian, Fernando de Alva, signifies Vanilla faced. M. Ternaux-Compans, in 1840, printed a French translation of this manuscript in Paris. The notice of the long elephant's hair which Cadamosto collected, is to be found in Ramusio, Vol. i. p. 109, and in Grynæus, cap. 43, p. 33.

(429) p. 277.—Clavigero, Storia antica del Messico (Cesena, 1780) T. ii. p. 153. The accordant testimonies of Hernan Cortes, in his reports to the Emperor Charles V., of Bernal Diaz, Gomara, Oviedo, and Hernandez, leave no doubt that at the time of the conquest of Montezuma's empire, there were in no part of Europe menageries and botanic gardens (collections of living animals and plants) which could be compared to those of Haaxtepec, Chapoltapec, Iztapalapon, and Tezcuco (Prescott, Vol. i. p. 178; Vol. ii. p. 66 and 117; Vol. iii. p. 42). Respecting the early attention stated in the text to have been paid to the fossil bones in the American "fields of giants," see Garcilaso, lib. ix. cap. 9; Acosta, lib. iv. cap. 30; and Hernandez (ed. of 1556), T. i. cap. 32, p. 105.

(430) p. 279.—Observations de Christophe Colomb sur le Passage de la Polaire par le Meridien, in my Relation hist. T. i. p. 506, and in the Examen crit. T. iii. p. 17—20, 44—51, and 56—61. (Compare also Navarrete, in Columbus's Journal of 16 to 30 Sept. 1492, p. 9, 15, and 254.)

(431) p. 282.—Respecting the singular differences of the Bula de concesion a los Reyes Catholicos de las Indias descubiertas y que se descrubieren of 3 May, 1493, and the Bula de Alexandro VI. sobre la particion del oceano of May 4, 1493 (elucidated in the Bula de estension of the 25th of September, 1493), see Examen crit. T. iii. p. 52—54. Very different from this line of demarcation is that settled in the Capitulacion de la Particion del Mar Oceano entre los Reyes Catholicos y Don Juan, Rey de Portugal, of the 7th June, 1494, 370 leguas (17½ to an equatorial degree) west of the Cape Verd Islands. (Compare Navarrete, Coleccion de los Viages y Descubr. de los Esp. T. ii. p. 28—35, 116—143, and 404; T. iv. p. 55 and 252). This last named line, which led to the sale of the Moluccas (de el Maluco) to Portugal, 1529, for the sum of 350000 gold ducats, had no connection with magnetical or meteorological fancies. The papal lines of demarcation, however, deserve more careful consideration in the present work, because, as I have mentioned

in the text, they exercised great influence on the endeavours to improve nautical astronomy, and especially the methods of finding the longitude. It is also very deserving of notice, that the capitulation of June 7, 1494, affords the first example of a proposal to fix a meridian in a permanent manner by marks graven in rocks, or by the erection of towers. It is commanded, "que se haga alguna señal ó torre" wherever the dividing meridian, in its course from pole to pole, whether in the eastern or the western hemisphere, intersects an island or a continent. In continents, the raya was to be marked, at proper intervals, by a series of such marks or towers; which would, indeed, have been no small undertaking.

([432]) p. 280.—It is a remarkable fact, that the earliest classical writer on terrestrial magnetism, William Gilbert, who we cannot suppose to have had any knowledge of Chinese literature, yet regards the mariner's compass as a Chinese invention, which had been brought to Europe by Marco Polo. "Illa quidem pyxide nihil unquam humanis excogitatum artibus humano generi profuisse magis, constat. Scientia nauticæ pyxidulæ traducta videtur in Italiam per Paulum Venetum, qui circa annum mcclx. apud Chinas artem pyxidis didicit" (Guilielmi Gilberti Colcestrensis, Medici Londinensis, de Magnete Physiologia nova, Lond. 1600, p. 4). There are, however, no grounds for the supposition that the compass was introduced by Marco Polo, whose travels were from 1271 to 1295, and who therefore returned to Italy after the mariner's compass had been spoken of by Guyot de Provins in his poem, as well as by Jacques de Vitry and Dante, as a long known instrument. Before Marco Polo set out on his travels in the middle of the 13th century, Catalans and Basques already made use of the compass (see Raymond Lully, in the treatise De Contemplatione, written in 1272).

([433]) p. 282.—For the anecdote respecting Sebastian Cabot, see Biddle's Memoirs of that celebrated navigator: a work written with a good historical and critical spirit (p. 222). "We know," says Biddle, "with certainty neither the date of the death nor the burying place of the great navigator who gave to Great Britain almost an entire continent, and without whom (as without Sir Walter Raleigh), the English language would perhaps not have been spoken by many millions who now inhabit America." Respecting the materials from which the variation-chart of Alonzo de Santa Cruz was compiled, as well as respecting the variation-compass, of which the construction was already such as to permit altitudes of the sun to be taken at the same time, see Navarrete, Noticia biografica del Cosmografo Alonso de Santa Cruz, p. 3—8. The first variation-compass was constructed before 1525, by an ingenious apothecary of Seville, Felipe Guillen. So earnest were the endeavours to learn more exactly the direction of the curves of magnetic declination, that in 1585 Juan Jayme sailed with Francisco Gali from Manila to Acapulco for the sole purpose of trying in the Pacific a declination instrument which he had invented. See my Essai politique sur la Nouvelle Espagne, T. iv. p. 110.

(434) p. 282.—Acosta, Hist. natural de las Indias, lib. i. cap. 17. These four magnetic lines of no variation led Halley, by the contests between Henry Bond and Beckborrow, to the theory of four magnetic poles.

(435) p. 282.—Gilbert de Magnete Physiologia nova, lib. v. cap. 8, p. 200.

(436) p. 283.—In the temperate and cold zones, the inflexion of the isothermal lines is general between the west coast of Europe and the east coast of America, but within the tropics the isothermal lines run almost parallel to the equator; and in the hasty conclusions into which Columbus suffered himself to be led, no account was taken of the difference between sea and land climates, or between east and west coasts, or of the influence of winds,—as in the case of winds blowing over Africa. Compare the remarkable considerations on climates which are brought together in the Vida del Almirante (cap. 66). The early conjecture of Columbus respecting the curvature of the isothermal lines in the Atlantic Ocean was well founded, if we limit it to the extra-tropical (temperate and frigid) zones.

(437) p. 283.—An observation of Columbus (Vida del Almirante, cap. 55; Examen crit. T. iv. p. 253; Kosmos, Bd. i. S. 479 (Engl. edit. Vol. i. note 388).

(438) p. 283.—The admiral, says Fernando Colon (Vida del Alm. cap. 58) ascribed the many refreshing falls of rain, which cooled the air whilst he was sailing along the coast of Jamaica, to the extent and denseness of the forests which clothe the mountains. He takes this opportunity of remarking, in his ship's journal, that "formerly there was as much rain in Madeira, the Canaries, and the Azores; but since the trees which shaded the ground have been cut down, rain has become much more rare." This warning has remained almost unheeded for three centuries and a half.

(439) p. 284.—Kosmos, Bd. i. S. 355 and 482 (Engl. edit. Vol. i. p. 327, and note 400); Examen crit. T. iv. p. 294; Asie centrale, T. iii. p. 235. The inscription of Adulis, which is almost fifteen hundred years older than Anghiera, speaks of "Abyssinian snow in which a man may sink up to the knees."

(440) p. 285.—Leonardo da Vinci says very finely of this proceeding, "questo è il methodo da osservarsi nella ricerca de' fenomeni della natura." See Venturi, Essai sur les Ouvrages physico-mathématiques de Léonard da Vinci, 1797, p. 31; Amoretti, Memorie storiche sù la Vita di Lionardo da Vinci, Milano, 1804, p. 143 (in his edition of the Trattato della Pittura, T. xxxiii. of the Classici Italiani); Whewell, Philos. of the Inductive Sciences, 1840, Vol. ii. p. 368—370; Brewster, Life of Newton, p. 332. Most of Leonardo da Vinci's physical works belong to the year 1498.

(441) p. 286.—The great attention paid by the early navigators to natural phenomena may be seen in the oldest Spanish accounts. Diego de Lepe, for example, (as we learn from a witness in the law-suit against the heirs of Columbus,) by means of a vessel provided with valves, which did not open until it had reached the bottom, found that at a distance from the mouth of the Orinoco, a stratum of fresh water of 6 fathoms depth flowed over the salt water (Navarrete, Viages y Descubrim. T. iii. p. 549). Columbus, on the south of the coast of Cuba, took up milk-white sea-water ("white as if meal had been mixed with it") to be carried to Spain in bottles (Vida del Almirante, p. 56). I was myself at the same spots, for the purpose of determining longitudes, and was surprised that the milk-white discolouration of sea-water, so common on shoals, should have appeared to the experienced admiral a new and unexpected phenomenon. In what relates to the gulf-stream itself, which must be regarded as an important cosmical phenomenon, various effects produced by it had been observed, long before the discovery of America, by the sea washing on shore at the Canaries and the Azores stems of bamboos, trunks of pines, corpses of foreign aspect from the Antilles, and even living men in canoes "which could not sink." But all this was then attributed solely to the strength of westerly tempests (Vida del Almirante, cap. 8; Herrera, Dec. i. lib. i. cap. 2, lib. ix. cap. 12); there was as yet no recognition of the movement of the waters which is independent of the direction of the wind, viz. the returning stream of the oceanic current, which brings every year tropical fruits from the West India Islands to the coasts of Ireland and Norway. Compare the Memoir of Sir Humphrey Gilbert, On the Possibility of a North-west Passage to Cathay, in Hakluyt, Navigations and Voyages, Vol. iii. p. 14; Herrera, Dec. i. lib. ix. cap. 12; and Examen crit. T. ii. p. 247—257, T. iii. p. 99—108.

(442) p. 287.—Examen crit. T. iii. p. 26 and 66—99; Kosmos, Bd. i. S, 328 and 330 (Engl. ed. Vol. i. p. 301 and 303).

([443]) p. 287.—Alonso de Ercilla has imitated the passage of Garcilaso in the Araucana: "Climas passè, mudè constelaciones;" see Kosmos, Bd. ii. S. 121, Anm. 62 (Engl. ed. Vol. ii. note 62).

([444]) p. 289.—Pet. Mart. Ocean. Dec. I. lib. ix. p. 96; Examen crit. T. iv. p. 221 and 317.

([445]) p. 289.—Acosta, Hist. natural de las Indias, lib. i. cap. 2; Rigaud, Account of Harriot's Astron. Papers, 1833, p. 37.

([446]) p. 289.—Pigafetta, Primo Viaggio intorno al Globo terracqueo, pubbl. da C. Amoretti, 1800, p. 46; Ramusio, Vol. i. p. 355 c; Petr. Mart. Ocean. Dec. III. lib. i. p. 217. (From the events to which Anghiera refers, Dec. II. lib. x. p. 204, and Dec. III. lib. x. p. 232, the passage of the Oceanica which speaks of the Magellanic clouds must have been written between 1514 and 1516.) Andrea Corsali (Ramusio, Vol. i. p. 177) also describes in a letter to Giuliano de Medici the movement of translation of "due nugolette di ragionevol grandezza." The star which he represents between Nubecula major and minor appears to me to be β Hydræ (Examen crit. T. v. p. 234—238). Respecting Petrus Theodor of Emden and Houtman, the pupil of the mathematician Plancius, see an historical article by Olbers, in Schumacher's Jahrbuch für 1840, S. 249.

([247]) p. 291.—Compare the researches of Delambre and Encke with Ideler, Ursprung der Sternnamen, S. xlix. 263 and 277; also my Examen crit. T. iv. p. 319—324; T. v. p. 17—19, 30 and 230—234.

([448]) p. 291.—Plin. ii. 70; Ideler, Sternnamen, S. 260 and 295.

([449]) p. 292.—I have attempted in another place to dispel the doubts which several distinguished commentators of Dante have expressed in modern times respecting the "quattro stelle." To take this problem in all its completeness, we must compare the passage, "Io mi volsi," &c. (Purgat. I. v. 22—24) with other passages:—Purg. I. v. 37; VIII. v. 85—93; XXIX. v. 121: XXX. v. 97; XXXI. v. 106; and Inf. XXVI. v. 117 and 127. The Milanese astronomer, De Cesaris, considers the three "facelle" ("Di che 'l polo di quà tutto quanto arde," and which set when the four stars of the Cross rise,) to be Canopus, Achernar and Fomalhaut. I have attempted to solve the difficulties by the following considerations:—"Le mysticisme philosophique et religieux qui pénétre et vivifie l'immense composition du Dante, assigne à tous les objets, à coté de leur existence réelle ou matérielle, une existence idéale. C'est comme deux mondes, dont l'un est le reflet de l'autre. Le groupe des quatres étoiles represente, dans l'ordre moral, les *vertus cardinales*, la prudence, la justice, la force et la temperance; elles méritent pour cela le nom de 'saintes lumières, *luci sante.*' Les trois étoiles 'qui éclairent le pole' representent les *vertus théologales*, la foi, l'espérance et la charité. Les premiers de ces êtres nous révelent eux-mêmes leur double nature; ils chantent: 'Ici nous sommes des nymphes, dans le ciel nous sommes des étoiles; *Noi sem qui Ninfe, e nel ciel semo stelle.*' Dans la *Terre de la verité*, le Paradis terrestre, sept nymphes se trouvent réunies: *In cerchio le facevan di se claustro le sette Ninfe.* C'est la réunion des vertus cardinales et théologales. Sous ces formes mystiques, les objets réels du firmament, éloignées les uns des autres, d'après les lois éternelles de la *Mécanique céleste*, se reconnaissent à peine. Le monde idéal est une libre création de l'ame, le produit de l'inspiration poétique." (Examen crit. T. iv. p. 324—332.)

([450]) p. 292.—Acosta, lib. i. cap. 5. Compare my Relation historique, T. i. p. 209. As the stars α and γ of the Southern Cross have almost the same right ascension, the Cross appears perpendicular when passing the meridian; but the natives too often forget that this celestial timepiece marks the hour each day 3′ 56″ earlier. I am indebted for all the calculations respecting the visibility of southern stars in the northern latitudes to the kind communications of Dr. Galle, by whom the planet of Le Verrier was first discovered in the heavens. "The uncertainty of the calculation according to which the star α of the Southern Cross, taking refraction into account, would have begun to be invisible in 52° 25′ N. lat. in the year 2900 before the Christian era, may possibly amount to more than 100 years, and according to the strictest formula of calculation could not altogether be removed, as the proper motion of the fixed stars cannot well be assumed to be uniform for such long intervals. The proper motion of α Crucis is about $\frac{1}{3}$ of a second annually, chiefly in right ascension. The uncertainty produced by neglecting this may be presumed not to exceed the above-mentioned limit.

([451]) p. 294.—Barros da Asia, Dec. I. liv. IV. cap. 2 (1778), p. 282.

([452]) p. 294.—Navarrete, Coleccion de los Viages y Descubrimientos que hiciéron por mar los Españoles, T. iv. p. xxxii. (in the Noticia biografica de Fernando de Magallanes).

([453]) p. 295.—Barros, Decad. III. Parte ii. p. 650 and 658—662.

([454]) p. 296.—The queen writes to Columbus: "Nosotros mismos y no otro alguno, habemos visto algo del libro que nos dejástes (a journal of his voyage in which the distrustful navigator had omitted all numerical data of degrees of latitude and of distances): quanto mas en esto platicamos y vemos, conocemos cuan gran cosa ha seido este negocio vu onestro y que habeis sabido en ello mas que nunca se pensó que pudiera saber ninguno de los nacidos.

Nos parece que seria bien que llevásedes con vos un buen Estrologo, y nos parescia que seria bueno para esto Fray Antonio de Marchena porque es buen Estrologo y siempre nos pareció que se conformaba con vuestro parecer." Respecting this Marchena, who is identical with Fray Juan Perez, the Guardian of the Convent de la Rabida where Columbus in his poverty in 1484 "asked the monks for bread and water for his child," see Navarrete, T. ii. p. 110; T. iii. p. 597 and 603 (Muñoz, Hist. del Nuevo Mundo, lib. iv. § 24).—Columbus, in a letter to the Christianissimos Monarcas from Jamaica, July 7, 1503, calls the astronomical Ephemerides "una vision profetica" (Navarrete, T. i. p. 306). The Portuguese astronomer Ruy Falero, a native of Cubilla, named by Charles V. 1519, Caballero de la Orden de Santiago, at the same time as Magellan, performed an important part in the preparations for Magellan's voyage of circumnavigation. He had prepared expressly for him a treatise on determinations of longitude, of which the great historian Barros possessed some chapters in manuscript (Examen crit. T. i. p. 276 and 302; T. iv. p. 315): probably the same which in 1535 were printed at Seville by John Exomberger. Navarrete (Obra postuma sobre la Hist. de la Nautica y de las ciencias matematicas, 1846, p. 147) could not find the book even in Spain. Respecting the four methods of finding the longitude which Falero had received from the suggestions of his "Demonio familiar," see Herrera, Dec. II. lib. ii. cap. 19; and Navarrete, T. v. p. lxxvii. Subsequently the cosmographer Alonso de Santa Cruz, the same who (like the apothecary of Seville, Felipe Guillen, 1525) attempted to determine the longitude by means of the variation of the compass-needle, made impracticable proposals for accomplishing the same object by the conveyance of time; but his chronometers were sand-and-water timepieces, wheelworks moved by weights, and even "wicks saturated with oil," which burnt out in very equal intervals of time! Pigafetta (Transunto del Trattato di Navigazione, p. 219) recommends altitudes of the moon on the meridian. Amerigo Vespucci speaking of the method of determining longitude by lunars, says with great naiveté and truth, that its advantages arise from the "corso più leggier de la luna" (Canovai, Viaggi, p. 57).

([455]) p. 298.—The American race, which is the same from 65° N. lat. to 55° S. lat., did not pass from the life of hunters to that of cultivators of the soil through the intermediate gradation of a pastoral life. This circumstance is the more remarkable, because the bison, enormous herds of which roam over the country, is susceptible of domestication, and yields much milk. Little attention has been paid to an account given in Gomara (Hist. gen. de las

Indias, cap. 214), of a tribe living in the 16th century to the north-west of Mexico in about 40° N. lat., whose greatest riches consisted in herds of tamed bisons (bueyes con una giba). From these animals the natives derived materials for clothing, food, and drink, probably the blood, (Prescott, Conquest of Mexico, Vol. iii. p. 416); for the dislike to milk, or at least its non use, appears, before the arrival of Europeans, to have been common to all the natives of the New Continent, as well as to the inhabitants of China and Cochin-china. It is true that there were from the earliest times in the mountainous parts of Quito, Peru, and Chili, herds of domesticated lamas; but these herds were in the possession of nations who led a settled life, and were engaged in the cultivation of the soil; in the Cordilleras of South America there were no "pastoral nations," and no such thing as a "pastoral life." What are the "tame deer," near the Punta de St. Helena, which I find spoken of in Herrera (Dec. II. lib. x. cap. 6, T. i. p. 471, ed. Amberes, 1728)? These deer are said to have yielded milk and cheese: "Ciervos que dan leche y queso y se crian en casa!" From what source is this notice derived? It may have arisen from a confusion with the lamas (which have neither horns nor antlers) of the cool mountainous region,—of which Garcilaso affirms that in Peru, and especially on the plateau of Collao, they were used for ploughing (Comment. reales, P. I. lib. v. cap. 2, p. 133). (Compare also Pedro de Cieça de Leon, Chronica del Peru, Sevilla, 1553, cap. 110, p. 264.) The employment of lamas for the plough would however appear to have been a rare exception, and a merely local custom. In general the want of domestic animals was a characteristic of the American race, and had a profound influence on family life.

(456) p. 298.—On the hopes which in the execution of his great and free-minded work, Luther placed especially on the younger generation, the youth of Germany, see the remarkable expressions in a letter of June 1518 (Neander de Vicelio, p. 7).

(457) p. 299.—I have shewn elsewhere how a knowledge of the period at which Vespucci was named Piloto mayor would alone be sufficient to refute the accusation, first brought against him in 1533 by the astronomer Schoner of Nuremberg, of having astutely inserted the words "Terra di Amerigo" in charts which he altered. The high esteem and respect which the Spanish court paid to the hydrographical and astronomical knowledge of Amerigo Vespucci, are clearly manifested in the instructions (Real titulo con extensas facultades) which were given to him when, on the 22d of March, 1508, he was appointed Piloto mayor (Navarrete, T. iii. p. 297—302). He was placed

at the head of a true Deposito hydrografico, and was to prepare for the Casa de Contratacion in Seville, (the central point of all Oceanic discoveries,) a general description of coasts and register of positions, in which all new discoveries were to be entered every year. But the name of "Americi terra" had been proposed for the New Continent as early as 1507, by a person whose existence even was assuredly unknown to Vespucci, the geographer Waldseemüller (Martinus Hylacomylus) of Freiburg in the Breisgau, the director of a printing establishment at St.-Dié in Lorraine, in a small work entitled, "Cosmographiæ Introductio, insuper quatuor Americi Vespucii Navigationes (impr. in oppido S. Deodati, 1507). Ringmann, professor of cosmography at Basle, (better known under the name of Philesius,) Hylacomylus and Gregorius Reisch, who published the "Margarita philosophica," were firm friends. In the last-named work there is a treatise by Hylacomylus on architecture and perspective written in 1509 (Examen crit. T. iv. p. 112). Laurentius Phrisius of Metz, a friend of Hylacomylus, and like him patronised by the Duke Renatus of Lorraine who corresponded by letter with Vespucci, speaks in the Strasburg edition of Ptolemy, 1522, of Hylacomylus as deceased. The map of the New Continent drawn by Hylacomylus and contained in this edition presents the first instance of the name of America "in the editions of Ptolemy's Geography:" but in the meanwhile, according to my investigations, there had appeared two years earlier a Map of the World by Petrus Apianus, which was inserted in Camer's edition of Solinus, and a second time in the Vadian edition of Mela, and which, like more modern Chinese maps, represents the Isthmus of Panama broken through (Examen crit. T. iv. p. 99—124; T. v. p. 168—176). It is a great error to regard the map of 1527 now in Weimar, obtained from the Ebner library at Nuremberg, and the map of 1529 of Diego Ribero, engraved by Güssefeld, as the oldest maps of the New Continent (Examen crit. T. ii. p. 184; T. iii. p. 191). Vespucci had visited the coasts of South America in 1499 (a year after Columbus's third voyage) in the expedition of Alonso de Hojeda, in company with Juan de la Cosa, whose map, drawn at the Puerto de Santa Maria in 1500 fully six years before Columbus's death, was first brought to light by myself. Vespucci could not even have had any motive for feigning a voyage in 1497, for he, as well as Columbus, was firmly persuaded until his death, that his discoveries were a part of Eastern Asia. (Compare the letter of Columbus, February 1502, to Pope Alexander VI., and another, July 1503, to Queen Isabella, Navarrete, T. i. p. 304, T. ii. p. 280, and Vespucci's letter to Pier Francesco de' Medici in Bandini's Vita e lettere di Amerigo Vespucci, p. 66 and 83.) Pedro de Ledesma, Columbus's pilot in his third voyage, still says in 1513, in the law-suit against the heirs, that Paria is considered a part of Asia, "la tierra firme que dicese que es de Asia:" Navarete, T. iii. p. 539. The frequent use of such periphrases as Mondo nuovo, alter Orbis, Colonus novi Orbis repertor, do not contradict this, as they only denote regions not before seen, and are used just in the same manner by Strabo, Mela, Tertullian, Isidore of Seville, and Cadamosto (Examen crit. T. i. p. 118; T. v. p. 182—184). For more than 20 years after the death of Vespucci, which took place in 1512, and indeed until the calumnious statements of Schoner in the Opusculum Geographicum, 1533, and of Servet in the Lyons edition of Ptolemy's Geography in 1535, we find no trace of any accusation against the Florentine navigator. Columbus himself a year before his death speaks of Vespucci in terms of unqualified esteem; he calls him "mucho hombre de bien,"—"worthy of all confidence," and "always inclined to render me service" (Carta a mi muy caro fijo D. Diego, in Navarrete, T. i. p. 351). The same goodwill towards Vespucci is displayed by Fernando Colon, who wrote the life of his father in 1535 in Seville four years before his death, and who with Juan Vespucci, a nephew of Amerigo's, was present at the astronomical junta of Badajoz, and at the proceedings respecting the possession of the Moluccas;—by Petrus Martyr de Anghiera, the personal friend of the Admiral, and whose correspondence goes down to 1525;—by Oviedo, who seeks for every thing which can lessen the fame of Columbus;—by Ramusio;—and by the great historian Guicciardini. If Amerigo had intentionally falsified the dates of his voyages, he would have brought them into agreement with each other, and not have made the first voyage terminate five months after the commencement of the second. The confusion of dates in the numerous versions of his voyages, is not to be attributed to him, as he did not himself publish any of these accounts; such mistakes and confusion of figures are moreover of very frequent occurrence in writings printed in the 16th century. Oviedo had been present, as one of the queen's pages, at the audience at which Ferdinand and Isabella received Columbus with much pomp on his return from his first voyage of diiscovery. Oviedo printed three times that this audience took place in the year 1496, and even that America was discovered in 1491. Gomara had the same printed not in figures but in words, and placed the discovery of the Terra firma of America in 1497, precisely therefore in the year so critical to Amerigo Vespucci's reputation (Examen crit. T. v. p. 196—202. The entire guiltlessness of the Florentine navigator, who never attempted to attach his name to the New Continent, but who had the misfortune by his magniloquence in the accounts addressed to the Gonfalionere Piero Soderini, to Pier Francesco de' Medici, and to Duke Renatus II. of Lorraine, to draw upon himself the attention of posterity more than he deserved, is most decisively shewn by the lawsuit which the fiscal authorities conducted from 1508 to 1527 against the heirs of Columbus, for the purpose of withdrawing from them the rights and privileges which had been ceded by the crown to the Admiral in 1492. Amerigo entered the service of the state as Piloto mayor in the same year that the lawsuit was commenced. He lived at Seville during four years of its proceedings, in which it was to be decided what parts of the New Continent were first seen by Columbus. The most miserable reports found a hearing, and were made matter of accusation by the fiscal; witnesses were sought for at St. Domingo and all the Spanish ports, at Moguer, Palos and Seville, and all this under the eyes of Amerigo Vespucci and his nephew Juan. The Mundus Novus, printed by Johann Otmer at Augsburg, 1504,—the Raccolta di Vicenza (Mondo novo e paesi novamente retrovati da Alberico Vespuzio Fiorentino, of Alessandro Zorzi, 1507,) usually attributed to Fracanzio di Montalboddo,—and the Quatuor Navigationes of Martin Waldseemüller (Hylacomylus) had already appeared; since 1520 maps were extant having in them the name of America, which had been proposed by Hylacomylus in 1507, and praised by Joachim Vadius in a letter addressed to Rudolphus Agricola from Vienna in 1512; and yet the person to whom extensively circulated writings in Germany, France, and Italy, attributed the discovery in 1497 of the Terra firma of Paria, was neither cited by the fiscal as a witness in the proceedings which had begun in 1508, and were continued for 19 years, nor was he even spoken of as opposed to Columbus, or as having preceded him. Why, after the death of Amerigo Vespucci (22d Feb. 1512 in Seville) was not his nephew Juan Vespucci called upon to give evidence, (as were Martin Alonso and Vicente Yañez Pinzon, Juan de la Cosa and Alonso de Hojeda,) that he might testify that the coast of Paria, to which great value was attached not as "part of the main land of Asia," but on account of the productive pearl fishery in its vicinity, had been already landed on before Columbus, before August (1498) by Amerigo? The disregard of this most important testimony would be inexplicable if Amerigo Vespucci had ever boasted of having made a voyage of discovery in 1497, or if any serious value had at that time been attached to the confused dates and misprints of the "Quatuor Navigationes." The different parts of the great and still unprinted work of a friend of Columbus, Fra Bartholomé de las Casas (the Historia general de las Indias), were as we know with certainty written at very different periods. It was not commenced until 1527, 15 years after the death of Amerigo, and was completed in 1559, 7 years before the death of the aged author in his 92d year. Praise and bitter censure are mingled in it in an extraordinary manner. We see that dislike and suspicion augmented progressively as the fame of the Florentine navigator spread. In the preface (Prologo) which was written first, Las Casas says, "Amerigo relates what he did in two voyages to our Indies, but he appears to me to have passed over many circumstances in silence, whether advisedly (á saviendas) or because he did not attend to them; his has led some to attribute to him that which is due to others, and which ought not to be taken from them." The sentence pronounced in the 1st book (chap. 140) is still equally moderate: "Here I must notice the injustice towards the Admiral which appears to have been committed by Amerigo, or perhaps by those who printed (lòs que imprimiéron) his Quatuor Navigationes. To him alone, without naming any other, the discovery of the continent is attributed. He is also said to have placed in maps the name of America, thereby sinfully failing towards the Admiral. As Amerigo was eloquent, and an elegant writer (era latino y eloquente), he makes himself appear in the letter to King Renatus like the leader of Hojeda's expedition: yet he was only one of the pilots, although experienced in seamanship and learned in cosmography (hombre entendido en las cosas de la mar y docto en cosmographia) . . . . In the world the belief prevails that he was the first at the main land. If he purposely gave currency to this belief, it was great wickedness; and if it was not really intentionally done, yet it looks like it (clara pareze la falsedad: y si fué de industria hecha, maldad grande fué; y ya que no lo fuese, al menos parezelo). . . . . Amerigo is represented as having sailed in the year 7 (1497): which seems indeed to have been only an error of the pen and not an intentional false statement (pareza aver avido yerro de pendola y no malicia), because he is made to have returned at the end of 18 months. The *foreign writers* call the country America. It ought to be Columba." This passage shews clearly that up to that time Las Casas had not accused Amerigo of having himself brought the name America into usage. He says, "an tomado los escriptores extrangeros de nombrar la nuestra Tierra firme America, como si Americo solo y no otro con él y antes que todos la oviera descubierto." Farther on in the work, lib. i. cap. 164—169, and lib. ii. cap. 2, violent animosity breaks out: nothing is now attributed to erroneous dates, or to the partiality of foreigners for Amerigo; all is intentional deceit of which Amerigo himself is guilty ("de industria lo hizo . . . persistió en el engaño . . . de falsedad està claramente convencido"). Bartholomé de las

Casas labours also in two passages to shew more particularly that Amerigo in his accounts falsified the true succession of the occurrences of his first two voyages, placing in the first voyage many things which belonged to the second, and vice versâ. It is strange that the accuser does not seem to have felt how much the weight of his accusations is diminished by what he himself says of the opposite opinion, and of the indifference of the person who would have been most interested in attacking Vespucci, if he had believed him guilty and adverse to his father and himself. "I cannot but wonder," says Las Casas (cap. 164), "that Hernando Colon, a clear-sighted man, who as I certainly know had in his hands Amerigo's accounts of his travels, should not have remarked in them any deceit or injustice towards the Admiral." Having had a few months ago a fresh opportunity of examining the rare manuscript of Bartholomé de las Casas, I have been led to embody in this long note what I had not already employed in 1839 in my Examen critique, T. v. p. 178—217. The conviction which I then expressed, in the same volume, p. 217 and 224, has remained unshaken. "Quand la dénomination d'un grand continent, généralement adoptée et consacrée par l'usage de plusieurs siècles, se présente comme un monument de l'injustice des hommes, il est naturel d'attribuer d'abord la cause de cette injustice à celui qui semblait le plus intérressé à la commettre. L'étude des documens a prouvé qu'aucun fait certain n'appuie cette supposition, et que le nom d'*Amérique* a pris naissance dans un pays éloigné (en France et en Allemagne), par un concours d'incidens qui paraissent écarter jusqu'au soupçon d'une influence de la part de Vespuce. C'est là que s'arrête la critique historique. Le champ sans bornes des causes *inconnues* ou des combinaisons morales *possibles*, n'est pas du domaine de l'histoire positive. Une homme qui pendant une longue carrière a joui de l'estime des plus illustre de ses contemporains, s'est élevé, par des connaissances en astronomie nautique, distinguées pour le temps où il vivait, à un emploi honorable. Le concours de circonstances fortuites lui a donné une célébrité dont le poids, pendant trois siècles, a pesé sur sa mémoire, en fournissant des motifs pour avilir son caractère. Une telle position est bien rare dans l'histoire des infortunes humaines : c'est l'exemple d'une flétrissure morale croissant avec l'illustration du nom. Il valait la peine de scruter ce qui, dans ce mélange de succès et d'adversités, appartient au navigateur même, aux hazards de la redaction précipitée de ses écrits, ou à de maladroits et dangereux amis." Even Copernicus contributed to this dangerous celebrity; for he also ascribes the discovery of the new part of the globe to Vespucci. In discussing the "centrum gravitatis" and "centrum magnitudinis" of the continent, he adds; "magis id erit clarum, si addentur insulæ ætate nostra sub Hispaniarum Lusitaniæque Principibus repertæ et præsertim America ab inventore denominata navium præfecto, quem, ob incompertam ejus adhuc magnitudinem, alterum orbem terrarum putant." (Nicolai Copernici de Revolutionibus orbium cœlestium, Libri sex, 1543, p. 2 a.)

(458) p. 300.—Compare my Examen crit. de l'Hist. de la Géographie, T. iii. p. 154—158, and 225—227.

(459) p. 302.—Compare Kosmos, Bd. i. S. 86 (Engl. trans. Vol. i. p. 73.)

(460) p. 303.—"The telescopes which Galileo constructed himself, and others which he used for observing Jupiter's satellites, the phases of Venus, and the solar spots, magnified 4, 7, and 32 times in linear dimensions, never more." (Arago, in the Annuaire du Bureau des Longitudes pour l'an 1842, p. 268).

(461) p. 304.—Westphal, in his Biography of Copernicus (1822, S. 33), dedicated to the great astronomer of Konigsberg, Bessel, like Gassendi, calls the Bishop of Ermland Lucas Watzelrodt von Allen. According to explanations very recently obtained, and for which I am indebted to the learned historian of Prussia, Archiv-Director Voigt, the family of the mother of Copernicus is called in original documents Weisselrodt, Weisselrot, Weisebrodt, and most usually Waisselrode. His mother was undoubtedly of German descent, and the family of Waisselrode, who were originally distinct from that of von Allen, which had flourished at Thorn from the beginning of the 15th century, probably took the name of von Allen in addition to their own, through adoption or connection. Sniadecki and Czynski (Kopernik et ses Travaux, 1847, p. 26) call the mother of the great Copernicus Barbara Wasselrode, married, in 1464, at Thorn, to his father, whose family they bring from Bohemia. The name of the astronomer, who Gassendi designates as Tornæus Borussus, is written by Westphal and Czynski, Kopernik, and by Krzyzianowski, Köpirnig. In a letter of the Bishop of Ermland, Martin Cromer of Heilsberg, dated Nov. 21, 1580, it is said, "Cum Jo. (Nicolaus) Copernicus vivens ornamento fuerit, atque etiam nunc post fata sit, non solum huic ecclesiæ, verum etiam toti Prussiæ patriæ suæ, iniquum esse puto, eum post obitum carere honore sepulchri sive monumenti."

(462) p. 304.—Thus Gassendi, in Nicolai Copernici Vita, appended to his biography of Tycho (Tychonis Brahei Vita, 1655, Hagæ-Comitum, p. 320: "eodem die et horis non multis priusquam animam efflaret." It is only Schubert, in his Astronomy, Th. i. S. 115, and Robert Small, in the very instructive Account of the Astronomical Discoveries of Kepler, 1804, p. 92, who state that Copernicus died "a few days after the appearance of his

work." This is also the opinion of the Archiv-Director Voigt at Konigsberg; because in a letter which George Donner, Canon of Ermland, wrote to the Duke of Prussia after the death of Copernicus, it is said, that "the estimable and worthy Doctor Nicolaus Koppernick sent forth his work, like the sweet song of the swan, a short time before his departure from this life of sorrows." According to the ordinarily received opinion (Westphal, Nikolaus Kopernikus, 1822, S. 73 and 82), the work was begun in 1507, and in 1530 was already so far completed that only a few corrections were subsequently added. The publication was hastened by a letter from Cardinal Schonberg, written from Rome in 1536. The cardinal wishes to have the manuscript copied and sent to him by Theodor von Reden. Copernicus himself, in his dedication to Pope Paul III. says, that the performance of the work has lingered on into the "quartum novennium." If we remember how much time was required for printing a work of 400 pages, and that the great man died in May 1543, we may presume that the dedication was not written in the last named year; which, reckoning backwards 36 years, would not give us a later but an earlier year than 1507.—Herr Voigt doubts whether the aqueduct and hydraulic works at Frauenburg, generally ascribed to Copernicus, were really executed according to his designs. He finds that so late as 1571, a contract was concluded between the Chapter and the "skilful Master Valentine Zendel at Breslau," to bring the water to Frauenburg, from the mill-ponds to the houses of the Canons. Nothing is said of any previous water-works, and therefore the existing ones cannot have been commenced until 28 years after the death of Copernicus.

(463) p. 305.—Delambre, Histoire de l'Astronomie moderne, T. i. p. 140.

(464) p. 304.—"Neque enim necesse est, eas hypotheses esse veras, imo ne verisimiles quidem, sed sufficit hoc unum, si calculum observationibus congruentem exhibeant," says the preface of Osiander. "The bishop of Culm, Tidemann Gise, a native of Dantzig, who had for years urged Copernicus to publish his work, at last received the manuscript, with permission to have it printed at his free pleasure. He sent it first to Rhæticus, Professor at Wittenberg, who had recently been living for a long time with his teacher at Frauenburg. Rhæticus regarded Nuremburg as the most suitable place for the publication, and entrusted the superintendence of the printing to the Professor Schoner and Andreas Osiander" (Gassendi, Vita Copernici, p. 319). The eulogium pronounced on the work at the close of the preface would suffice to shew, without the express testimony of Gassendi, that the preface was by another hand. Also on the title of the first edition (that of Nurem-

berg, 1543), Osiander has made use of an expression which is always carefully avoided in Copernicus's own writing: "motus stellarum novis insuper ac admirabilibus hypothesibus ornati," together with the very ungentle addition, "igitur, studiose lector, eme, lege, fruere." In the second Bale edition of 1566, which I have very carefully compared with the first Nuremberg edition, there is no longer mention in the title of the book of the "admirable hypothesis;" but Osiander's "Præfatiuncula de Hypothesibus hujus Operis," as Gassendi calls the interpolated preface, is preserved. It is also evident that Osiander, without naming himself, meant to shew that the præfatiuncula was by a different hand from the work itself, as he designates the dedication to Paul III. as the "Præfatio Authoris." The first edition has only 196 leaves; the second has 213, on account of the added Narratio Prima of the astronomer George Joachim Rhæticus, and a letter directed to Schoner, which, as I have remarked in the text, being printed in 1541 by the intervention of the mathematician Gassarus of Basle, gave to the learned world the first correct knowledge of the Copernican system. Rhæticus had given up his professorship at Wittenberg for the sake of enjoying the instructions of Copernicus at Frauenberg itself. (Compare, on these subjects, Gassendi, p. 310—319.) The explanation of what Osiander was induced to add from timidity, is given by Gassendi: "Andreas porro Osiander fuit, qui non modo operarum inspector (the superintendent of the printing) fuit, sed Præfatiunculam quoque ad lectorem (tacito licet nomine) de Hypothesibus operis adhibuit. Ejus in ea consilium fuit, ut, tametsi Copernicus Motum Terræ habuisset, non solum pro Hypothesi, sed pro vero etiam placito; ipse tamen ad rem, ob illos, qui heinc offenderentur, leniendam, excusatum eum faceret, quasi talem Motum non pro dogmate, sed pro Hypothesi mera assumpsisset."

([465]) p. 307.—"Quis enim in hoc pulcherrimo templo lampadem hanc in alio vel meliori loco poneret, quam unde totum simul possit illuminare? Siquidem non inepte quidam lucernam mundi, alii mentem, alii rectorem vocant. Trimegistus visibilem Deum, Sophoclis Electra intuentem omnia. Ita profecto tanquam in solio regali Sol residens circumagentem gubernat astrorum familiam: Tellus quoque minime fraudatur lunari ministerio, sed ut Aristoteles de animalibus ait, maximam Luna cum terra cognationem habet. Concepit interea a Sole terra, et impregnatur annuo partu. Invenimus igitur sub hac ordinatione admirandam mundi symmetriam ac certum harmoniæ nexum motus et magnitudinis orbium: qualis alio modo reperiri non potest (Nicol. Copern. de Revol. Orbium Cœlestium, lib. i. cap. 10, p. 9 b). In this passage, which is not without poetic grace and elevation of style, we recognise, as was the case with all the astronomers of the 17th century, traces of long and intimate acquaintance with classical antiquity. Copernicus had in his mind Cic. Somn. Scip. c. 4; Plin. ii. 4; and Mercur. Trismeg. lib. v. (ed. Cracov. 1586), p. 195 and 201. The allusion to the Electra of Sophocles is obscure, as the sun is not termed any where "all-seeing," as it is in the Iliad and the Odyssey, and also in the Chœphoræ of Æschylus (v. 980), which yet Copernicus would not probably have called Electra. According to Böckh's conjecture, the allusion is to be ascribed to a vague remembrance of verse 869 of Sophocles' Œdipus Coloneus. It is singular that quite lately, in an otherwise instructive memoir (Czynski, Kopernik et ses Travaux, 1847, p. 102), the Electra of the tragedian is confounded with "electric currents." The passage of Copernicus quoted above is thus translated: "Si on prend le soleil pour le flambeau de l'univers, pour son ame, pour son guide, si Trimegiste le nomme un Dieu, si Sophocle le croit une puissance électrique qui anime et contemple l'ensemble de la création.". . . .

([466]) p. 307.—"Pluribus ergo existentibus centris, de centro quoque mundi non temere quis dubitabit, an videlicet fuerit istud gravitatis terrenæ, an aliud. Equidem existimo, *gravitatem* non aliud esse, quam appetentiam quandam naturalem partibus inditam a divina providentia opficis universorum, ut in unitatem integritatemque suam sese conferant in formam globi coëtuntes. Quam affectionem credibile est etiam Soli, Lunæ, cæterisque errantium fulgoribus inesse, ut ejus efficacia in ea qua se repræsentant rotunditate permaneant, quæ nihilominus multis modis suos efficiunt circuitus. Si igitur et terra faciat alios, ut pote secundum centrum (mundi), necesse erit eos esse qui similiter extrinsecus in multis apparent, in quibus invenimus annuum circuitum.—Ipse denique Sol medium mundi putabitur possidere, quæ omnia ratio ordinis, quo illa sibi invicem succedunt, et mundi totius harmonia nos docet, si modo rem ipsam ambobus (ut ajunt) oculis inspiciamus" (Copern. de Revol. Orb. Cœl. lib. i. cap. 9, p. 7, b).

([467]) p. 308.—Plut. de Facie in Orbe Lunæ, p. 923, c (compare Ideler, Meteorologia Veterum Græcorum et Romanorum, 1832, p. 6). In the passage of Plutarch, Anaxagoras is not named; but that the latter applied the same theory of "falling if the force of rotation intermitted" to all the material celestial bodies, we learn from Diog. Laert. ii. 12, and the many passages which I have collected (Kosmos, Bd. i. S. 139, 397, 401, and 408; Engl. trans. Vol. i. p. 123-124, Notes 62, 69, 89). Compare also Aristot. de Cœlo, ii. 1, p. 284, a 24, Bekker and a remarkable passage of Simplicius, p. 491, b, in the Scholia, according to the edition of the Berlin Academy, where the "not falling of heavenly bodies" is spoken of "when the force of rotation predominates over the proper falling force or downward attraction." We may connect with these ideas, which also partially belong to Empedocles and Democritus as well as to Anaxagoras, the instance adduced by Simplicius, (*l. c.*) "that water in a phial is not spilt when the phial is swung round with a movement of rotation more rapid than the downward movement of the water" (*τῆς επι το κατω του υδατος φορας*).

(468) p. 308.—Kosmos, Bd. i. S. 139 and 408; Engl. trans, p. 124 and note 89. (Compare Letronne, des Opinions cosmographiques des Pères de l'Eglise, in the Revue des deux Mondes, 1834, T. i. p. 621.)

(469) p. 308.—For all that relates to attraction, gravity, and the fall of bodies, as regarded in antiquity, see a collection of passages from the ancients, made with great industry and discrimination, by Th. Henri Martin, Etudes sur le Timée de Platon, 1841, T. ii. p. 272–280, and 341.

(470) p. 308.—Joh. Philoponus de Creatione Mundi, lib. i. cap. 12.

(471) p. 308.—He afterwards gave up the correct opinion (Brewster, Martyrs of Science, 1846, p. 211); but that there dwells in the central body of the planetary system, the Sun, a power which governs the movements of the planets, and that this solar force decreases either as the square of the distance or in direct ratio, was expressed by Kepler, in the Harmonice Mundi, completed in 1618.

(472) p. 308.—Kosmos, Bd. i. S. 30 and 58 (Engl. trans. Vol. i. p. 31 and 52.)

(473) p. 309.—Kosmos, Bd. ii. S. 139 and 209 (Eng. trans. Vol. ii. p. 105 and 175). The scattered passages in the work of Copernicus, relating to the Ante-Hipparchian system of the structure of the universe, exclusive of the dedication, are the following:—lib. i. cap. 5 and 10; lib. v. cap. 1 and 3 (ed. princ. 1543, p. 3, b; 7, b; 8, b; 133, b; 141 and 141, b; 179 and 181, b). Every where Copernicus shews a predilection for, and a very accurate acquaintance with, the views entertained by the Pythagoreans, or which, to speak more circumspectly, were attributed to the most ancient among them. For example, as we see by the beginning of the dedication, he was acquainted with the letter of Lysis to Hipparchus; which, indeed, shews that the mystery loving Italic school only designed to communicate their opinions to friends, "as had at first been the purpose of Copernicus likewise." The period to which Lysis belonged is somewhat uncertain; he is sometimes termed an immediate disciple of Pythagoras himself, sometimes, and with more probability, a teacher of Epaminondas (Böckh, Philolaos, S. 8—15).

The letter of Lysis to Hipparchus (an old Pythagorean, who had disclosed the mysteries of the sect), is, like so many other writings, a forgery of later times. Copernicus had probably become acquainted with it from the collection of Aldus Manutius, Epistolæ diversorum Philosophorum (Romæ, 1494), or from a Latin translation by Cardinal Bessarion (Venet. 1516). In the prohibition of Copernicus' work, De Revolutionibus, in the famous decree of the Congregazione dell' Indice of the 5th of March, 1616, the new system of the universe is expressly designated as "falsa illa doctrina Pythagorica, Divinæ Scripturæ omnino adversans." The important passage on Aristarchus of Samos, of which I have spoken in the text, is in the Arenarius, p. 449 of the Paris edition of Archimedes of 1615 by David Rivaltus. But the editio princeps is the Basle edition of 1544, apud Jo. Hervagium. The passage in the Arenarius says very distinctly, that "Aristarchus had confuted the Astronomers who imagined the earth to be immoveable in the centre of the universe; that this centre was occupied by the sun, which was immoveable, like other stars, while the earth revolved round it." Copernicus, in his work, twice names Aristarchus, p. 69 b and 79, but without any allusion to his system. Ideler, in Wolf and Buttmann's Museum der Alterthums-Wissenschaft (Bd. ii. 1808, S. 452), asks whether Copernicus was acquainted with Nicolaus von Cusa's work, De Docta Ignorantia. The first Paris edition of it was indeed published in 1514, and the expression, "jam nobis manifestum est terram in veritate moveri," from a platonising cardinal, might have been expected to make some impression on the Canon of Frauenberg (Whewell, Philosophy of the Inductive Sciences, Vol. ii. p. 343); but a fragment of Cusa's writing discovered very recently (1843) in the library of the Hospital at Cues, sufficiently proves, as does the work De Venatione Sapientiæ, cap. 28, that Cusa imagined the earth not to move round the sun, but to move together with it, though more slowly, "round the constantly changing pole of the universe" (Clemens, in Giordano Bruno, and Nicol. von Cusa, 1847, S. 97—100).

(474) p. 309.—See the profound treatment of this subject in Martin, Etudes sur Timée, T. ii. p. 111 (Cosmographie des Egyptiens), and p. 129—133 (Antécédents du Système de Copernic). The statement of this learned philologist, according to which the original system of Pythagoras himself differed from that of Philolaos, and placed the earth at rest in the centre, does not appear to me quite convincing (T. ii. p. 103 and 107). Respecting the remarkable statement of Gassendi mentioned in the text, of the similarity of the systems of Tycho Brahe and Apollonius of Perga, I here add

further explanation. In Gassendi's biographies, he says, "Magnam imprimis rationem habuit Copernicus duarum opinionum affinium, quarum unam Martiano Capellæ, alteram Apollonio Pergæo attribuit.—Apollonius Solem delegit, circa quem, ut centrum, non modo Mercurius et Venus, verum etiam Mars, Jupiter, Saturnus suas obirent periodos, dum Sol interim, uti et Luna, circa Terram, ut circa centrum, quod foret affixarum mundique centrum, moverentur, quæ deinceps quoque opinio Tychonis propemodum fuit. Rationem autem magnam harum opinionum Copernicus habuit, quod utraque eximie Mercurii ac Veneris circuitiones repræsentaret, eximieque causam retrogradationum, directionum, stationum in iis apparentium exprimeret et posterior (Pergæi) quoque in tribus planetis superioribus præstaret" (Gassendi, Tychonis Brahei Vita, p. 296). My friend the astronomer Galle, from whom I sought information, like myself finds nothing which could justify Gassendi's decided statement. He writes to me, "In the passages which you refer to in Ptolemy's Almagest (in the commencement of Book XII.), and in the works of Copernicus (lib. v. cap. 3, p. 141, a; cap. 35, p. 179, a and b; cap. 36, p. 181, b), there is only question of explaining the retrogressions and stationary appearances of the planets, in which there is indeed a reference to Apollonius's assumption of the revolution of the planets round the sun (and Copernicus himself mentions expressly the assumption of the earth's standing still); but it does not appear possible to determine where he obtained what he supposes to have been derived from Apollonius. I can only therefore conjecture, that some late writer gave a system attributed to Apollonius of Perga which resembled that of Tycho; although I do not find, even in Copernicus, any clear exposition of such a system, or any quotations of ancient passages respecting it. If the source from whence the complete Tychonic view is attributed to Apollonius should be merely lib. XII. of the Almagest, we may consider that Gassendi went too far in his suppositions, and that the case resembled that of the phases of Mercury and Venus, which Copernicus spoke of indeed (lib. i. cap. 10, p. 7, b, and 8, a), but without decidedly applying them to his system. Apollonius, perhaps, in a similar manner may have treated mathematically the explanation of the retrogressions of the planets under the assumption of a revolution round the sun, without subjoining any thing decided and general as to the truth of this assumption. The difference of the Apollonian system described by Gassendi from that of Tycho would only be, that the latter explained the inequalities of the movements as well. The remark of Robert Small, that the fundamental idea of the Tychonian system was by no means a stranger to the mind of Copernicus, but had rather served him as a point of transition to his own system, appears to me well founded."

([475]) p. 310.—Schubert, Astronomie, Th. i. S. 124. Whewell has given, in the Philosophy of the Inductive Sciences, Vol. ii. p. 282, an Inductive Table of Astronomy, which presents an exceedingly good and complete tabular view of the astronomical contemplation of the structure of the universe, from the earliest times to Newton's system of gravitation.

([476]) p. 311.—Plato inclines, in the Phædrus, to the system of Philolaus; but in the Timæus, to that which represents the earth as immoveable in the centre, subsequently called the Hipparchian or the Ptolemaic system (Böckh, de Platonico Systemate Cœlestium Globorum, et de vera indole Astronomiæ Philolaicæ, p. xxvi.—xxxii.; also the same author in the Philolaos, S. 104—108. Compare also Fries, Geschichte der Philosophie, Bd. i. S. 325—347, with Martin's Etudes sur Timée, T. ii. p. 64—92). The astronomical vision in which the structure of the universe is veiled, at the end of the Book of the Republic, reminds one at once of the planetary systems of intercalated spheres, and of the concord of tones (the music of the spheres)—"the voices of the Sirens winging their flight with the revolving orbs." (See, on the discovery of the true system of the universe, the fine and comprehensive work of Apelt, Epochen der Gesch. der Menscheit, Bd. i. 1845, S. 205—305 and 379—445.)

([477]) p. 311.—Kepler, Harmonices, Mundi libri quinque, 1619, p. 189. "On the 8th of March, 1618, Kepler, after many unsuccessful attempts, came upon the thought of comparing the squares of the times of revolution of the planets with the cubes of the mean distances; but he made an error of calculation, and rejected the idea. On the 15th of May, 1618, he came back upon it, and calculated correctly. The third Keplerian law was now discovered." This discovery, and those related to it, coincide with the distressing period when this great man, exposed from early childhood to the hardest strokes of fate, was labouring, in a trial for witchcraft which lasted six years, to save his aged mother, 70 years old, accused of poison-mixing, incapacity of shedding tears, and sorcery, from the torture and the stake. The suspicion was strengthened by her own son, the wicked Christopher Kepler a worker in tin, being his mother's accuser; and by her having been brought up by an aunt who was burnt at Weil as a witch. See an exceedingly interesting work, but little known in foreign countries, drawn from newly discovered manuscripts by Baron von Breitschwert, entitled, "Johann Keppler's Leben und Wirken," 1831, S. 12, 97—147, and 196. According to this work,

Kepler, who in German letters always signed his name Keppler, was not born on the 21st of December, 1571, in the imperial town Weil, as is usually supposed, but on the 27th of December, 1571, in the Wurtemberg village of Magstatt. Of Copernicus it is uncertain whether he was born on the 19th of January, 1472, or on the 19th February, 1473 (as Möstlin supposes), or (according to Czynski) on the 12th February of the same year. The year of Columbus's birth was long uncertain within 19 years. Ramusio places it in 1430, Bernaldez, the friend of the discoverer, in 1436, and the celebrated historian Muñoz in 1446.

([478]) p. 312.—Plut. de Plac. Philos. ii. 14; Aristot. Meteorol. xi. 8, De Cœlo, ii. 8. On theories of the spheres generally, and on the retrograding spheres of Aristotle in particular, see Ideler's Vorlesung über Eudoxus, 1828, S. 49—60.

([479]) p. 313.—A better insight into the free movement of bodies, and into the independence of the direction of the axis of the earth once given, and of the rotatory and progressive movement of the terrestrial globe in its orbit, has freed the original system of Copernicus from the assumption of a declination-movement, or a so-called third movement of the earth (De Revolut. Orb. Cœl. lib. i. cap. 11, triplex motus telluris). In the annual revolution round the sun, the parallelism of the earth's axis is maintained in conformity with the law of inertia, without the application of a "correcting" epicycle.

([480]) p. 314.—Delambre, Hist. de l'Astronomie ancienne, T. ii. p. 381.

([481]) p. 314.—See Sir David Brewster's judgment on Kepler's optical works, in the "Martyrs of Science," 1846, p. 179—182. (Compare Wilde, Gesch. der Optik, 1838, Th. i. S. 182—210.) If the law of the refraction of rays of light belongs to the Leyden Professor Willebrord Snellius (1626), who at his decease left it behind him buried in his papers, on the other hand the publication of the law in a trigonometrical form was first made by Descartes. See Brewster, in the North British Review, Vol. vii. p. 207; Wilde, Gesch. der Optik, Th. i. S. 227.

([482]) p. 314.—Compare two excellent memoirs on the discovery of the telescope, by Professor Moll of Utrecht, in the Journal of the Royal Institution, 1831, Vol. i. p. 319, and by Wilde at Berlin, in his Gesch. der Optik, 1838, Th. i. S. 138—172. The work of Moll, written in the Dutch language, is entitled, "Geschiedkundig Onderzoek naar de eerste Uitfinders der Verrekykers, uit de Aantekenningen van wyle den Hoogl. van Swinden zamengesteld door G. Moll," Amsterdam, 1831. Olbers has given an extract from this interesting treatise in Schumacher's Jahrbuch für 1843, S. 56—65. The optical instruments which Prince Maurice of Nassau and the Archduke Albert had from Jansen (the Archduke gave his to Cornelius Drebbel), were, (as is evident from the letter of the ambassador Boreel, who had been often in Jansen's house when a child, and at a later period saw the instruments in the shop,) microscopes 18 inches long, "through which small objects, when one looked down at them from above, appeared wonderfully magnified."—The confusion between the microscope and the telescope has contributed to obscure the history of the invention of both instruments. The letter of Boreel (Paris, 1655), above alluded to, notwithstanding the authority of Tiraboschi, renders it improbable that the first invention of the compound microscope belonged to Galileo. Compare, on this obscure history of optical inventions, Vincenzio Antinori, in the Saggi di Naturali Esperienze fatte nell' Accademia del Cimento, 1841, p. 22—26. Even Huygens, who was born scarcely twenty-five years after the supposed date of the invention of the telescope, did not venture to decide with certainty respecting the name of the first inventor (Opera reliqua, 1728, Vol. ii. p. 125). According to the researches which Van Swinden and Moll have made in Archives, not only was Lippershey, as early as the 2d of October, 1608, in possession of a telescope made by himself, but the French Ambassador at the Hague, President Jeannin, wrote, on the 28th of December of the same year, to Sully, "that he was in treaty with the Middleburg spectacle-maker for a telescope which he wished to send to the king" (Henry IV.) Simon Marius (Mayer of Gunzenhausen, one of the two independent discoverers of Jupiter's satellites) even relates that his friend Fuchs of Bimbach, Privy Councillor of the Margrave of Ansbach, was offered a telescope for sale in the autumn of 1608 at Frankfort-on-Maine, by a Belgian. Telescopes were made in London in February 1610, or a year after Galileo had completed his telescope (Rigaud on Harriot's Papers, 1833, p. 23, 26, and 46). They were at first called cylinders. Porta, the inventor of the camera obscura, as well as, at earlier periods, Fracastoro the cotemporary of Columbus, Copernicus, and Cardanus, had merely spoken of the possibility "of seeing every thing larger and nearer" by looking through convex and concave glasses placed on each other (duo specilla ocularia alterum alteri superposita); but we cannot ascribe to them the invention of the telescope (Tiraboschi, Storia della Letter. ital. T. xi. p. 467; Wilde, Gesch. der Optik, Th. i. S. 121). Spectacles had been known in Haarlem since the beginning of the 14th century; and an epitaph in the Church of Maria Maggiore at Florence names as the inventor (inventore degli occhiali) Salvino degli Armati, deceased in 1317. Separate and apparently authentic notices of the use of spectacles by aged persons occur even as early as 1299 and 1305. The passages of Roger Bacon relate to the magnifying power of spherical segments of glass. See Wilde, Gesch. der Optik, Th. i. S. 93—96; and above, note 284.

([483]) p. 315.—In like manner, the above named physician and mathematician of the Margravate of Ansbach, Simon Marius, as early as 1608, after receiving a description of the action of a Dutch telescope, is believed to have constructed one himself. On Galileo's earliest observation of the mountains in the moon, referred to in the text, compare Nelli Vita di Galilei, Vol. i. p. 200—206; Galilei Opere, 1744, T. ii. p. 60, 403, and (Lettera al Padre Cristoforo Grienberger, in Materia delle Montuosità della Luna) p. 409—424. Galileo found in the moon some circular districts, surrounded on every side by mountains, like the form of Bohemia. "Eundem facit aspectum lunæ locus quidam, ac faceret in terris regio consimilis Boemiæ, si montibus altissimis, inque peripheriam perfecti circuli dispositis occluderetur undique" (T. ii. p. 8). The measurements of the altitudes of the mountains were made by the method of the tangent of the solar ray. Galileo, as Helvetius still later, measured the distance of the summit of the mountains from the boundary of the illuminated portion, at the moment when the mountain summit first caught the solar ray. I find no observation of the lengths of the shadows of the mountains. He found the summits "incirca miglia quattro" in height, and "much higher than our terrestrial mountains." The comparison is curious, because, according to Riccioli, very exaggerated ideas of the height of our mountains then prevailed; and one of the principal or most celebrated amongst them, the Peak of Teneriffe, was first measured with some degree of exactness, trigonometrically, by Feuillée, in 1724. Galileo, like all other observers up to the end of the 18th century, believed in the existence of many seas in the moon, and of a lunar atmosphere.

([484]) p. 316.—I again find occasion (Kosmos, Bd. i. S. 434; Engl. trans. Vol. i. note 159) to recal here the proposition laid down by Arago, "Il n'y a qu'une maniere rationelle et juste d'écrire l'histoire des sciences, c'est de s'appuyer exclusivement sur des publications ayant date certaine: hors de là, tout est confusion et obscurité." The singularly delayed publication of the Fränkischen Kalender's oder der Practica (1612), and of the astronomically important memoir entitled "Mundus Jovialis anno 1609 detectus ope perspicilli Belgici (Feb. 1614)," may indeed have given occasion to the suspicion that Marius had derived information from the Nuncius Sidereus of Galileo, the dedication of which bears date in March 1610, or even from earlier communications by letter. Galileo, excited by the not forgotten law-suit against Balthasar Bapra, calls him a pupil of the Marius, "usurpatore del Sistema di Giove"; Galileo even reproaches the heretical protestant astronomer of Gunzenhausen with founding the apparently earlier date of his observation on a confusion between the calendars. "Tace il Mario di far cauto il lettore, come essendo egli separato della chiesa nostra, ne avendo accettato l'emendatione gregoriana, il giorno 7 di Gennaio del 1610 di noi cattolici (the day on which Galileo discovered the satellites) è l'istesso, che il dì 28 di Decembre del 1609 di loro cretici, e questa è tutta la precedenza delle sue finte osservationi" (Venturi, Memorie e Lettere di G. Galilei, 1818, P. i. p. 279; and Delambre, Hist. de l'Astr. mod. T. i. p. 696). According to a letter which Galileo wrote, in 1614, to the Academia di Lincei, he, somewhat unphilosophically, thought of addressing his complaint against Marius to the Marchese di Brandeburgo. On the whole, however, Galileo continued well disposed towards the German astronomers. He writes, in March 1611, "Gli ingegni singolari, che in gran numero fioriscono nell' Alemagna, mi hanno lungo tempo tenuto in desiderio di vederla" (Opere, T. ii. p. 44). It has always appeared to me remarkable, that if, in a conversation with Marius, Kepler was playfully cited as a sponsor for the bestowal of the mythological denominations of Io and Callisto, there should not occur any mention of his countryman, either in the Commentary to the Nuncius Sidereus, nuper ad mortales a Galilæo missus, published in Prague, in April 1610, or in his letters to Galileo or to the Emperor Rudolph in the autumn of the same year; instead of which, Kepler every where speaks of "the glorious discovery of the Medicean stars by Galileo." In publishing his own observations of the satellites, made from the 4th to the 9th of September, 1610, he gives to a little memoir which appeared at Frankfort in 1611, the title, "Kepleri Narratio de Observatis a se quatuor Jovis Satellitibus erronibus quos Galilæus Mathematicus Florentinus jure inventionis Medicea Sidera nuncupavit. "A letter addressed to Galileo from Prague, Oct. 25, 1610, concludes with the words "neminem habes, quem metuas æmulam." Compare Venturi, P. i. p. 100, 117, 139, 144, and 149. Baron von Zach, misled by a mistake, and after a by no means careful examination of the valuable manuscripts preserved at Petworth, the seat of Lord Egremont, stated that the distinguished astronomer and Virginian traveller, Thomas Harriot, had discovered the satellites of Jupiter at the same time as Galileo, and even earlier. A more close examination of Harriot's manuscripts, by Rigaud, has shewn that they began, not on the 16th of January, but only on the 17th of October, 1610, nine months after Galileo and Marius. (Compare Zach, Corr. Astron. Vol. vii. p. 105; Rigaud, Account of Harriot's Astron. Papers, Oxford, 1833, p. 37; Brewster, Martyrs of Science, 1846, p. 32. The earliest observations of Jupiter's satellites by Galileo and his pupil Renieri, were only discovered two years ago.

(485) p. 317.—It ought to be 73 years; for the prohibition of the Copernican system by the Congregation of the Index was given on the 5th of March, 1616.

(486) p. 317.—Freiherr von Breitschwert, Keppler's Leben, S. 36.

(487) p. 317.—Sir John Herschel, Astron. S. 465.

(488) p. 318.—Galilei, Opere, T. ii. (Longitudine per via de' Pianeti Medicei), p. 435—506; Nelli, Vita, Vol. 2, p. 656—688; Venturi, Memorie e Lettere di G. Galilei, P. i. p. 177. As early as 1612, or scarcely two years after the discovery of Jupiter's satellites, Galileo boasted, somewhat prematurely, of having completed tables of those satellites to such a degree of exactness, that the phenomena could be computed by them to 1′ of time. A long diplomatic correspondence, which did not lead to the desired object, was commenced with the Spanish ambassador in 1616, and with the Dutch ambassador in 1636. The telescopes were to magnify 40 to 50 times. In order to find the satellites more easily when the ship is in motion, and (as he imagined) to keep them in the field, he invented, in 1617 (Nelli, Vol. ii. p. 663), the binocular telescope, which has usually been attributed to the Capucine monk, Schyrleus de Rheita, who had much experience in optical matters, and was seeking to find the means of constructing telescopes magnifying 4000 times. Galileo made experiments with his binocular (to which he also gave the name of celatone or testiera) in the harbour of Leghorn, during a strong wind and much motion of the ship. He also had a contrivance prepared in the arsenal at Pisa, for protecting the observer of the satellites from the motion of the ship, by seating him in a kind of boat, which was to float in another boat filled with water or with oil (Lettera al Picchena de 22 Marzo, 1617; Nelli, Vita, Vol. i. p. 184; Galilei, Opere, T. ii. p. 473; Lettera a Lorenzo Realio del 5 Giugno, 1637). The proofs which Galileo assigns of the advantages for the naval service of his method over Morin's method of lunar distances, are very remarkable. (Opere, T. ii. p. 454)

(489) p. 319.—Arago, in the Annuaire for 1842, pp. 460—476 (Decouvertes des taches Solaires et de la Rotation du Soleil), and Brewster (Martyrs of Science, pp. 36 and 39) place the first observation of Galileo in October or November 1610. Compare Nelli, Vita, Vol. i. pp. 324—384; Galilei, Opere, T. i. p. lix. T. ii. pp. 85—200, T. iv. p. 53. On Harriot's observations, see

Rigaud, pp. 32 and 38. The Jesuit Schoner, who was summoned from Gratz to Rome, has been accused of seeking to revenge himself of Galileo on account of the literary contest respecting the discovery of the solar spots, by getting it whispered, through another Jesuit, Grassi, to Pope Urban VIII. that he (the Pope) was the person represented by the foolish and ignorant Simplicius in the Dialoghi delle Scienze Nuove (Nelli, Vol. ii. p. 515).

(490) p. 320.—Delambre, Hist. de l'Astronomie moderne, T. i. p. 690.

(490) p. 320.—In Galileo's Letters to the Principe Cesi (May 25, 1612) the same opinion is expressed; Venturi, P. i. p. 172.

(492) p. 321.—See on this subject some ingenious and interesting considerations by Arago, in the Annuaire pour l'An 1842, pp. 481—488. (The experiments with Drummond's light projected on the sun's disk are mentioned by Sir John Herschel in his Astronomy, S. 334.)

(493) p. 321.—Giordano Brano und Nic. von Cusa verglichen, von J. Clemens, 1847, S. 101. On the phases of Venus, see Galileo, Opere, T. ii. p. 53, and Nelli, Vita, Vol. i. pp. 213—215.

(494) p. 322.—Compare Kosmos, Bd. i. S. 160 and 416; Eng. translation, Vol. i. p. 144, Note 120.

(495) p. 323.—Laplace says of Kepler's theory of the measurement of casks (Stereometria doliorum, 1615, which, like Archimedes, contains the development of elevated ideas in reference to an insignificant subject):—Kepler présente dans cet ouvrage des vues sur l'infini qui ont influé sur la révolution que la géométrie a eprouvée à la fin du 17me siècle; et Fermat, que l'on doit regarder comme le véritable inventeur du calcul différentiel, a fondé sur elles sa belle méthode *de maximis et minimis* (Précis de l'hist. de l'Astronomie, 1821, p. 95). On the geometrical acuteness manifested by Kepler in the five books of his Harmonices Mundi, see Chasles, Aperçu hist. des Méthodes en Géométrie, 1837, pp. 482—487.

(496) p. 323.—Sir David Brewster says well in the account of Kepler's method of investigating truth:—"The influence of imagination as an instrument of research has been much overlooked by those who have ventured to give laws to philosophy. This faculty is of greatest value in physical inquiries: if we use it as a guide, and confide in its indications, it will infallibly deceive us; but if we employ it as an auxiliary, it will afford us the most invaluable aid" (Martyrs of Science, p. 215).

(497) p. 324.—Arago, in the Annuaire, 1842, p. 434 (De la transformation des Nébuleuses et de la matière diffuse en étoiles). Compare Kosmos, Bd. i. S. 148 and 158 (English translation, Vol. i. pp. 132 and 142).

([498]) p. 324.—Compare the ideas of Sir John Herschel on the position of our planetary system, Kosmos, Bd. i. S. 157 and 415; and Struve, Etudes d'Astronomie Stellaire, 1847, p. 4.

([499]) p. 324.—Apelt says (Epochen der Geschichte de Menschheit, Bd. i. 1845, S. 233): "The remarkable law of the distances, which usually passes under the name of Bode's law (or that of Titius), was a discovery of Kepler's, who, after many years of persevering industry, first deduced it by calculation from the observations of Tycho de Brahe." See Harmonices Mundi, libri quinque, cap. 3. Compare also Cournot, Additions to a Translation of Sir John Herschel, Traité d'Astronomie, 1834, S. 434, p. 324, and Fries, Vorlesungen über die Sternkunde, 1813, S. 325 (Law of the distances in the secondary planets or satellites). The passages from Plato, Pliny, Censorinus, and Achilles Tatius, in the Prolegomena to the Aratus, are carefully collected in Fries, Geschichte der Philosophie, Bd. i. 1837, S. 146—150; in Martin, Etudes sur le Timée, T. ii. p. 38; and in Brandis, Geschichte der Griechisch-Römischen Philosophie, Th. ii. Abth. i. 1844, S. 364.

([500]) p. 325.—Delambre, Hist. de l'Astronomie moderne, T. i. p. 360.

([501]) p. 325.—Arago, in the Annuaire for 1842, pp. 560—564 (Kosmos, Bd. i. S. 102; English translation, Vol. i. p. 88).

([502]) p. 326.—Compare Kosmos, Bd. i. S. 142—148, and 412 (English translation, Vol. i. p. 127—133, Notes, 91—93.)

([503]) p. 327.—Annuaire du Bureau des Longitudes pour l'an 1842, p. 312—353 (Etoiles changeantes ou périodiques). In the seventeenth century there were recognised as variable stars, besides Mira Ceti (Holwarda, 1638) and α Hydræ (Montanari, 1672), β Persei or Algol, and χ Cygni (Kirch, 1686). On what Galileo calls nebulæ, see his Opere, T. ii. p. 15, and Nelli, Vita, Vol. ii. p. 208. Huygens, in the Sistema Saturninum, points in the clearest manner to the nebula in the sword of Orion, in saying of nebulæ generally:—"Cui certe simile aliud nusquam apud reliquas fixas potui animadvertere. Nam ceteræ nebulosæ olim existimatæ atque ipsa via lactea, perspicillis inspectæ, nullas nebulas habere comperiuntur neque aliud esse quam plurium stellarum congeries et frequentia." This passage shews that Huygens (as previously Galileo) had not attentively considered the nebula in Andromeda which Marius had first described.

([504]) p. 329.—On the important law, discovered by Brewster, of the connection between the angle of complete polarisation and the refractive power of bodies, see Philosophical Transactions of the Royal Society for the year 1815, pp. 125—159.

(505) p. 329.—See Kosmos, Bd. i. S. 35 and 48; English translation, Vol. i. p. 37, Note 16.

(506) p. 329.—Sir David Brewster, in Berghaus and Johnson's Physical Atlas, 1847, Part vii. p. 5 (Polarisation of the Atmosphere).

(507) p. 329.—On Grimaldi's and Hooke's attempt to explain the polarisation (?) of soap-bubbles by the interference of the rays of light, see Arago, in the Annuaire for 1831, p. 164 (Brewster's Life of Newton, p. 53).

(508) p. 330.—Brewster, Life of Sir Isaac Newton, p. 17. The year 1660 has been assumed for the date of the invention of the method of fluxions, which, according to the official explanations of the Committee of the Royal Society of London, April 24, 1712, is "one and the same with the differential method, excepting the name and mode of notation." For the whole unhappy contest with Leibnitz on the subject of priority, in which, extraordinary to say, accusations against Newton's veracity were even interspersed, see Brewster, pp. 189—218. That all colours are contained in white light was already maintained by De la Chambre, in his work entitled "La Lumière" (Paris, 1657), and by Isaac Vossius, who was afterwards a Canon at Windsor, in a remarkable memoir, entitled "De Lucis natura et proprietate" (Amstelod. 1662), for the communication of which I was indebted two years ago to M. Arago, at Paris. This memoir is treated of by Brandes in the new edition of Gehler's physikalischen Wörterbuch, Bd. iv. (1827), S. 43, and very circumstantially by Wilde, in his Gesch. der Optik, Th. i. (1838), S. 223, 228, and 317). Isaac Vossius, however, regarded sulphur, which forms, according to him, a component part of all bodies, as the fundamental substance of all colours (cap. 25, p. 60). In Vossii Responsum ad objecta Joh. de Bruyn, Professoris Trajectini, et Petri Petiti, 1663, it is said, p. 69—Nec lumen ullum est absque calore, nec calor ullus absque lumine. Lux, sonus, anima (!) odor, vis magnetica, quamvis incorporea, sunt tamen aliquid (De Lucis Nat. cap. 13, p. 29).

(509) p. 331.—Kosmos, Bd. i. S. 427 and 429, Bd. ii. S. 482, Anm. 92; Engl. trans. Vol. i. Notes 141 and 144, Vol. ii. Note 432.

(510) p. 331.—Lord Bacon, whose comprehensive and, generally speaking, free and methodical views were unfortunately accompanied by very limited mathematical and physical knowledge, even for the period at which he lived, therefore did Gilbert the greater injustice. "Bacon showed his inferior aptitude for physical research in rejecting the Copernican doctrine which William Gilbert adopted (Whewell, Philosophy of the Inductive Sciences, Vol. ii. p. 378.)

([511]) p. 331.—Kosmos, Bd. i. S. 194 und 435, Anm. 31 and 32; Engl. trans. Vol. i. p. 176, Notes 161 and 162.

([512]) p. 332.—The first observations of the kind were made on the tower of the Augustine's Church, at Mantua (1590.) Grimaldi and Gassendi were acquainted with similar instances, all in geographical latitudes where the inclination of the magnetic needle is very considerable. On the subject of the first measurements of the magnetic intensity by the oscillation of a needle, compare my Relation hist. T. i. pp. 260—264, and Kosmos, Bd. i. S. 432—434 (Engl. transl. Vol. i. Note 159).

([513]) p. 334.—Kosmos, Bd. i. S. 436—439, Anm. 36 (Engl. trans. Vol. i. Note 166).

([514]) p. 334.—Kosmos, Bd. i. S. 189 (Engl. trans. Vol. i. p. 171.)

([514] *bis*) p. 335.—[*Additional note by the Editor.*—The desire so earnestly expressed by the author in the text, pp. 334 and 335, that "the laws of terrestrial magnetism should be thoroughly sought out by naval expeditions, which should examine, as nearly as possible at the same time the state of magnetism over all the accessible parts of the globe which are covered by the ocean," that "such expeditions should be combined with land surveys," and that "the year 1850 might deserve to be marked as the first normal epoch in which the materials of a magnetic map of the world should be assembled," is much nearer its fulfilment than M. de Humboldt seems to have been aware of when the second volume of Kosmos was published in Germany (October 1847). The antarctic expedition of Sir James Clark Ross, referred to in the text, has been followed by that of Lieuts. Moore, R.N. and Clerk, Royal Artillery (1845), by which the magnetic survey of the accessible portions of the high latitudes of the southern hemisphere has been completed; by the voyages of Lieuts. Smith and Dayman, R.N. (1844 and 1845) between the Cape of Good Hope and Van Diemen Island; of Lieut. Moore, R.N. (1846) to Hudson's Bay; and by the land expedition of Lieut. Lefroy, Royal Artillery (1843-44), by which the whole of British North America east of the Rocky Mountains, from the frontiers of the United States to the shores of Hudson's Bay and the Polar ocean, has been magnetically surveyed. These were all special surveys, undertaken by the British Government expressly for the magnetical purposes which they accomplished; and, with the exception of the observations in Lieutenant Moore's voyage to Hudson's Bay, which are now (February 1848) in process of reduction, their results have been deduced and published. In addition to special expeditions to parts of the globe which are either remote or difficult of access, the British Government has availed itself of the services of Her Majesty's ships and vessels employed in Hydrographical Surveys, by directing that determinations of the three magnetic elements should be made by them at the several ports and harbours which they may visit, as well as at sea *daily*, as often as the weather permits, in their passages from port to port. Such determinations have been executed, in whole or in part, by the surveying expeditions of Sir Edward Belcher (1837—1840, and 1843—1847) to the north-west coast of America, the islands of the Pacific, and the Indian and Chinese seas; of Captain Sullivan (1838—1839) to the Falkland Islands; of Captain Allen (1841—1842) to the western coast of Africa; of Captain Blackwood (1842—1846) to Australia and Torres Strait; of Captain Barnett (1843—18 *) to Bermuda and the West Indies; of Captain Kellett (1845—18 *) to the Pacific; of the Arctic Expedition under Sir John Franklin (1845—18 *); of Captain Stanley (1847—18 *) to Australia and New Guinea; of Captain Moore (1848—18 *) to Kamptschatka and Behring's Strait; and of Captain Stokes (1848—18 *) to New Zealand. To these should be added, as a special undertaking at the expense of the East India Company, a magnetic survey of the islands of the Indian Archipelago, by Lieut. Elliot, of the Madras Engineers, commenced in 1846, and still in progress. When it is remembered that several of the above-named surveys include periods of three or four years, and in some instances not only determinations at the several ports and harbours which may have been visited, but also daily observations, weather permitting, of the three magnetic elements at sea in passages from port to port, the accumulation of materials, and their already extensive distribution over the surface of the globe, may in some degree be judged of. These surveys, with others which may be expected to be made under the present favourable disposition of Her Majesty's Government towards scientific researches, and, taken in conjunction with extensive magnetic surveys which are in progress on the continent of Europe (particularly in the Austrian dominions), give a full promise of the speedy realisation of M. de Humboldt's wish so earnestly expressed, that the materials of the first general magnetic map of the globe should be assembled, and even permit the anticipation, that the first normal epoch of such a map will be but little removed from the year 1850.]

([515]) p. 335.—On the oldest thermometers, see Nelli, Vita e Commercio litterario di Galilei (Losanna, 1793), Vol. i. p. 68—94; Opere di Galilei (Padovo, 1744), T. i. p. lv.; Libri, Histoire des Sciences mathematiques en

* When the concluding date is not filled up, the observations are still in progress.

Italie, T. iv. 1841, p. 185—197. Evidence respecting the first comparative observations of temperature may be found in the letters of Gianfrancesco Sagredo and Benedetto Castelli, in 1613, 1615, and 1633, in Venturi, Memorie e Lettere inedite di Galilei, P. i. 1818, p. 20.

(516) p. 335.—Vincenzio Antinori, in the Saggi di Naturali Esperienze, fatte nell'a Academia del Cimento, 1841, p. 30—44.

(517) p. 335.—On the determination of the thermometric scale of the Academia del Cimento, and on the meteorological observations continued for 16 years by Father Raineri, a pupil of Galileo, see Libri, in the Annales de Chimie et de Physique, T. xlv. 1830, p. 354; and a more recent similar work by Schouw, in his Tableau du Climat et de la Végétation de l'Italie, 1839, T. 99—106.

(518) p. 336.—Antinori, Saggi dell' Accad. del Cim. 1841, p. 114, and in the Aggiunte at the end of the book, p. lxxvi.

(519) p. 337.—Antinori, p. 29.

(520) p. 337.—Ren. Cartesii Epistolæ (Amstel. 1682), P. iii. Ep. 67.

(521) p. 337.—Bacon's Works, by Shaw, 1733, Vol. iii. p. 441 (see Kosmos, Bd. i. S. 338 and 479, Anm. 58; Engl. trans. Vol. i. p. 310, and note 388).

(522) p. 338.—Hooke's Posthumous Works, p. 364. (Compare my Relat. historique, T. I. p. 199.) Hooke however, unhappily, like Galileo, assumed a difference in the velocity of rotation of the earth and of the atmosphere: see Posth. Works, p. 88 and 363.

(523) p. 338.—Although Galileo also speaks of the remaining behind of the particles of air as a cause of the Trade Winds, yet his view ought not to be confounded with that of Hooke and Hadley as it has recently been. Galileo, in the Dialogo quarto (Opere, T. iv. p. 311) makes Salviati say: "Dicevamo pur' ora che l' aria, come corpo tenue, e fluido, e non saldamente congiunto alla terra, pareva, che non avesse necessità d' obbedire al suo moto, se non in quanto l' asprezza della superficie terrestre ne rapisce, e seco porta una parte a se contigua, che di non molto intervallo sopravanza le maggiori altezze delle montagne; la qual porzion d' aria tanto meno dovrà esser renitente alla conversion terrestre, quanto che ella è ripiena di vapori, fumi, ed esalazioni, materie tutte participanti delle qualità terrene: e per conseguenza atte nate per lor natura (?) ai medesimi movimenti. Ma dove mancassero le cause del moto, cioè dove la superficie del globo avesse grandi spazi piani, e meno vi fusse della mistione dei vapori terreni, quivi cesserebe in parte la causa, per la qule l'aria ambiente dovesse totalmente obbedrie al rapimento della conversion terrestre; si che in taliluoghi, mentre che la terra

si volge verso Oriente, si dovrebbe sentir continuamente un vento, che ci ferisse, spirando da Levante verso Ponente; e tale spiramento dovrebbe farsi piu sensibile, dove la vertigine del globo fusse piu veloce: il che sarebbe ne i luoghi piu remoti da i Poli, e vicini al cerchio massimo della diurna conversione. L'esperienza applaude molto a questo filosofico discorso, poichè ne gli ampi mari sottoposti alla Zona torrida, dove anco l' evaporazioni terrestri mancano (?) si sente una perpetua aura muovere da Oriente. . . . ."

(524) p. 338.—Brewster, in the Edinburgh Journal of Science, Vol. ii. 1825, p. 145. Sturm has described the Differential Thermometer in a little work, entitled, Collegium experimentale curiosum, (Nuremberg, 1676,) p. 49. On the Baconian law of the rotation of the wind, which Dove first extended to both zones, and recognised in its intimate connection with the causes of all aerial currents, see the detailed treatise of Muncke in the new edition of Gehler's Physikal. Worterbuch, Bd. x. S. 2003—2019 and 2030—2035.

(525) p. 339.—Antinori, p. 45, and even in the Saggi, p. 17—19.

(526) p. 339.—Venturi, Essai sur les ouvrages physico-mathématiques de Léonard de Vinci, 1797, p. 28.

(527) p. 339.—Bibliothèque universelle de Genève, T. xxvii. 1824, p. 120.

(528) p. 340.—Gilbert de Magnete, lib. ii. cap. 2—4, p. 46—71. In interpreting the nomenclature employed he already said: "Electrica quæ attrahit eadem ratione ut electrum, versorium non magneticum ex quovis metallo, inserviens electricis experimentis." In the text itself we find it said; "Magneticè ut ita dicam, vel electricè attrahere (vim illam electricam nobis placet appellare . . . .) (p. 52); "effluvia electrica, attractiones electricæ." He neither employed the abstract expression *electricitas*, nor the barbarous term *magnetismus* introduced in the 18th century. On the derivation of ηλεκτρον, the "attracter or drawer, and the drawing or attracting stone," from ελξις and ελκειν, already indicated in the Timæus of Plato, p. 80 c, and the probable transition through a harder ελκτρον, see Buttmann, Mythologus, Bd. ii. (1829), S. 357. Among the theoretical propositions put forward by Gilbert (which are not always expressed with equal clearness), I select the following: "Cum duo sint corporum genera, quæ manifestis sensibus nostris motionibus corpora allicere videntur, Electrica et Magnetica; Electrica naturalibus ab humore effluviis; Magnetica formalibus efficientiis seu potius primariis vigoribus, incitationes faciunt. . . . . Facile est hominibus ingenio acutis, absque experimentis et usu rerum labi, et errare. Substantiæ proprietates aut familiaritates, sunt generales nimis, nec tamen veræ designatæ causæ, atque, ut ita dicam, verba quædam sonant, re ipsâ nihil in specie ostendunt. Neque

ista succini credita attractio, a singulari aliquâ proprietate substantiæ, aut familiaritate assurgit; cum in pluribus aliis corporibus eundem effectum, majori industria invenimus, et omnia etiam corpora cujusmodicunque proprietatis, ab omnibus illiis alliciuntur." (De Magnete, p. 50, 51, 60, and 65.) Gilbert's principal labours appear to belong to the interval from 1590 to 1600. Whewell justly assigns him an important place among those whom he terms "practical Reformers of the physical sciences." Gilbert was surgeon to Queen Elizabeth and James I. and died in 1603. A second work, entitled "De Mundo nostro sublunari Philosophia Nova," was published after his death.

(529) p. 341.—Brewster, Life of Newton, p. 307.

(530) p. 344.—Rey, strictly speaking, only mentions the access of air to the oxides; he did not know that the oxides themselves (which were then called metallic calxes) are only combinations of metals and air. According to him, the air makes "the calx heavier, as sand increases in weight when water hangs about it." The calx is susceptible of being saturated with air. "L'air espaissi s'attache à la chaux, ainsi le poids augmente du commencement jusqu'à la fin: mais quand tout en est affablé, elle n'en sçauroit prendre d'avantage. Ne continuez plus votre calcination soubs cet espoir, vous perdriez vostre peine." Rey's work thus contains the first approximation to the better explanation of a phenomenon, the more complete understanding of which was afterwards influential in reforming the whole of chemistry. See Kopp, Gesch. der Chemie, Th. iii. S. 131—133. (Compare also in the same work, Th. i. S. 116—127, and Th. iii. S. 119—138, as well as S. 175—195.)

(531) p. 345.—Priestley's last complaint of that which "Lavoisier is deemed to have appropriated to himself," makes itself heard in his little memoir entitled, "The Doctrine of Phlogiston established," 1800, p. 43.

(532) p. 346.—Sir John Herschel, Discourse on the Study of Natural Philosophy, p. 116.

(533) p. 346.—Humboldt, Essai géognostique sur le Gisement des Roches dans les deux Hémisphères, 1823, p. 38.

(534) p. 347.—Steno de Solido intra Solidum naturaliter contento, 1669, p. 2, 17, 28, 63, and 69 (fig. 20—25).

(535) p. 347.—Venturi, Essai sur les Ouvrages physico-mathématiques de Léonard de Vinci, 1797, S. 5, No. 124.

(536) p. 347.—Agostino Scilla, La vana Speculazione disingannata dal Senso, Nap. 1670, Tab. xii. fig. 1. Compare Joh. Müller, Bericht über die

von Herrn Koch, in Alabama gesammelten fossilen Knochenreste seines Hydrachus (the Basilosaurus of Harlan, 1835; the Zeuglodon of Owen, 1839; the Squalodon of Grateloup, 1840; the Dorndon of Gibbes, 1845), read in the Royal Academy of Sciences at Berlin, April—June 1847. These valuable fossil remains of an animal of the ancient world, which were collected in the state of Alabama (in Washington County, not far from Clarksville), have become by the munificence of our King, since 1847, the property of the Zoological Museum at Berlin. Besides the remains found in Alabama and South Carolina, parts of the Hydrachus have been found in Europe, at Leognan near Bordeaux, not far from Linz on the Danube, and, in 1670, in Malta.

(537) p. 348.—Martin Lister, in the Philosophical Transactions, Vol. vi. 1671, No. lxxvi. p. 2283.

(538) p. 348.—See a luminous exposition of the earlier progress of palæontological studies, in Whewell's History of the Inductive Sciences, 1837, Vol. iii. p. 507—545.

(539) p. 349.—Leibnizen's geschichtliche Aufsätze und Gedichte, herausgegeben von Pertz, 1847 (in the gesammelten Werken: Geschichte, Bd. iv.) On the first, Protogæa of 1691, and the subsequent revisions, see Tellkampf, Jahresbericht der Burgerschule zu Hannover, 1847, S. 1—32.

(540) p. 350.—Kosmos, Bd. i. S. 172 (Engl. trans. Vol. i. p. 155).

(541) p. 350.—Delambre, Hist. de l'Astronomie mod. T. ii. p. 601.

(542) p. 351.—Kosmos, Bd. i. S. 171 (Engl. trans. Vol. i. p. 154). The contest respecting priority relative to the knowledge of the earth's compression, in reference to a memoir read by Huygens, in 1669, before the Paris Academy, was first cleared up by Delambre, in his Hist. de l'Astr. mod. T. i. p. lii. and T. ii. p. 558. Richer's return to Europe took place indeed in 1673, but his work was not printed until 1679; and as Huygens left Paris in 1682, he did not write the Additamentum to the Memoir of 1669, the publication of which was very late, until the period when he had already before his eyes the results of Richer's Pendulum Experiments, and of Newton's great work, Philosophiæ Naturalis Principia Mathematica.

(543) p. 351.—Bessel, in Schumacher's Jahrbuch für 1843, S. 32.

(544) p. 352.—Wilhelms von Humboldt, gesammelte Werke, Bd. i. S. 11.

(545) p. 358.—Schleiden, Grundzüge der wissenschaftlichen Botanik, Th. i. 1845, S. 152, Th. ii. S. 76; Kunth, Lehrbuch der Botanik, Th. i. 1847, S. 91—100, and 505.

INDEX TO VOL. II.

END OF VOL. II.

Printed in the United States
By Bookmasters